DRY CELL BATTERIES

CHEMISTRY AND DESIGN

DRY CELL BATTERIES

CHEMISTRY AND DESIGN

Louis F. Martin

NOYES DATA CORPORATION

Park Ridge, New Jersey   London, England

1973

Published in the United States of America by
Noyes Data Corporation
Noyes Building, Park Ridge, New Jersey 07656

FOREWORD

The detailed, descriptive information in this book is based on U.S. patents since 1969 relating to the chemistry and design of dry cell batteries.

This book serves a double purpose in that it supplies detailed technical information and can be used as a guide to the U.S. patent literature in this field. By indicating all the information that is significant, and eliminating legal jargon and juristic phraseology, this book presents an advanced, commercially oriented review of dry cell chemistry and technology.

The U.S. patent literature is the largest and most comprehensive collection of technical information in the world. There is more practical, commercial, timely process information assembled here than is available from any other source. The technical information obtained from a patent is extremely reliable and comprehensive; sufficient information must be included to avoid rejection for "insufficient disclosure."

The patent literature covers a substantial amount of information not available in the journal literature. The patent literature is a prime source of basic commercially useful information. This information is overlooked by those who rely primarily on the periodical journal literature. It is realized that there is a lag between a patent application on a new process development and the granting of a patent, but it is felt that this may roughly parallel or even anticipate the lag in putting that development into commercial practice.

Many of these patents are being utilized commercially. Whether used or not, they offer opportunities for technological transfer. Also, a major purpose of this book is to describe the number of technical possibilities available, which may open up profitable areas of research and development. One should have to go no further than this condensed information to establish a sound background before launching into research in this field.

Advanced composition and production methods developed by Noyes Data are employed to bring these durably bound books to you in a minimum of time. Specialized techniques are used to close the gap between "manuscript" and "completed book." Industrial technology is progressing so rapidly that time-honored, conventional typesetting, printing, binding and shipping methods can render a technical or scientific book quite obsolete before the potential user gets to see it.

The Table of Contents is organized in such a way as to serve as a subject index. Other indexes by company, inventor and patent number help in providing easy access to the information contained in this book.

15 Reasons Why the U.S. Patent Office Literature Is Important to You—

(1) The U.S. patent literature is the largest and most comprehensive collection of technical information in the world. There is more practical commercial process information assembled here than is available from any other source.

(2) The technical information obtained from the patent literature is extremely comprehensive; sufficient information must be included to avoid rejection for "insufficient disclosure."

(3) The patent literature is a prime source of basic commercially utilizable information. This information is overlooked by those who rely primarily on the periodical journal literature.

(4) An important feature of the patent literature is that it can serve to avoid duplication of research and development.

(5) Patents, unlike periodical literature, are bound by definition to contain new information, data and ideas.

(6) It can serve as a source of new ideas in a different but related field, and may be outside the patent protection offered the original invention.

(7) Since claims are narrowly defined, much valuable information is included that may be outside the legal protection afforded by the claims.

(8) Patents discuss the difficulties associated with previous research, development or production techniques, and offer a specific method of overcoming problems. This gives clues to current process information that has not been published in periodicals or books.

(9) Can aid in process design by providing a selection of alternate techniques. A powerful research and engineering tool.

(10) Obtain licenses – many U.S. chemical patents have not been developed commercially.

(11) Patents provide an excellent starting point for the next investigator.

(12) Frequently, innovations derived from research are first disclosed in the patent literature, prior to coverage in the periodical literature.

(13) Patents offer a most valuable method of keeping abreast of latest technologies, serving an individual's own "current awareness" program.

(14) Copies of U.S. patents are easily obtained from the U.S. Patent Office at 50¢ a copy.

(15) It is a creative source of ideas for those with imagination.

CONTENTS AND SUBJECT INDEX

INTRODUCTION

Of the several billion batteries manufactured throughout the world each year, the conventional dry cell ($Zn-MnO_2$, Leclanché) still accounts for well over 70% of the market. This cell, first described by George Leclanché in 1868, evolved into the "dry cell" by immobilizing the electrolyte on a porous material and was mass produced beginning in 1890. The alkaline dry cell, known in the early 1900's did not reach commercial production until 1949 due to need for better sealing techniques for the strong alkaline electrolytes. Indeed, much of the very recent battery design literature is devoted to sealing techniques which are adaptable to modern manufacturing methods.

Magnesium dry cell batteries have demonstrated almost double the capacity of the conventional zinc dry cell batteries at moderate discharge rates and have significantly increased the storage capability of such batteries. These batteries commonly employ either magnesium perchlorate or magnesium bromide as the electrolyte. Magnesium dry batteries are finding increasing use in military communication equipment and the cost of the batteries has dropped significantly over the past few years.

Alkaline zinc-mercury oxide cells were developed during World War II. By 1945, approximately one million cells were being produced each day. These cells, while expensive, do provide a relatively constant voltage during discharge and exhibit a good energy-to-volume ratio.

Alkaline silver oxide-zinc batteries are used in many military applications such as guided missiles, portable radios and radar equipment. Either the divalent silver oxide or monovalent silver oxide may be used as the cathode. These batteries exhibit high capacity per unit of weight, coupled with excellent high rate capabilities. Cells containing organic electrolytes are able to use high energy light metal electrodes such as lithium, which are incompatible with aqueous electrolytes. Typically, a lithium-cupric chloride cell in a propylene carbonate electrolyte will provide high operating voltages and energy densities.

In recent years, much research effort has been devoted to solid state devices employing solid electrolytes. Various metal-halogen compounds and charge transfer

1

complexes have been developed as electrolytes and for use in electrodes. These solid electrolyte batteries are of considerable interest because of their exceptional stability, physical ruggedness, and ease of miniaturization. The efficiency of these cells is determined by the solid electrolyte which serves as an ionic conductor, and most of the recent effort has been directed to enhancing this ionic conductivity while minimizing cost.

The heat activated batteries (fused salt electrolyte) work on the principle that fused salts are ionically conductive and thus perform as an electrolyte. Much of the work on thermal batteries has been conducted for the military and little has been published to date.

The accelerating pace of technology in battery electrochemistry and design is nowhere more evident than in the patent literature. This book describes over 130 processes relating to recent developments in dry cell batteries employing zinc, magnesium, cadmium-nickel, lithium, solid state devices, organic electrolytes, and thermal activation. It includes many modern manufacturing techniques.

BASIC ZINC DRY CELLS

ELECTRODES

Wet Amalgamation Bonding

R.E. Ralston and Y.L. Ko; U.S. Patent 3,669,754; June 13, 1972; assigned to
P.R. Mallory & Co. describe a negative electrode which is fabricated by a process
which might be called "wet amalgamation bonding". A mixture of the following
composition has been used to prepare electrodes of 65% porosity:

> 46.6 grams zinc
> 6.0 grams mercurous chloride
> 5.0 grams ammonium chloride
> 1 drop glycerine

In mixing, one drop of glycerine is added to the proper weight of zinc particles of
-100, +200 mesh, and mixed thoroughly to wet and cover the surfaces of the zinc
particles. Mercurous chloride is added, again with complete mixing, to coat the
surface of the zinc with amalgamating material. Ammonium chloride is then added
and dispersed evenly. The mix is then ready for use.

The molded electrode is made by amalgamation-bonding zinc particles to a conductor
of brass, or zinc plated brass, or amalgamated zinc plated brass. Preamalgamation
not only cleans the surface but also decreases the tendency of the zinc plate on the
conductor to be stripped off or go into solution in the ammonium chloride solution.
This is accomplished by placing the conductor in a mold conveniently made of plastic, and applying the correct amount of electrode mix in contact with the surface to
be bonded. Excess air is expelled and good particle-to-particle relationship established by vibration compacting.

While still confined in the mold, a quantity of saturated ammonium chloride solution
is added to wet the electrode material and to initiate the reaction process. After
approximately 15 minutes the electrode can be removed from the mold and placed in
a solution of ammonium chloride, for 12 hours or more, to allow the bonding reaction

3

to reach completion; and then the electrode is washed and dried. The bonding action
by this technique might be called solution sintering by liquid mercury. The initial
reaction is the chemical displacement of zinc by mercury, depositing liquid mercury
over the entire zinc surface. Immediately the mercury begins to alloy with the clean
zinc, since the ammonium chloride acts as a flux to clean the zinc surface of oxide,
and other unwanted matter, and the surface changes from 100% mercury to about 10%
mercury as the homogeneous alloy forms.

However, during the early stages of the alloy formation, zinc dissolves in the liquid
mercury which is concentrated at the points or areas of contact between particles,
and forms a solid solution bond at these contact points or areas when the alloy formation is complete. The same type of bonding is also accomplished at the surface
of the anode collector. The superiority of the molded anode has been demonstrated
by the performance of alkaline manganese cells at minus 40°C. The following
service was obtained with "C" cells on an intermittent duty cycle of 4 minutes
on load per hour, 10 hours per day, 5 days per week to an end voltage of 0.93
volts on a 7.5 ohm load:

	Minutes
Conventional Brand A	3
Conventional Brand B	11
This process Molded Anode	*190-200

Leached Amalgamated Zinc

R.R. Clune and H. Field; U.S. Patent 3,427,204; February 11, 1969; assigned to
P.R. Mallory & Co., Inc. describe a battery system in which the efficiency is greatly enhanced, especially for low temperature usage. This increase in efficiency is
among other features due to the presence, in the battery, of an anode consisting of
porous zinc amalgam and utilizing ammonium chloride, mercurous chloride and magnesium sulfate. The anode is formed as a uniform, porous pellet by means primarily
of reactive cementation of the mixed materials.

The extent of the anodic surface is greatly enhanced by the high porosity of the
structure thus formed. After the chemical displacement reaction and cementation,
excess unused materials are carefully leached away resulting in a highly porous and
uniformly dense amalgamated zinc pellet of great surface area. The anode, moreover, is of much greater porosity and more consistent surface area, affording better
electrolytic conductivity and lower impedance, and has a greater ability to hold
more electrolyte than anodes priorly available. The preferred pelleted mixture consists primarily of about 55.4% zinc, 37.7% ammonium chloride, 6.9% mercurous
chloride, and the rest an aliphatic solvent. The final anode structure has a porosity
of about 77%, and consists essentially of about 90% zinc and 10% mercury.

This process enables alkaline cells to function at low temperatures (-30°C.) with up
to 85% of their efficiency at normal temperatures. It is an improvement over common
pellet, dispersed pellet or wound anode structures, which deliver at best 20% of room
temperature performance at -30°C. and over such devices as heaters, frangible vials,
antifreezes, etc., employed to obtain low temperature operation. These have added
greatly to the complexity of the structure, while decreasing space available for active
elements; e.g., depolarizer, anode and electrolyte.

The leached amalgamated anode of great porosity is produced as follows:

Fabrication and Reactive Cementation — Place 300 grams of ammonium chloride and 25 grams of magnesium sulfate into a blender. Introduce 916 grams of zinc and mix. Introduce 0.080 gram (approximately 50 drops) of a nonpolar, low volatility substance such as deodorized kerosene. Mix until uniform. The kerosene acts as a coating which prevents premature reaction of the material and as a blending agent to coat the pellets and cohere them in order to prevent stratification and segregation. This step assures uniform blending so that homogenity is obtained in the mix.

Introduce 236 grams of mercurous chloride and mix until uniform. It is required that all mixing be done at a temperature of 65°F. maximum and at a relative humidity of 40% maximum. The mix is then pelletized at a pressure of about 12,800 pounds per square inch.

Leaching — The pellets are placed in washing trays and are then immersed in a saturated solution of ammonium chloride. The pellets must be soaked for a minimum of 1/2 hour in the saturated solution of ammonium chloride. The pellets are then leached until free of chloride ion. Test to determine an absence of chloride ion by the use of silver nitrate added to the water drippings from the washed pellets.

Drying — After leaching is completed by showing that all chloride ion is absent from the leaching water, place the anodes in a solution of methanol for about 10 minutes. The methanol is used to dry the anodes. Remove the pellets from methanol and place them in a vacuum oven preheated at 100°F. until the pellets are completely dried. The pellets are then stored in a dry atmosphere in a container which is air tight and which may include a desiccant.

Potassium Hydroxide Impregnation

R.W. Blossom and A. Charkey; U.S. Patent 3,655,451; April 11, 1972; assigned to Yardney International Corp. describe a process for manufacturing an electrode impregnated with potassium hydroxide. Potassium hydroxide in powder form having a low content of water is mixed with the anode metal in powder form and blended to form a coherent mixture. The mixture is spread over a conductive grid, subjected to a pressing operation to form a unitary structure of the desired electrode shape and then heated in the absence of air. The electrode is then stored in an airtight container. A strong KOH-impregnated metal anode capable of long storage is obtained.

In the manufacture of primary cells where quick activation is needed, several methods have been proposed for impregnating a zinc electrode with a potassium hydroxide electrolyte by immersing the completely formed zinc electrode in an alkaline solution. Thereafter, it is dried and placed in either cold storage or some closed environment devoid of water vapor to obtain a long-lasting electrode. However, it has been found that, during manufacture, these electrodes tend to be pyrophoric and will spontaneously burst into flame.

An alternate method for manufacturing such electrodes has been described where potassium hydroxide, in powder form is mixed with zinc, in powder form, spread over a conductive grid and subjected to a pressing operation to form the desired electrode shape. This method avoided the pyrophoric stage and provided usable electrodes.

The potassium hydroxide powder useful in the process may be purchased on the open market and should contain a very low amount of water. A preferable range is between 5 and 7% of H_2O by weight. It has been found that moisture contents of less than 3.5% should be avoided. Such substantially anhydrous KOH tends to cause caking and is difficult to repulverize. On the other hand, KOH powders with water contents of 10% or greater render the zinc/KOH mixture too pyrophoric. When handling this potassium hydroxide powder and the zinc/KOH mixture, safety precaution such as safety glasses and dust masks are recommended. Since KOH is very hygroscopic, it should be carefully stored to maintain the moisture range within the limits of 3.5 to 10% by weight.

In the preparation of the electrode, KOH in powder form is mixed with zinc in powder form. The ratio of KOH to zinc powder may be from 0.15 to 0.25 grams of KOH for each gram of zinc powder and preferably is above 0.23 gram. The zinc and KOH powders are well blended either by handmixing, by tumbling the powders or by large-scale blending equipment. The amount of KOH mixed with the zinc powder depends to some extent upon the desired concentration of the electrolyte after the addition of of water in a battery system. In a battery of given configuration, the space available for the electrolyte also influences the amount of KOH. Too little KOH will prevent proper functioning of the electrolyte.

On the other hand, too much admixed KOH tends to render the electrode too brittle and of course reduces the amount of zinc available in the anode. A compromise must then be reached, keeping in mind these considerations. The amount of dry KOH added to the plate is such that the final concentration of electrolyte in the assembled cell of any configuration will be between 20 and 45% aqueous KOH. This means that the ratio of 0.015 to 0.25 grams of KOH per gram of zinc set forth above is useful for zinc-air cell having limited quantity of free electrolyte, e.g., a cell having 32 cc of electrolyte volume with 29.5 cc of zinc-electrode volume. In other configurations suitable adjustments are required.

The electrodes are fabricated by cold-pressing the composite Zn/KOH mixture on a supporting grid at various pressures. The grid may be made of expanded-metal meshes of copper, silver, or cold-rolled steel having 40 to 60% open area and 5 to 10 mil thickness. The mixture is evenly distributed into the grid and the composite is compressed at a pressure ranging, in a typical situation, from 4 to 10 tons per square inch. The preferable range is about 4.4 to 5.0 tons per square inch.

The final density of the pressed electrode is approximately 2.2 grams per cc. Since this compacted mass includes KOH, the addition of water which will dissolve the KOH renders the density of the zinc plate in the activated cell substantially lower. Thus, the zinc density is computed to provide the optimum of 1.8 grams per cc in situ. KOH impregnated zinc plates having thicknesses from 0.020 to 0.0200 inches have been made for the above described cells having a 32 cc electrolyte volume.

After compression, the dry plates are placed in an oven provided with a nonoxidizing atmosphere such as vacuum, nitrogen, hydrogen or argon, and are heated at temperatures in the range of 200° to 400°F., for a period of time sufficient to agglomerate the particles and bind them together with the water contained in the KOH powder acting as a binder. At the higher temperature, 10 minutes is sufficient to develop the strength of the electrode. At lower temperatures correspondingly longer times are necessary, but below 200°F. either the reaction of the interstitial water is not started or the time required is too long for practical purposes, (in excess of 1 1/2 hours). Following the heating treatment, the plates are sealed in moistureproof composite bags made from craft paper, aluminum or polyethylene depending upon the desired storage time between electrode manufacture and complete battery fabrication or electrode insertion. It has been found that electrodes prepared from such Zn/KOH powder mixtures can be left exposed in air from 1 to 2 hours depending on the relative humidity. As a practical matter, however, pyrophoric conditions may arise when the KOH absorbs in excess of 10% of water.

Since oxidation of the plates is initiated after they have been exposed for more than the stated periods, and proceeds rapidly until the electrodes may ignite, sealing of the electrode must be completed before the specified time. The exact correlation between the rate of moisture pick-up leading to pyrophoric electrode oxidation and the relative ambient humidity has not been established. However, it has been observed that on days when the humidity is less than 40%, the plates can be exposed for about 2 hours. On days when the humidity is high, for instance greater than 60%, the plates should not be exposed to the air for more than 1 hour. Plate oxidation is best minimized by drying the atmosphere in the locale where the mixing, pressing, heat-treating and sealing operations are conducted.

The performance of the above described composite electrodes was evaluated with good results. In a typical example, the electrode for a cell was made by blending 30 grams of zinc powder with 7.4 grams of KOH powder, mold-pressing to a density of approximately 2.2 grams per cc and a thickness of 0.2 inches and subsequently heating at 300°F. for 20 minutes in a nitrogen-flushed oven. Sample electrodes were compared with electrodes prepared according to a previous process. It was found that the heat-treated electrodes of this process had much greater tensile strength, compressive strength and resistance to flexure.

In addition, when subjected to vibration tests, the electrodes survived frequencies and amplitudes that destroyed the referenced samples. The electrodes were then inserted into 12 mils pellon separator bags and placed in an air cell having a 12-square inch cathode area. The cell was activated with 14 cc of distilled water and discharged immediately at rates of 2.5 amp. and 5.0 amp. At these discharges the respective plateau voltages were 1.06 and 1.02 volts, yielding further respective storage capacities of 13.7 and 12.9 amp. hours. The zinc utilization was respectively 2.33 and 2.45 grams per amp. hour. Statistically these performace data were identical with the unheated electrodes of the prior disclosure referred to.

Sintered Zinc Powder

In a process described by R.D. Weller; U.S. Patent 3,663,297; May 16, 1972; assigned to the U.S. Secretary of the Navy sintered zinc powder battery electrodes are prepared by (1) rinsing zinc powder with a solvent that dissolves zinc oxide at a faster rate than it dissolves zinc and which is saturated with a water-soluble salt having a higher melting point than zinc; (2) pouring off the solvent; (3) mixing the zinc with the water-soluble salt used in step (1); (4) packing the mixture around a metal screen that is made of a metal which is an electrical conductor having a higher melting point than zinc; (5) heating the packed article at a temperature above 419°C. but below the melting point of the water-soluble salt and the metal screen; (6) cooling the article; and (7) soaking the article in water to remove the water-soluble salt.

Example: 20 grams of zinc powder was rinsed with 1 N HCl which was saturated with NaCl. Contacting was maintained for about 30 seconds in order to remove the zinc oxide coating from the zinc particles. Most of the HCl was then poured off leaving the zinc powder damp. 8 grams of NaCl powder was added to the wet zinc making a damp mixture of NaCl and zinc powders. This mixture was packed into a die while it was still damp and pressed to 7,000 psi into an electrode which was approximately 1 1/2 x 1 1/2 x 1/8 inches. A silver metal screen was included in the middle for a current collector.

The pressed electrode, still damp, was held between two discs of sintered glass and placed into an argon atmosphere furnace which was heated up to 500°C. in approximately 15 minutes. This temperature was then held for about 1 more minute. After cooling the electrode was removed and soaked in water to dissolve the NaCl thus leaving the zinc electrode.

Addition of Stainless Steel Filaments to Cathode

C.K. Ching, W. Cohen and S. Tseng; U.S. Patent 3,660,167; May 2, 1972 describe an alkaline dry cell of high energy density in which a cathode of powdered manganese dioxide, carbon, and graphite has dispersed throughout it, short filaments of stainless steel, long filaments of stainless steel, and stainless steel powder.

Referring to Figure 1.1, the anode (10) has a steel jacket (11) which is generally tubular with a closed bottom (9) and an open top (8). The jacket (11) is a protective shell or covering that protects the internal parts of the alkaline dry cell from impact damage or damage due to deformation of the cell. The steel jacket is approximately 0.010 inch thick. A preformed liner (12) of nonconductive material having an external diameter substantially equal to the internal diameter of the steel jacket (11) is inserted into the jacket (11) to cover the side wall and the bottom. It is preferred that the liner (12) be made of some plastic material such as polyethylene or polyvinyl chloride 0.006 inch thick. The liner (12) has a bottom center opening to which a contact button (13) is inserted.

Since the liner (12) is of dielectric material the contact button (13) establishes conductivity between the exterior of the liner (12) and the interior. A deposit of amalgamated zinc powder is now placed in the liner (12) to form the bottom (14) of the anode. The bottom (14) of the anode is a layer originally 0.150 inch thick. The thickness is not critical. The bottom (14) is in electrical conductive contact with the contact button (13) and in turn with the jacket (11). The contact (13) is preferably a stainless steel button formed of the 400 series. To complete the anode from the bottom (13) to the top, an expandable split mold is positioned in the liner (12), seated on the bottom (14).

The split mold has an external diameter of 0.437 inch leaving an interstitial space. The interstitial space is filled with amalgamated zinc powder to within approximately 1/16 of an inch from the top. A tapered pin is then inserted in the split mold. This expands the mold and compresses the amalgamated zinc powder to a thickness of approximately 0.030 inch. The tapered pin is removed and the split mold is thereupon withdrawn leaving a central cavity in the anode (15) for the insertion of a cathode assembly. Two generally vertical slots are formed in the vertical wall of the anode (15) by reason of the expansion of the split mold.

FIGURE 1.1: ADDITION OF STAINLESS STEEL FILAMENTS TO CATHODE

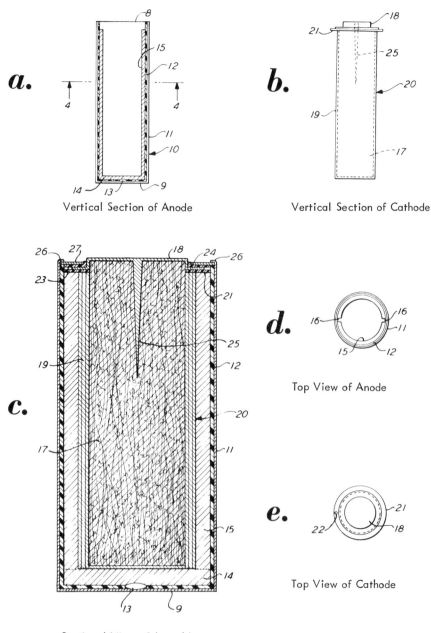

a. Vertical Section of Anode

b. Vertical Section of Cathode

c. Sectional View of Assembly

d. Top View of Anode

e. Top View of Cathode

Source: C.K. Ching, W. Cohen and S. Tseng; U.S. Patent 3,660,167; May 2, 1972.

The composition of the anode (15) and the bottom of the anode (14) is 99.9% pure zinc metal which is melted; to this, 10% of mercury by weight is added and thoroughly mixed in. The mixture is then cooled and it is ground into a powder having a particle size of 50 to 100 mesh. Although the bottom (14) of the anode is not critical as to thickness, it is preferred that it be compressed to a layer 0.10 inch thick overlying the contact button (13). The powder comprising the anode may be compacted by vibrating the jacket until the mass is dense. The densification of the amalgamated zinc is performed before the mold is expanded. The bottom (14) should be compressed before the vertical portion (15) of the anode is formed.

The cathode (17) is separately prepared and is later inserted into the jacket (11) of the anode (10). A liquid electrolyte is first prepared, consisting of 30% potassium hydroxide and 70% by weight of water. To this zinc oxide is added until 80% saturation at 70°F. is attained. Then approximately 3% by weight of the entire solution of sodium carboxymethylcellulose, high viscosity grade is added. Alternatively, other thickening agents such as "sodium starch glycolate" (Dupont) or "Kelzan" (Kelco Company, Clark, New Jersey) may be used to increase viscosity of the mixture. The electrolyte is now complete.

A mold is provided which has an internal diameter of 0.375 inch and a height substantially greater than the interior cavity of the anode (10). Into this mold a mixture of manganese dioxide, carbon, and graphite is inserted. It is preferred that the carbon have low ohmic resistance, such as "Conductex SC" (Colombian Carbon Co., New York). The manganese dioxide should be battery grade. The mixture has the following composition: 5.5 to 6.5 grams manganese dioxide, 0.3 to 0.5 grams carbon, 0.5 to 1.5 grams graphite, 0.5 to 1.5 grams stainless steel powder (0.020 x 0.006 x 0.006 inch), 0.2 to 0.8 grams stainless steel filaments (0.002 inch thick, 0.004 inch wide, 0.060 inch long), 0.2 to 0.8 grams longer stainless steel filaments (0.002 inch thick, 0.004 inch wide and 3.0 inches long).

Stainless steel powder and filaments are of the 400 series and preferably 416 series. This provides a cathode composition which is 5 to 17% by weight of stainless steel powder, 3.4 to 8% by weight short filaments, and 2.0 to 8.0% by weight long filaments. The manganese dioxide, carbon, graphite, stainless steel powder and short filaments are mixed by adding the ingredients together (except the long filaments) and agitating until the mixture is visibly determined to be complete. Twelve percent of water by weight of the mass of powder and filaments, including the as yet unadded long filaments is added to the mixture, and the mixture proceeds until the mass has a granular texture, resembling in appearance coarse beach sand.

The long filaments are then bunched arranged in a square area approximately 3 inches by 3 inches and having a thickness of perhaps no greater than a 1/16 of an inch. The mixture previously obtained is then applied to this bunch or mat and the mat is invested with the mixture, and it is rolled into a cylindrical bundle, all the while kneading the bundle to thoroughly disperse the powder throughout the mat. The roll, mat or bunch produced should be approximately 3/8 inch in diameter. It is loosely inserted into a cylindrical mold having an internal diameter of 0.375 inch. A cap or closure is then applied to the bottom of the mold. A cap (18) is positioned across the bottom of the bore of the mold, and the mold is closed at the bottom of the base plate. A plug is inserted in the top, and pressure of 40,000 psi applied. The base plate from the mold is removed. The cap (18) has become bonded to the compacted mat so formed and the cathode (17) so formed is discharged from the mold and is allowed to dry for 24 hours or more. The cylindrical cathode (17) takes on the appearance of a shiny, dense, homogeneous rod. After the cathode is dried, it is inserted in a separator (19). The separator is a tubular body closed at the bottom and open at the top and fitting snugly over the compressed cathode (17). The separator is made of high wet strength paper which is ionically permeable and has a low ohmic resistance. The separator (19) is sufficiently strong and dense to prevent any anode or cathode particles from invading the cathode or anode by penetrating the separator (19). Suitable for this purpose is a paper such as Flex Pak Crepe (Hollingsworth and Vose Company of East Walpole, Massachusetts). It is preferred that the separator (19) be formed of double layers of paper, each of which is 0.0075 inch thick.

The cathode assembly (20) is now ready to be inserted into the anode assembly (10), but first a lower seal (21) is positioned underneath the cap (18). The lower seal (21) extends beyond the cap. The cathode assembly (20) is then applied to the anode (10) and the lower seal (21) limits the insertion of the cathode by resting upon the top of the vertical portion (15) of the anode. The lower seal (21) is provided with a passage (22) and is arranged in registration with the slot (16) left by the mold which has formed and compressed the vertical portion (15) of the anode.

Through the hole (22) in the lower seal (21), the reserved electrolyte previously prepared is injected into the interface between the liner (12) and the separator (19) until thorough saturation has been achieved. When sufficient electrolyte has been added to accomplish this purpose, the aperture (22) is sealed by a suitable inert seal such as an epoxy resin, and an upper seal (23) is applied. The upper seal (23) is an annular member having an upstanding flange (24) which abuts on the side of the cap (18).

This seal (23) is made of dielectric material such as polyethylene or polyvinyl chloride. The lower seal (21) is likewise made of polyvinyl chloride. The cap (18) is made of an electrically conductive material such as stainless steel. The cap is provided with an axial pin (25) of 0.075 inch in diameter which penetrates into the cathode. The distance of penetration is preferably 1/2 of the height of the cathode, although this may vary with differing sizes and configurations of the cell. On top of the upper seal, a steel ring (27) is provided to fill the space between the flange (24) and the jacket (11). The top (26) of the jacket (11) is formed or rolled over to retain the ring (27) in place and the assembly is complete.

It is believed that dense mat of long filaments throughout the cathode, in combination with the shorter filaments and the powder coact to conduct the electrical energy so generated through readily accessible, highly conductive paths extending into each and every corner of the cathode, providing throughout the structure a continuing of conductivity that is not achieved in conventional constructions.

Manganese Fibers for Self-Supporting Electrodes

D.V. Louzos; U.S. Patent 3,685,983; August 22, 1972; assigned to Union Carbide Corporation describes stable, high surface area manganese fibers having a central spine portion with poly-directional side growths. The fibers are prepared by the electrolysis of a soluble Mn salt-containing electrolyte solution under conditions of high cathode current density. Galvanic cell electrodes are fabricated using the manganese fibers by compression molding techniques. The process for preparing the manganese fibers may be carried out at room temperature in a typical electrolysis cell using an anode suitably of manganese and a thin cathode, e.g., manganese, suspended in the electrolyte bath. The fibers electroform at the cathode and may be broken off and collected at the bottom of the bath, or if the fibers are not removed and the electrolysis is allowed to proceed, the fibers tend to electrodeposit in the form of an interconnected skeletal manganese fibrous mat which resembles "finger coral" in appearance. This interconnected skeletal fibrous mat consists basically of multiple fibers joined to one or more neighboring fibers throughout the mat. By the term "electroform" or "electroformation" is meant the production of manganese fibers by electrodeposition.

Galvanic cell electrodes can be readily fabricated using the manganese fibers by conventional compression molding techniques. The manganese fibers prepared as described above are placed within the mold and then compression molded to form an electrode compact of the desired size and configuration. In forming the compact, the interconnected skeletal manganese fibrous mats are preferably used. If the individual fibers are used, it is essential that they should be thoroughly intermingled when placed within the mold. When compression is applied, the fibers readily interlock or interknit producing a highly cohesive electrode body which is capable of supporting its own weight and retaining the shape in which it is molded.

Galvanic cell cathodes can be fabricated using an oxidized form of the manganese fibers. The manganese fibers may be first subjected to an electro-oxidizing treatment to convert the surfaces of the fibers to manganese dioxide. The oxidized fibers are then placed within the mold and compressed to form the cathode body in the same manner as described above.

The galvanic cell cathodes are most advantageously used in both primary and secondary alkaline dry cells. The principal advantage of such cathodes is that they are self-supporting and can be fabricated without the need for an electro-chemically inert binder for the active material such as employed in cathodes of the prior art.

Referring to Figure 1.2a, there is shown a photomicrograph of a typical interconnected, skeletal manganese fibrous mat under ten-fold magnification. As will be seen from the photomicrograph, the fibrous mat closely resembles the "finger coral" in appearance. It will be further seen from a close inspection of the photomicrograph that the mat is basically a skeletal structure of interconnected fibers joined to one or more neighboring fibers. Fibure 1.2b shows a photomicrograph of the same skeletal fibrous mat under 100-fold magnification. It will be seen from this photomicrograph that the individual manganese fibers consist essentially of a central spine portion with some polydirectional side growths.

Figure 1.2c shows schematically a typical electrolysis call for preparing manganese fibers in accordance with the process. The cell consists of an open tank (10) which is approximately 3/4 filled with a soluble manganese salt-containing electrolyte bath (12). Suspended in the electrolyte bath (12) is an anode (14), suitably a manganese rod. A cathode (16), e.g., of manganese, is dipped just below the surface of the electrolyte bath (12). In a practical cell, an array of multiple cathodes suspended within the electrolyte bath from a common bus bar may be used, there being only one cathade shown here for the purpose of illustration. The anode (14) and the cathode (16) are connected respectively through means of wires (18), (20) into an external circuit. The circuit includes a source of direct electrical current and means such as a rheostat for controlling the flow of electrical current through the cell.

FIGURE 1.2:   PRODUCTION OF MANGANESE FIBERS FOR CELL ELECTRODES

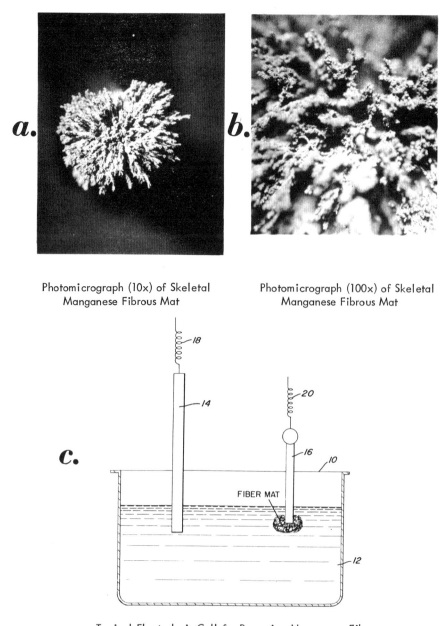

Photomicrograph (10x) of Skeletal
Manganese Fibrous Mat

Photomicrograph (100x) of Skeletal
Manganese Fibrous Mat

Typical Electrolysis Cell for Preparing Manganese Fibers

Source: D.V. Louzos; U.S. Patent 3,685,983; August 22, 1972

To carry out the electroformation process, the external circuit is closed suitably by means of a switch and electrical current is allowed to flow through the cell. Manganese is deposited at the cathode in accordance with the following reaction:

$$Mn^{++} + 2e^- \longrightarrow Mn$$

Essentially all of the electrical current flowing through the cell is utilized in forming the manganese deposit. From the earliest experimental work leading to the process, it was recognized that one of the essential requirements for carrying out the electroformation process is the maintenance of extremely high cathode current densities. It has been found that the cathode current density should be at least about 500 amp. per square foot. This is considerably higher than that used in the conventional electroplating art for depositing smooth coatings of manganese from a manganese sulfate bath where a cathode current density of from about 5 to 45 amp. per square foot has been reported (A Comprehensive Treatise on Inorganic and Theoretical Chemistry, vol. 12, p. 166, by J.W. Mellor).

Since the current density is inversely proportional to the cathode surface area for a given current, it is advantageous to employ a cathode of the smallest practical surface area exposed to the electrolyte and preferably a thin wire or rod cathode suitably of manganese is used. During the electroformation process, manganese first deposits at the cathode in the form of individual fibers which may be easily broken off and then collected at the bottom of the electrolyte bath.

If the process is allowed to proceed without removing the individual fibers, the electroformation will continue with more and more fibers being deposited from the initial growth at the cathode surfaces. More and more fibers will continue to grow in this manner so long as sufficient electrical current is flowing through the cell, and eventually an interconnected skeletal manganese fibrous mat will be formed. This skeletal fibrous mat consists basically of multiple fibers joined to one or more neighboring fibers in the mat.

Once the electroformation process has been started and the formation of the interconnected skeletal fibrous mat has begun, it may be necessary to periodically increase the flow of electrical current through the cell, such as by means of the rheostat, in order to meet the increased current requirements due to the increasingly greater number of fibers being deposited. It is virtually impossible during this period of the electrodeposition process to determine the cathode current density with any degree of accuracy due to the rapidly changing surface area of the manganese deposits.

However, the electroformation process may be expediently carried out by properly controlling the amount of electrical current flowing through the cell to provide an estimated cathode current density which is above the minimum requirement for the electroformation of the fibers. The proper range of cathode current density can be estimated simply by visual observation of the type of deposit or reaction occuring at the cathode. If the cathode current density is too low, no fiber deposit can be observed. The deposit in this instance will be of the level, adherent type or the powdery type. If the cathode current density is too high, gas evolution (hydrogen) will be readily observed.

To illustrate the process, manganese fibers have been prepared using an electrolysis cell similar to that shown in Figure 1.2c. The cathode was a single manganese chip with about 0.003 square inch of its surface in contact with the electrolyte. The anode was another manganese chip having about 2 square inches of its area in contact with the electrolyte. The electrolyte was a saturated solution of manganese chloride in water and the anode-to-cathode distance was about 1.5 inches. The electrolyte was maintained at about room temperature. Upon closing the electrical circuit, a current of about 10 milliamperes flowed through the cell and manganese was observed to electrodeposit at the cathode surfaces in the form of individual fibers. The fibers formed initially at the high current density edges of each cathode and could be easily broken off immediately as they were formed by scraping the surfaces of each cathode, the fibers then falling to the bottom of the electrolyte bath.

When the electrolysis was allowed to proceed without removing the fibers, more and more fibers were observed to electrodeposit from the surface of the fibers initially formed at each cathode and this process continued with each of the fibers joining to one or more neighboring fibers until an interconnected skeletal manganese fiber mat was produced. Eventually, the weight of the fiber mat so produced caused it to be broken off from the cathode surfaces and the mat then fell to the bottom of the electrolyte bath. The process was continued to produce more fiber mat.

The manganese fibers prepared in the above example were relatively short fibers having a length of about 1/10 of an inch. Shorter fibers of about 0.01 inch in length have also been prepared. One advantage of the manganese fibers is that they possess a relatively high specific surface area as evidenced by the rather rough-surface or "fuzzy" appearance of the fibers shown in the photomicrographs of Figures 1.2a and 1.2b. Although the manganese fibers possess a high specific surface area, they are not so highly developed as to be pyrophoric and subject to rapid oxidation when exposed to the atmosphere.

The fibers are composed essentially of pure manganese metal. The electroformation process is accompanied by electropurification and the fibers so prepared are probably one of the purest forms of manganese obtainable within a reasonable economic framework. It has been found that manganese fibers prepared in the manner as described above and then oxidized suitably by an electrooxidizing treatment to convert the surfaces of the fibers to manganese dioxide can be advantageously used for fabricating galvanic cell cathodes for use in both primary and secondary galvanic dry cells.

Manganese Dioxide Production

In a process described by H. Yamagishi and M. Tanaka; U.S. Patent 3,702,287; November 7, 1972; assigned to Nippon Kokan Kabushiki Kaisha, Japan manganese dioxide is produced by direct current electrolysis of aqueous sulfuric acid solution of manganese sulfate under specified conditions. Mn^{3+} ions are produced and are hydrolyzed with the formation of MnO_2.

In the process, the raw material and aqueous sulfuric solution of manganese sulfate to be used as a circulating acid are introduced in a leaching tank to leach manganese component from the raw material. The manganese sulfate solution is supplied continuously to an electrolytic cell, and the DC electrolysis is carried out under agitation at a specified temperature. Mn^{2+} is oxidized electrolytically to Mn^{3+}, and is hydrolyzed gradually, and MnO_2 is deposited in the electrolyte. Slurry containing MnO_2 and Mn^{3+} ions is continuously discharged from the electrolytic cell, and the slurry is filtered by a filter to separate it into MnO_2 and the filtrate. The filtrate contains Mn^{3+} at high concentration together with H_2SO_4 and Mn^{2+}, the major part of which are returned to the cell, and a portion is heated in a heat-treatment tank, hydrolyzing Mn^{3+} to produce MnO_2, filtering it in a second filter, returning the produced MnO_2 to the electrolytic cell, which is used as seed, and the filtrate is passed to a second heat-treatment tank. In some cases the first filtrate is passed directly to the second heat-treatment tank. MnO_2 obtained on the first filter is passed to the second heat-treatment tank and (the solution containing the filtrate from the second filter which contains H_2SO_4, Mn^{2+} and whose Mn^{3+} concentration is decreased to about the equilibrium state by heating) is heated together with the first filtrate or after which filtering by a third filter to separate it into MnO_2 and the filtrate.

MnO_2 is subjected to treatments such as neutralizing and washing, drying, etc., and is made into a product. The final filtrate contains free sulfuric acid, Mn^{2+} and low concentration Mn^{3+}, but it is returned to the leaching tank, and is used for leaching manganese from the raw material as a circulating acid. Considered from the thermo-economical point of view, the heat required for heating of the filtrate containing high concentration Mn^{3+} after separation of MnO_2 discharged from the electrolytic cell can be utilized for heating of MnO_2 in the following heat-treatment tank, and the heat required for heating the latter may be utilized effectively for leaching manganese portion from the raw material in the next leaching tank.

Example 1:   Electrolysis —

Electrolytic cell: 4 liters
Electrode:
        Anode: Pure PbO_2 formed by electrodeposition
        Cathode: Graphite
Distance between electrodes: 9 mm.
Composition of electrolyte:
        Concentration of free sulfuric acid: 400 g./l.
        Concentration of Mn^{2+}:   20 g./l.
Temperature: 37°C.
Current density: 45 amp./dm.^{2}
Mean bath voltage: 3.7 v.
Raw material: $MnSO_4$ solution containing 120 g./l. as Mn^{2+}
Quantity of raw material supplied: 200 ml./hr., continuous feeding

Heating After Electrolysis —

Heating temperature: 95°C.
Heating time: 2 hours
Electrolyte used for heating contained 400 g./l. of H_2SO_4, 20 g./l. of Mn^{2+} and 0.3 g./l. of Mn^{3+}

Current Efficiency and Composition of MnO_2 Produced —

Current efficiency: 76.0%, 6 hours of electrolysis
Product:
        Total Mn: 58.6%
        MnO_2:   87.9%
        Combined water: 7.0%
        Pb: 0.009%
Crystal structure: γ-MnO_2

Example 2:   Electrolysis —

> Electrolytic cell, electrodes and the distance between electrodes are the
> same as in Example 1.
> Composition of electrolyte:
> > Concentration of free sulfuric acid: 500 g./l.
> > Concentration of Mn^{2+}:   16 g./l.
>
> Temperature: 45°C.
> Current density: 60 amp./dm.^{2}
> Mean bath voltage: 4.3 v.
> Material: $MnSO_4$ solution containing 110 g./l. as Mn^{2+}, obtained
> by leaching $MnCO_3$
> Quantity of raw material supplied: 240 ml./hr. continuous feeding

Heating After Electrolysis —

> Heating temperature: 70°C.
> Heating time: 3 hours
> Electrolyte used for heating contained 500 g./l. of H_2SO_4, 16 g./l.
> of Mn^{2+} and 9.5 g./l. of Mn^{3+}

Current Efficiency and Composition of MnO_2 Produced —

> Current efficiency: 71.2%, 24 hours of electrolysis
> Product:
> > Total Mn: 58.0%
> > MnO_2:   86.4%
> > Combined water: 9.8%
>
> Crystal structure: γ-MnO_2 and a small amount of α-MnO_2

Example 3:   As the result of discharge performance test by 10 ohm continuous discharge based on JIS K 1467–1965, it was found that MnO_2 produced by the process
the "duration of closed-circuit at a voltage of more than 1.0 v." is increased about
15 to 20% as compared with the conventional products, and that it has remarkably
excellent characteristics as depolarizer.

Discharge Performance

	Minutes*
Commercial electrolytic MnO_2	215
MnO_2 produced according to Japanese Patent	
No. 1696/66	200
MnO_2 produced according to this process	250

*Duration of closed-circuit at a voltage of more than 1.0 v.

Since the temperature selected is high such as 30° to 50°C., no cooling device is
needed for electrolyte, and it is easy to maintain the temperature constant. By
heating MnO_2, produced by the electrolysis of the electrolyte having the composition of the process, there were obtained a decrease in the quantity of combined water,
increase in purity of MnO_2 and a decrease in lower oxide. "The duration of closed-
circuit at a voltage of more than 1.0 v." is increased about 15 to 20% as compared
with the conventional products, thus MnO_2 having excellent characteristics as a depolarizer was obtained.

SEPARATORS

Paper-Polyvinyl Alcohol Paste

A process described by S. Yamamoto, J. Watanabe, S. Hosoi, M. Kuwazaki, A. Ota, T. Takata and J. Asaoka; U.S. Patent 3,655,449; April 11, 1972, assigned to Matsushita Electrical Industrial Co., Ltd., Japan involves improving the discharge performance, preservability and electrolyte leakage resistability of a dry cell and simplifying the production process by interposing between a positive electrode and a negative zinc electrode an integral separator layer composed of three layers consisting of a barrier membrane formed of polyvinylalcohol alone or having incorporated one or a plurality of materials having at least one of water absorbing property, water retaining property, swelling property and adhesive property with respect to an electrolyte liquid, a paper having excellent water absorbing and water retaining property, and a paste layer of a natural or synthetic paste.

The separator layer can sufficiently attain the purpose of separation in a thickness only 1/10 of that of the conventional paste type separator layer. Therefore, the amount of the positive electrode can be increased by an amount corresponding to the decrease in thickness and the discharge performance of the cell can be improved accordingly. The barrier membrane used in the separator layer is formed of a polyvinylalcohol having a polymerization degree of 1,200 to 1,600 and a saponification degree of 75 to 95 and capable of being dissolved into a pasty state during discharge of the dry cell, or of such polyvinylalcohol incorporating one or a plurality of materials having at least one of water absorbing property, water retaining property, swelling property and adhesive property with respect to an electrolyte liquid and also capable of being dissolved into a pasty state during discharge of the dry cell.

The materials to be incorporated in the polyvinylalcohol include colloidal silica, pulp, agar, polyacrylamide, hydroxypropyl cellulose, starch, wheat flour, corn starch, karaya gum, methylcellulose, carboxymethylcellulose, etc. These materials may be contained in an amount of 1 to about 50%. Such a barrier membrane can be simply produced by adding to a liquid polyvinylalcohol the abovementioned material or materials in powdered form or in solution in a solvent and shaping the resultant mixture into the shape of a film. The barrier membrane thus produced excels in water absorbing property, water retaining property, swelling property and adhesive property, and hence a large amount of the electrolyte liquid can be contained in the barrier membrane proper.

Therefore, the separator layer will create a very small internal resistance when incorporated in the dry cell. Further, since the diffusion of the ions through the barrier membrane takes place utilizing the water as the conductive medium which is present in the vicinity of the adjacent material, the diffusion velocity of the ions, for example, through a polyvinylalcohol barrier membrane containing 10% of colloidal silica is higher than 5 times that through a barrier membrane formed solely of polyvinylalcohol and, therefore, a sharp increase in zinc ion concentration at the paste layer caused by dissolution of the negative zinc electrode during discharge of the cell can considerably be alleviated.

The most important feature of the barrier membrane according to the process is that it is dissolved into a pasty state, during discharge of the cell, due to the increasing zinc ion concentration caused by the solution of the negative zinc electrode.

Furthermore, the barrier membrane can be caused to dissolve at any point of capacity, by changing the polymerization degree and the saponification degree of the membrane, the amount and the type of a material to be incorporated and the mixing ratio of the three components of the electrolyte liquid used in the dry cell, i.e., ammonium chloride, zinc chloride and water, if the mode of discharge of the cell is previously known. Once the barrier membrane has been dissolved into a pasty state, the resultant paste exhibits properties which are less inferior or even superior to those of the paste layer in the conventional paste-type separator layer, and the diffusion velocity of ions is further increased. Before dissolution, the barrier membrane of course serves its intended purpose sufficiently by preventing the transfer of the paste material and water from the paste layer to the positive electrode and the penetration of the depolarizer from the positive electrode to the paste layer.

Swellable Anion Exchange Compound

A process described by W. Krey; U.S. Patent 3,558,364; January 26, 1971; assigned to Varta GmbH, Germany is directed to a leakproof electric cell which comprises a positive electrode, a negative electrode, a depolarizer and a separator layer disposed between the negative electrode and the depolarizer where the separator layer comprises granules of at least one swellable anion exchange compound and may also consist of a mixture of granular swellable anion and cation exchange compounds. The following example illustrates the process.

Example 1: In this example an anion exchange compound was used which consisted of a styrene polymer that had been cross-linked with 0.5 mol percent divinylbenzene and contained quaternary ammonium exchange groups as the active groups. The exchange material had a particle size between 20 and 50μ. Five grams of this dry pulverulent compound were added to 5 grams of acetone and thereafter 4 grams of a 20% solution of a binding agent consisting of a butadiene acrylonitrile copolymerizate were introduced into the mass.

An electric cell was formed in a conventional manner. A small amount of the suspension formed as just described, that is 1 ml. was sprayed onto the interior wall of a rotating zinc cup which had a utilizable cylinder surface of 43 cm.^{2} and a cylinder thickness of 0.45 mm. After the rapid evaporation of the solvent, there was formed a uniform firmly adhering separator coating on the interior wall of the zinc cup.

To produce the completed cell, the bottom of the thus-coated cup was then covered with a disk of synthetic material and a depolarizer mass was directly pressed into the interior of the cup, or it could also have been inserted as a preshaped pressed body. The depolarizer body was then covered at its upper end with an annular disk of a synthetic fiber fleece whereupon a carbon rod was inserted into the depolarizer body. Thereby the depolarizer mass was firmly and uniformly pressed against the separator coating.

The cell was then closed in conventional manner by a packing and a terminal cap. The expansion space provided in conventional cells could be dispensed with in this case, with the exception of the provision of a relatively small space to permit accomodation of the depolarizer body which, within established tolerances, may have slight variations in size. The depolarizer used in this example had the following composition, all parts by weight. Manganese dioxide, 87 parts; carbon black, 13 parts; zinc oxide, 0.5 part; 30% concentration zinc chloride solution (density

1.29), 70 parts. The electric cell was completely leakproof even under conditions of great stress, for instance in case of an electrical circuit resulting in complete discharge. No liquid escaped from the element even under these particularly strenuous conditions.

Examples 2 through 4: The following examples will illustrate electric cells in which mixtures of anion and cation exchange compounds were used. The anion exchange compounds were the same as described in Example 1. The cation exchange compounds consisted of a cross-linked Na-polystyrenesulfonate. The following mixtures were used.

Example 2: 95% of anion exchange compound and 5% of cation exchange compound

Example 3: 80% of anion exchange compound and 20% of cation exchange compound

Example 4: 55% of anion exchange compound and 45% of cation exchange compound

The exchange capacity of the anion exchange compound in these cases was 4.4 mval. per gram and the exchange capacity of the cation exchange compound was 5.3 mval. per gram. All electric cells made with these mixtures showed more or less the same type of complete leak-proofness as exhibited by the cell described in Example 1.

The usual additional structure features which are necessary in conventional electric cells to prevent leaking, such as use of absorbent materials or use of steel or plastic envelopes can be dispensed with in the electric cells of the process. Likewise the volume of the expansion space for the electrolyte can be drastically reduced as already indicated in favor of more electrical capacity.

In addition there is still a further advantage when the cells are compared with the prior art elements in which a paper lining was used or which had a thickening agent of for instance starch and methylcellulose. The cells of the process have a capacity yield which is higher by about 15 to 25% in case of harsh continuous discharges. The reason for this is that during the discharge no electrolyte solution is withdrawn from the depolarizer body.

With the cell of the process the particular thickener or, rather, separator layer, the capacity yield, the short-circuit current and the storability of the cells are strongly influenced by the swelling properties and thus the degree of cross-linking and the exchange capacity of the exchange material that has been used; see Helfferich, "Ionenaustauscher", vol. I, pages 92 through 96. The criterion for the usefulness of an anion exchanger of the process may therefore be found, apart from the cross-linking degree which is not easily determined, in the water content $W°$ of the exchanger that is in equilibrium with distilled water and charged with Cl^- ions and also in the total weight capacity GK. The water content $W°$ is defined as gram water per 1 gram dry weight of the exchanger that has been charged with Cl^- ions. The total weight capacity GK is equal to the quotient as follows:

$$\frac{\text{the reagent consumed during titration in mval.}}{\text{dry weight of specimen in grams}}$$

All anion exchange compounds which have high swelling properties and are not easily oxidizable may be used. As already stated, the preferred compound is based on a cross-linked polystyrene matrix. The water content indicating the swelling properties of the preferred material should be W° no less than 3 and the weight capacity GK should be no less than 2 mval./g. This requirement is met by the described anion exchange material consisting of styrene and divinylbenzene with ring-attached substituents comprising quaternary ammonium compounds where the degree of cross-linking amounts to less than 3 mol-percent divinylbenzene.

Fused Paste Powder and Thermoplastic Resin

J. Watanabe, S. Hosoi, M. Kuwazaki, T. Sawai and H. Ueno; U.S. Patent 3,513,033; May 19, 1970; assigned to Matsushita Electric Industrial Co., Ltd., Japan describe a dry cell comprising a zinc casing serving as the negative electrode, a cathodic mixture disposed in the zinc casing, a separator interposed between the zinc casing and the cathodic mixture, and an electrolyte retained in the separator. The separator is composed of a paste powder and a thermoplastic resin powder which will form, upon being fused with heat, a net structure capable of retaining the paste powder, and the electrolyte is retained by the paste powder gelatinized within the net structure.

In a preferred form of the process, the separator is formed of a mixture of a paste material consisting of a cellulose derivative powder, such as methylcellulose or carboxymethylcellulose, and a thermoplastic resin powder, such as polyethylene powder, powder of polyethylene-polyvinyl acetate copolymer or polyvinyl acetate powder, which has a melting point of 90 to 180°C. and a size of 80 to 350 meshes. The mixed powder is melt-bonded to the inner surface of a zinc casing, by filling the zinc casing with the mixed powder and heating the zinc casing to a temperature higher than the melting point of the thermoplastic resin, whereby a net structure of thermoplastic resin is formed with the paste powder retained.

Methylcellulose or carboxymethylcellulose is a paste material which dissolves or swells on absorption of water, without being heated. The powder of polyethylene, polyethylene-polyvinyl acetate copolymer or polyvinyl acetate is insoluble in water or electrolyte and stable chemically. This thermoplastic resin is preferably mixed with the paste powder in an amount of 50 to 80 parts per 50 to 20 parts of the latter, in view of the fact that an amount greater than that specified will result in a high internal resistance of the separator, whereas an amount smaller than that specified will result in an insufficient amount of the paste powder being retained by the separator.

The mixture of the paste powder and the thermoplastic resin powder can be attached to the inner surface of the zinc casing by previously heating the zinc casing at a temperature higher than the melting point of the thermoplastic resin powder, e.g., a temperature higher than 180°C., filling the zinc casing with the mixed powder and retaining the same for 2 to 3 seconds and then removing the excess mixed powder from the zinc casing, or alternatively by filling the zinc casing with the mixed powder, heating the zinc casing at a temperature of 90° to 180°C. for 5 to 15 seconds and removing the excess powder not having been attached to the inner surface of the zinc casing from the zinc casing.

Upon contacting the heated inner surface of the zinc casing, the thermoplastic resin powder in the mixed powder is softened and the individual particles of the resin are melt-bonded with each other forming a net structure. Simultaneously, these particles are bonded to the inner surface of the zinc casing at the sides which are in contact with the surface of the zinc casing. The paste powder and unmelted particles of the thermoplastic resin powder are retained in the net structure thus formed, whereby a separator is produced. Immediately after the mixed powder is attached to the inner surface of the zinc casing, the mixed powder is rolled under pressure or heated again, so as to eliminate the pin holes and thereby produce a separator of thin, uniform thickness.

The rolling of the mixed powder, attached on the inner surface of the zinc casing, under pressure or reheating of the same causes the fear of the powder being formed into a film or the meshes in the net structure being reduced to an excessively small size, thereby degrading the electrolyte absorbing and retaining property of the final separator. Such fear may be eliminated entirely by adding to the mixed powder 10 to 20% of a 80 to 150 mesh pulp powder having good liquid-absorbing property as an electrolyte retaining material. Besides the pulp powder, a powder processed starch, such as cross-linking starch, or a powder of natural starch, such as corn starch, wheat flour or glutinous rice, may also be used as an electrolyte retaining material.

Although the separator can be formed on the inner surface of the zinc casing by using a thermoplastic resin having a melting point higher than 150°C., the melting point of the thermoplastic resin used is preferably not higher than 150°C. so as to avoid deterioration of the paste powder. The zinc casing serving as the negative electrode may be heated in a heating drier or the like, or by a steel which is contacted with the outer surface of the zinc casing. When the zinc casing is molded from zinc pellets by impact molding, the temperature of the molded zinc casing drops from the molding temperature of 220° to 260°C. to a temperature of 140° to 150°C. about 15 to 25 seconds after molding, so that this residual heat in the molded zinc casing may also be used.

In order to attach the mixed powder uniformly on the inner surface of the zinc casing, it is essential to effect pouring and removal of the mixed powder into or from the zinc casing smoothly as well as to impart a good fluidity to the mixed powder. For this purpose, it is recommended to add to the mixed powder from about 0.1 to 3% of a fine silicon dioxide powder having a particle size of 5 to 50 mμ. By the addition of such silicon dioxide powder, it is possible to improve the fluidity of the mixed powder, to effect the operation of pouring and removing the mixed powder into and from the zinc casing smoothly and thereby to attach the mixed powder uniformly on the inner surface of the zinc casing.

Besides the addition of silicon dioxide powder, a uniform layer of mixed powder may also be obtained on the inner surface of the zinc casing by using a thermoplastic resin, e.g., polyethylene, copolymer of polyethylene and polyvinyl acetate or polyvinyl acetate, in the form of a short fiber, preferably a short fiber of 0.1 to 0.5 mm. in length, whereby the mixed powder can be attached on the inner surface of the zinc casing evenly.

The cathodic mixture to be inserted into the zinc casing, with the mixed powder attached to the inner surface, is composed primarily of manganese dioxide, acetylene black, graphite and ammonium chloride, and molded into a columnar shape with a carbon rod embedded centrally thereof, the carbon rod serving as positive electrode. For producing a dry cell using a cathodic mixture having a carbon rod embedded in the center, the electrolyte composed of ammonium chloride, zinc chloride and water is poured in a zinc casing having a separator formed on the inner surface by attaching the mixed powder, and the cathodic mixture is inserted into the zinc casing forcing the electrolyte to rise between it and the separator and thereafter dissolving or swelling the paste powder in the separator.

A mixture of 50 parts of manganese dioxide, 5 parts of acetylene black, 5 parts of graphite, 3 parts of zinc chloride, 17 parts of ammonium chloride and 20 parts of water is molded to form a columnar cathodic mixture having a diameter of 28.0 mm., a height of 42.0 mm. and a net weight of 48 g. On the other hand, a mixture of 30 parts of methylcellulose powder, 20 parts of polyvinyl alcohol powder, 10 parts of pulp powder and 40 parts of polyethylene-polyvinyl acetate copolymer powder, is filled in the zinc casing and heated at 150°C. for 10 seconds.

After removing excessive mixed powder, the zinc casing is again heated at 150°C. for 10 seconds and the layer of the mixed powder on the inner surface of the zinc casing is rolled to form a separator of 0.3 mm. in thickness, 2.7 cc of electrolyte composed of 15 parts of ammonium chloride, 15 parts of zinc chloride and 70 parts of water is poured into the zinc casing and then the cathodic mixture is inserted into the zinc casing, whereby a UM-1 type dry cell is produced.

The dry cell produced in the manner described and the so-called paste-type dry cell which is of the same size as the former but in which the paste layer is thick and accordingly the diameter of the cathodic mixture is about 10% smaller than that in the former, were connected to a resistance of 4 ohms and discharged at 20°C. for 30 minutes per day and 6 days per week, until the voltages dropped to 0.85 volt. The relationships between the accumulated discharge time and the voltage, on the respective dry cells are shown in Figure 1.3. In Figure 1.3, curve (D) represents a discharge curve of this dry cell and curve (E) represents a discharge curve of the paste-type dry cell. It will be seen from the characteristics shown in Figure 1.3 that the dry cell according to this process is superior to the conventional dry cell in respect of discharge capacity.

FIGURE 1.3:   INTERMITTENT DISCHARGE CURVES FOR DRY CELLS EMPLOYING DIFFERENT SEPARATORS

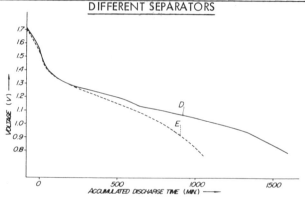

Source: J. Watanabe, S. Hosoi, M. Kuwazaki, T. Sawai and H. Ueno; U.S. Patent 3,513,033; May 19, 1970

Thick Wall Paste Separator

In a process described by B. Schumm, Jr.; U.S. Patent 3,615,859; October 26, 1971; assigned to Union Carbide Corporation substantial reduction in the formation and movement of spew in a Leclanche dry cell is obtained by the provision of a thick wall gelatinous electrolyte paste separator interposed between the consumable zinc anode and the cathode mix cake without seriously reducing the service life of the cell.

Basically, the process is predicated on the findings that if, contrary to previous practice, the gelatinous electrolyte paste separator wall thickness is significantly increased, the formation of spew can be substantially reduced without seriously reducing the service life. It has been found that the use of a thick wall paste separator brings about a distinct change in the character and location of the by-products of the cell reaction while at the same time increasing the electrical efficiency of the cell.

Typical of the process, D-size Leclanche round dry cells are provided with a gelatinous electrolyte paste separator having a wall thickness of about 0.125 inch or more whereas prior dry cells of the same size employed a paste wall thickness of between about 0.06 to 0.09 inch. In further contrast, other D-size round dry cells of the prior art have employed thin paper or film separators having a thickness of only about 0.005 inch.

SEALING TECHNIQUES

Two Separate Exudate-Impervious Plastic Zones

A process described by S.J. Angelovich; U.S. Patent 3,694,267; September 26, 1972; assigned to P.R. Mallory & Co., Inc. relates to electrochemical cells which are widely utilized in flashlights, portable radios and electronic devices, photoflashes and other devices. A common problem that has been encountered in the use of these cells is that of leakage of liquid. This problem is a particularly vexing one since such liquid leakage is corrosive and damaging to the battery itself as well as to the device in which the battery cells are used.

This process comprises the utilization of two separated plastic zones created by upper and lower layers of exudate-impervious plastic materials separated by an exudate-impervious adhesive-backed plastic disc. An annular absorbent washer located in the lower zone traps and holds mobile liquids, vapor, gases and fluids present in that zone while the adhesive-backed plastic disc and the lower plastic layer prevent the absorbent washer from discharging any fluids outside of this zone onto the metal terminals of the battery as exudate where evaporation would create the unwanted deposits. The upper zone is established by the dual pressure clamping of the upper plastic layer between the negative terminal of the cell and an inner metal container wall and an outer metal container wall.

The function of the upper zone is to electrically insulate the outer can from the negative terminal plate and to seal the cell against the inward diffusion of contaminants and against the outward creepage of electrolyte exudate. Referring to Figures 1.4a and 1.4b, reference numeral (10) denotes the side wall of an outer cylindrical metal container can which extends vertically along the full length of the battery from the crimped upper portion (22). The anode collector assembly which is shown and described in great detail in U.S. Patent 3,663,301, will be discussed here only to the extent necessary to adequately describe the improvements associated with this process.

This anode collector assembly includes a nylon plastic insert cap (12) which closes the upper end of the inner metal cylindrical container can (18). This cap has a central axial bore into which is tightly fitted the nail-shaped shank (14) of an anode collector element of nickel-silver, which has a flat head (15) spaced above the top surface (46) of the plastic insert cap. To hold the insert cap in place in the upper portion of the cell, there is provided a shaped notched peripheral border rim (42) confined between a head constriction (44) in the wall (18) of the inner cylindrical metal can and between the flat open upper edge and end (35) of the inner can.

An annular washer (16) constituted by a liquid absorbent cellulosic material, such as "kraft" paper, is set on the flat open upper edge end (35) of the inner can wall. In order to accommodate the placement of this absorbent washer, the flat open upper edge end of the inner can is bent inwardly through approximately 90 degrees to point toward the anode collector element. This flat edge end is substantially perpendicular to both the inner and the outer metal can walls and is substantially parallel to both the central circular area (11) and the annular moat ring (33) of the negative terminal plate.

To lock the absorbent washer in place on the flat upper edge, the inner end of the flat upper edge (35) tapers to a point in a direction toward the anode collector element (14). Furthermore, opposite the annular moat ring there is a portion of the seated annular absorbent washer shaped in a concave downward curvature to correspondingly and securely engage the taper of the upper edge end. The washer has a central circular cut out portion being of a diameter great enough to permit the washer to fit around and to snugly contact the hub (48) of the anode collector assembly.

According to the prior art shown in Figure 1.4a a single layer of an insulating material (28) such as a paper tube extends continuously from the crimped upper end (22) of the outer steel metal can downward past portion (24) surrounding the outer surface of the inner steel metal can. It has been found that by having two separate plastic layers, a lower layer (40) and an upper layer (38) separated by an adhesive-backed vinyl plastic disc (34), that two leakage-stopping zones can be created in a tandem relationship. The first lower layer (40) of insulating plastic surrounds substantially all of the inner cylindrical metal can and is folded flat to surround part of the annular washer (16).

As shown in Figure 1.4b the length of the lower layer (40) and the length of the upper layer (38) need not be equal. It is only required that each extend in length from the inner surface of the outer can wall radially inward past the annular moat ring (33) and the flat open upper edge end (35) of the inner can wall, so that these two plastic layers, the annular washer, and the vinyl disc can be pressure clamped between the ring (33) and the open edge (35) to create two hermetically sealed zones. The materials constituting both the upper and lower layers are high strength, flexible, liquid-impermeable plastics such as polyethylene, polyvinylidene, polyvinyl chloride, polypropylene, polystyrene and mixtures. Positioned between these two layers is an adhesive-backed vinyl disc (34) simultaneously engaging the lower plastic layer (40), the absorbent washer (16) and the top surface (46) of the plastic insert cap with the disc positioned between the top surface of the insert cap and the lower surface of the flat head (15) of anode collector element.

FIGURE 1.4:   LEAKPROOF CLOSURE FOR BATTERY

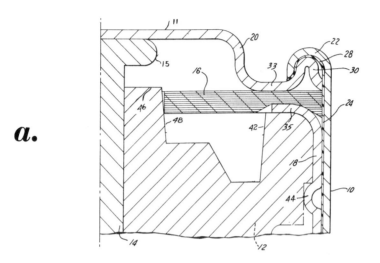

a.

Prior Art Cell Design

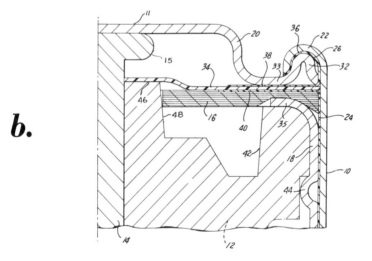

b.

Improved Leakproof Design

Source: S.J. Angelovich; U.S. Patent 3,694,267; September 26, 1972

Thus, the adhesive-backed vinyl disc extends radially from the shank portion (14) of the anode collector element to the inner surface of the outer cylindrical metal can wall in order to establish a first leakproof zone. The anode terminal of the battery serves a dual function in providing electrical contact to an outside circuit and in providing in combination with the flat upper end the pressure sealing necessary to establish the two leakproof zones. This preformed negative end terminal plate is made of steel and has a central circular area portion (11) with a downwardly dependent wall (20) having an annular moat ring (33) connected to an encircling pinched rim (30) in Figure 1.4a and (32) in Figure 1.4b.

The upper wall (22) of the outer can (10) is crimped downwardly over the pinched rim of the negative terminal to trap therebetween a portion (36) of the upper layer of plastic (38), so as to electrically insulate the outer can from the negative terminal plate and so as to seal the cell against the inward diffusion of contaminants and against the outward creepage of electrolyte exudate. As shown in Figure 1.4a the upper layer of plastic (38) extends radially linearly and horizontally from beneath the central circular area portion of the terminal plate outwardly past the downwardly dependent wall and past the annular moat ring; then layer (38) extends in an upward arc (26) to (36) around the pinched rim (32) in a direction back toward the central area in order to establish a second leakproof zone.

It is to be pointed out that in order to establish the two zone leakproof closure seal, that both of the plastic layers (38) and (40) are each subjected to a double pressure grip seal. The lower plastic layer is pressure gripped once between walls (10) and (18) and secondly between ring (33) and (35), while the upper plastic layer is pressure gripped once between ring (33) and end (35) and then secondly between crimp (22) and rim (32) with the applied pressure being maximum against portion (36) to prevent the ingress of contaminants and the egress of exudate due to shifting of the negative terminal plate from mishandling of the cell during shipping, installation, etc.

The flat head (15) of the anode collector element prior to assembling the cell is welded to the central portion (11) of the negative terminal plate which serves as a handle for manipulating this element. In the process of assembling together the component parts of the cell, annular rings of a positive cathode depolarizer material are comprised of manganese dioxide, mercuric oxide, or a combination of both, intermixed with a suitable inert and electronically conductive filler material such as graphite.

One or more annular self-supporting depolarizer rings are pressed from a mixture of 91% by weight electrolytic manganese dioxide and 9% by weight of graphite although other proportions can be used, for example 5 to 20% graphite. The dimensions of the pressed depolarizer are such that tight fit and good electrical contact is made with the inner surface of the side wall and the closed bottom surface of the inner metal can. The negative anodic electrode is made of amalgamated zinc powder which is sufficiently pressed to permit the insertion of anode collector element (14).

An absorbent separator material such as dexter paper and/or parchment paper is disposed within the cathode and this separator comprises a concentrically fitted electrically insulating porous lining, the center of which defines a central chamber into which is inserted the anode electrode, as shown and described in U.S. Patent 3,663,301.

This absorbent separator lining maintains the presence of a suitable electrolyte such as 35 to 40% KOH, 3.5 to 6.5% ZnO, the balance water, in contact with the electrodes while providing a suitable ionically permeable barrier material. After inserting the anode electrode into the central chamber, the plastic insert cap (12) is positioned within the inner metal can with the peripheral border rim (42) resting upon bead constriction (44) in wall (18), this bead having been formed by metal working wall (18) after the cathode rings had been inserted into the inner metal can.

After the nylon plastic disc (12) is seated on bead (44), the open upper edge end (35) of the inner can is bent inwardly to confine the peripheral border rim (42) and to form a compression seal within. Annular absorbent washer (16) is seated on the flat open upper edge end (35) with the hub (48) projecting above the washer and snugly contacting the washer. A lower layer of plastic tubing (40) is shrunk on over the absorbent washer and over the outer wall of the inner metal can.

An adhesive-backed vinyl disc (34) blanked from electrical tape is pressed onto the lower plastic layer, the absorbent washer (16), and the hub (48) with the adhesive surface in direct contact with hub surface (46). Since the shrinkable tubing (40) may be polyvinyl chloride, PVC, the adhesion between the disc and the PVC tubing is excellent in a caustic medium. Alternatively the tape layer (46) may be laminated to the absorbent washer (16) before the washer is inserted onto the upper edge (35). The concept of shrinking the PVC plastic refers to stretch pressing the plastic around another structure of greater diameter such that the elastic nature of the plastic will establish an impermeable seal.

The upper layer of plastic (38) is then laid on top of disc (34). At this point the shank (14) of the nail-shaped anode collector element is inserted into the central axial bore of the insert cap (12) and forced into the anodic electrode to make contact therewith. The preformed negative end terminal plate being welded to the flat head (15) is used to manipulate the collector element during this insertion. Finally, the annular moat ring (33) will rest upon the upper surface of the upper plastic layer (38), which layer is placed in tension by extension of portion (38) radially linearly from beneath the moat ring outwardly past the pinched rim (32) and then by extending it in an upward arc to point (26) where the rim is again contacted.

The upper layer is then folded around the pinched rim in a direction back toward the central area (11), with extra material being utilized to maintain the tensile forces while the outer metal can (10) is slipped over the upper and the lower plastic layers and then while the upper end (22) of tube (10) is crimped downwardly over the pinched rim of the negative terminal to trap therebetween a portion of the negative end terminal plate. This crimp of end (22) over rim (32) not only holds portion (36) in a compression grip but also holds elements (38), (34), (40), and (16) in a compression grip between ring (33) and end (35) to create respectively the upper and lower leakproof sealing zones.

When a group of test cells were constructed according to the process and subjected to a temperature cycling test for accelerating leakage, these cells went 40 to 80 days with no leakage evident on the exterior. Prior art batteries lasted a maximum of 4 days before leakage appeared.

Further similar design considerations for leakproof primary cells are described by R.E. Ralston and Y.L. Ko; U.S. Patent 3,660,168; May 2, 1972; and U.S. Patent

3,663,301; May 16, 1972 both assigned to P.R. Mallory & Co., Inc.

Plastic Diaphragms and Closures

R.J. Bosben, R.H. Feldhake and P.J. Spellman; U.S. Patent 3,617,386; Nov. 2, 1971; assigned to ESB Incorporated describe a cylindrical cell assembled in a metal case which has a metallic terminal and cover piece insulated from the metal case. A flexible, nonconductive diaphragm member is located beneath the cover piece and its lip forms the insulating means between the metal case and the cover piece. A metallic contact member is mounted upon and traverses the diaphragm and provides an electrical path between one of the cell electrodes and a flexible metallic member located between the diaphragm and the cover piece.

The flexible metallic member provides an electrical path between the contact member and the metallic terminal and cover piece. The shapes of the flexible member and of the flexible diaphragm are such that when the diaphragm is pressed outward by internal gas pressure, the contact between the flexible metallic member and the metallic contact member mounted on the flexible diaphragm is broken. The flexible diaphragm is chosen from a class of materials that are readily permeable to hydrogen but not readily permeable to water vapor.

The diaphragm is provided with a weakened section so that it will rupture at a pressure safely below the rupture strength of the metal battery case. By these means, a sealed battery design is provided that nondestructively protects the user against rupture caused by the slow evolution of hydrogen gas, and the rapid evolution of gas from electrochemical action. If, for other reasons, dangerous gas pressure should build up in the cell, the user is protected against violent cell rupture by the fracture of the weak portions of the diaphragm. These features are provided by a simple but reliable construction not requiring additional space in normal cell designs.

In Figure 1.5, (10) represents a metal container in which the cell parts are located. A tight fitting tube of positive depolarizer mix (11) is forced into container (10), makes electrical contact with it and forms the positive electrode. The depolarizer mix is comprised of compacted manganese dioxide and graphite powder. A separator tube (12) made of a porous material such as propylene felt wet with electrolyte is located within the bore of the positive mix. The space within the separator tube is filled with a suspension (13) of amalgamated zinc powder, alkaline electrolyte and a jelling agent such as carboxymethycellulose to form the negative electrode.

A metallic current collector (14) is embedded in the jelled zinc. It leads from the zinc component (13) to a first contact means (15), in this case a rivet passing through a flexible diaphragm (16), and sealing the opening. The current collector (14) is flexible and has one or more bends in it so that it does not hinder the motion of flexible diaphragm (16). The flexible diaphragm (16) has a reinforced lip (17) which encloses the circumference of cell terminal and cover (18). Lip (17) is compressed between a bead (19) and a flange (20) sealing the cell cavity and electrically insulating cover (18) from can (10).

The flexible diaphragm (16) is made of a material readily permeable to hydrogen but not readily permeable to water vapor, in this case nylon is used. A V-shaped circular convolution (21) is formed on diaphragm (16) to increase its area and flexibility. The convolution (21) penetrates into the end of separator (12) as is shown at (27) and serves to prevent zinc from bridging from negative to positive electrodes over the edge of separator (12). The cell terminal and cover piece (18) has one or more holes (22) so that the entire space (23) between diaphragm (16) and piece (18) is kept at atmospheric pressure.

A resilient metallic second contact means (24) normally touching near its center the first contact means (15) and at its extremities the cell terminal and cover piece (18) electrically bridges and completes the internal electrical circuit of the cell from first contact means (15) to cell terminal and cover piece (18). The second contact means (24) may be a single strip of spring material or it may take the form of a star having three or more arms. Preferred materials for this part are copper spring alloy or spring steel.

FIGURE 1.5:   SEALED CELL CONSTRUCTION

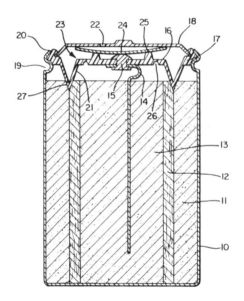

Source: R.J. Bosben, R.H. Feldhake and P.J. Spellman; U.S. Patent 3,617,386;
          November 2, 1971

The resilient second contact means (24) bears upon an annular projection (25) formed upon the surface of flexible diaphragm (16). The bearing point of projection (25) on the resilient second contact means (24) is located between the points where re-silient second contact means (24) touches the cell terminal and cover piece (22) and the first contact means (15) fastened to the diaphragm (16). The resilient second contact means therefore forms a lever of the third class. By this mechanism, the motion of the center portion of the second contact means in the vicinity of first con-tact means (15) will be greater than the motion of projection (15) on diaphragm (16). This insures that when diaphragm (16) moves upward from internal gas pressure a given distance, the resilient second contact means (24) will move a greater distance and separate from first contact means (15) thereby breaking the electrical circuit within the cell. Protruberances (not shown) on cell terminal and cover piece (18) and on annular projection (25) serve to position the resilient second contact piece (24) and prevent it from slipping out of its proper location. Groove (26) formed in diaphragm (16) provides a calculated weak point in the diaphragm. By making the groove concentric with the center of the diaphragm, a desirable degree of flexibility can be given to the diaphragm.

Under normal conditions, the small amount of hydrogen evolved by the zinc electrode (13) will diffuse through diaphragm (16). If through mischance the cell should be electrically connected so as to produce large amounts of gas by electrolysis, diaphragm (16) will flex upward forcing flexible second contact means (24) out of contact with first contact means (15) breaking the electrical circuit within the cell and preventing further formation of gas by electrical means. After pressure has been relieved by diffusion through diaphragm, the cell will again be useable. If, however, the cell should evolve gas in large quantity from a nonelectrical cause, such as being overheated, diaphragm (16) will rupture at weak area (26) before external damage can be done.

Thus, a safety feature for use on many forms of sealed cell has been described. It protects the cell against a slow pressure buildup from normal chemical action as well as rapid pressure buildup from electrical abuse without harm to the cell. It also protects the user from the dangers of cell pressure buildup from any other reason. Thus in this cell construction, a single member, namely diaphragm (16), serves as a gas diffusion membrane, a cell sealing member, a cell terminal insulator, a switch operating diaphragm, and a frangible safety diaphragm.

In related work, M.J. Terlecke; U.S. Patent 3,615,860; October 26, 1971; assigned to ESB Incorporated describes a dry cell construction having a one-piece plastic closure covering the open end of the negative electrode can which contains a depolarizer mix and a current collector centrally imbedded in the depolarizer mix. The plastic closure rests on the top edge of the negative electrode can and has a tubular projection extending from the bottom thereof through which the current collector passes.

The tubular projection covers the current collector throughout an airspace located above the depolarizer mix. The plastic closure contains at least one venthole extending from its top surface into contact with the airspace above the depolarizer mix and at least one groove in the top surface of the closure extending from the outer edge of the closure into contact with the airspace between the terminal cap on top of the current collector and the plastic closure. It is preferred that the plastic closure have a plurality of ventholes and grooves which are offset from each other. The cell construction of this process is particularly adapted for rechargeable dry cells which utilize a depolarizer mix containing an azodicarbonamide depolarizer material.

A process described by S. Ohki; U.S. Patent 3,556,859; January 19, 1971; assigned to Toshiba Ray-O-Vac Co., Ltd., Japan relates to a dry battery where a cylindrical member of plastic material housing a zinc can, a power generating component, filled with other power generating components is improved in its sealing construction, the plastic cylindrical member is expanded by the pressure of gases evolved in the battery thereby to prevent the electrical contact between the zinc can and anode seal plate from being loosened and also render the battery completely free from the leakage of the electrolyte. Complete details of construction are provided.

Synthetic Molded Electrode Casing

A process described by J.F. Jammet and F.M. Joyeux; U.S. Patent 3,575,724;
April 20, 1971; assigned to Societe des Accumulateurs Fixes et de Traction, SA,
France relates to a cylindrical primary dry cell of the type comprising a cylindrical
negative electrode cooperating with a depolarizing mass provided with a conductive
rod used as a positive electrode, the end of the rod protruding from the cell and
being optionally covered with a metal cap constituting the positive terminal characterized by the fact that the negative electrode constituted by a curved metal sheet
is coated on its outer surface except for an annular zone at one end with a synthetic
casing which latter also constitutes a bottom for the cell, the noncoated zone of the
electrode serving as the cell's negative output terminal. Advantageously the synthetic casing is molded over the negative electrode.

According to one form of the process, the noncoated zone of the negative electrode
is closely pressed against a plastic cover, thus hermetically closing the cell between
the rod and the negative electrode, as by means of a metal ring with an L-shaped
half-section. Advantageously, the metal ring is fixed in its place after mounting
by a diameter reducing operation, the part of the negative electrode on which it has
been applied having an outer diameter slightly smaller than that of the remainder
of the electrode.

According to a further characteristic of the process, the longitudinal edges of the
curved metal sheet constituting the negative electrode do not quite meet. Such a
form is particularly advantageous when the negative electrode is made of magnesium.
The manufacture of magnesium cups raises many problems since the impact extrusion
of magnesium is a difficult operation; graphite powder must be used in such operating
and thereafter be washed off, e.g., by means of acetic acid. However, when according to the process the negative electrode is constituted by a curved metal sheet
whose longitudinal edges do not quite meet, no extrusion operation is necessary.

Referring to Figure 1.6a, reference numeral (1) denotes a metallic sheet that has
been curved or shaped into generally cylindrical tubular form with open ends and
whose longitudinal edges (1a) and (1b) are unjoined and somewhat spaced apart.
This shaped tube (1) is intended to serve as the negative electrode of the cell and
may be of any suitable metal such as zinc, magnesium or their alloys. After an
eventual cleaning of conventional nature, the shaped tube (1) is mounted on the
central core (2) of an injection mold (M) as seen in Figure 1.6b.

After the mold (M) has been closed by cover part (3), the synthetic material (P) which
will constitute the casing of the cell by partly coating the outer surface of the negative electrode tube is injected into the annular spaces (S) and (B) within the closed
mold (M) through the injection nozzle (4).

The molding operation is shown diagrammatically in Figure 1.6b during the injection
of the synthetic material (P). It fills the spaces (S) and (B) forming a plastic coating
(S) on the major length of the tube (1), also spanning one end of the tube by portion
(B) so that according to the process the molded casing (5) includes also a bottom for
the cell and also covers the negative electrode (1) for its entire length except for a
zone (Z) at the other end of the tubular electrode (1). As shown in Figures 1.6c

and 1.6d, the synthetic material also fills the space or gap between the longitudinal edges (1a) and (1b) of sheet (1) on the whole height of the tube (1) at (5a), thus creating a resilient seal which enables the cell to bear some dilatations.

Figure 1.6c is a vertical sectional view taken through the filled gap between the edges (1a) and (1b) of tubular sheet (1). As seen in Figures 1.6e and 1.6f, the tubular electrode (1) coated with synthetic plastic casing (5) after removal from the mold is then provided with a tubular separator (6) and a centering shallow cup (7). Separator (6) as may be seen clearly in Figures 1.6e and 1.6g terminates short of the zone (Z), may be constituted by a rolled-up sheet already impregnated or coated with conventional electrolyte. The center ring cup (7) made, e.g., of paper or the like, is more especially intended to hold the separator (6) in its mounted position internally of tube (1).

Thereafter, a sausage-shaped depolarizer mass or mix (8) is inserted in the thus created space within the tube (1). A disk (10) perforated at its center and made, e.g., of cardboard or synthetic material is then positioned over the depolarizer mass. A conductive rod (9) of carbon is driven into the mass along its axis. This operation expands the mass outwardly, thus pushing it into intimate contact with the inner wall of separator (6) and further pushes separator (6) into intimate contact with the inner wall of the negative tubular electrode (1). After the carbon rod (9) has been driven home to the position shown in Figure 1.6e, a selected pressure is applied to disk (10) thus effecting a further compression of the depolarizer mass (8).

Subsequently, a plastic cover (11) (see Figure 1.6g) is mounted in the uncoated still open end of tubular electrode (1) to hermetically close the cell (C) between the rod (9) and the negative electrode (1). This cover (10) is provided with an annular internally extending flange (12) which closely engages the carbon rod (9) whose protruding end is then provided optionally with a metal cap (13) which will constitute the positive output terminal of the cell. The uncoated portion (Z) of the negative electrode (1) which was not coated with the synthetic plastic of casing (5) is then pressed against the side wall of plastic cover (11) as by means of a metal ring (14) which as shown in Figure 1.6g has an L-shaped half-section.

After mounting of the ring (14) according to the process, it is set and held in place by reduction of its diameter. As a consequence, the part (Z) of the electrode on which it has been applied possesses thereafter a diameter slightly smaller than that of the remainder of the tubular electrode. The plastic cover (11) is provided with at least one aperture (15) through which gases evolved in the cell during discharge may escape when they have reached a given overpressure in the cell. This aperture (15) constitutes a valve since air cannot come into the cell through it from outside. The lips (15a) and (15b) of this aperture are closed by the compression applied to the upper part of cover (11) by the ring (14) so that they can play the part of a valve, preventing ingress of air while permitting egress of gases evolved in the cell during use. It would also be possible to provide the cover (11) with a thinner wall on at least part of its top so that gases evolved in the cell could diffuse outwards on occurrence of undesirable internal overpressures.

The ring (14) which is in intimate contact with the uncoated zone (Z) of the tubular electrode (1) constitutes the negative output terminal of the cell and may be made, e.g., of tin-plated steel sheet. If then several cells are to be series- or parallel-connected, suitable connections can easily be welded or soldered to this ring (14).

This is very advantageous when the negative electrode is made of magnesium since welding or soldering is always difficult with magnesium.

FIGURE 1.6:   CYLINDRICAL PRIMARY DRY CELLS

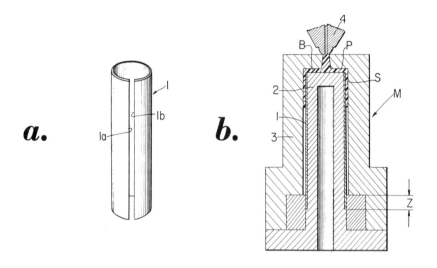

a.   b.

Negative Electrode   Molding Device for Synthetic Casing

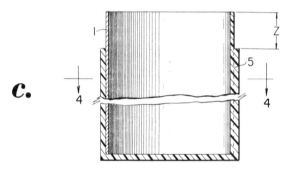

c.

Sectional View of Electrode and Casing
Along Line (3—3) of Figure 1.6d

(continued)

FIGURE 1.6:   (continued)

d.

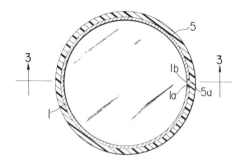

Sectional View of Electrode and Casing
Along Line (4—4) of Figure 1.6c

e.

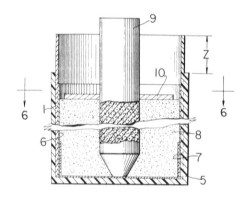

Sectional View of Cell After Filling

f.

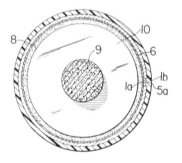

Sectional View Along Line (6—6)
of Figure 1.6e

(continued)

FIGURE 1.6:   (continued)

g.

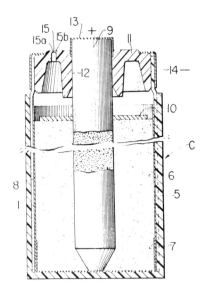

Sectional View of Completed Cell

Source: J.F. Jammet and F.M. Joyeux; U.S. Patent 3,575,724; April 20, 1971

In related work J.F. Jammet; U.S. Patent 3,627,586; December 14, 1971; assigned to Societe des Accumulateurs Fixes et de Traction, SA, France describes sealed electrochemical cells, preferably of cylindrical shape comprising a rodlike positive electrode surrounded by a depolarizing mix and separator enclosed within an overall cylindrically shaped cuplike negative electrode formed from sheet metal which is mechanically folded and bowed to provide a bottom and cylindrically shaped side walls having at least one longitudinal slot. A synthetic plastic casing is molded in situ about the negative electrode.

After assembly of the positive electrode, depolarizer mix and separator and other required cell components with this negative electrode bearing the synthetic casing, an insulative closure member is mounted on the assembly to seal the contents of the casing surrounded negative electrode. The positive electrode projects through this cover and optionally a metal cap covers the protruding end of the positive electrode. The bottom of the negative electrode is either directly left uncovered by the casing to constitute the negative terminal of the generator, or in the alternative, a metallic cup in electrical contact with the bottom is left partially uncovered by the casing to constitute the negative terminal. This cup may be ribbed, if desired. Complete process and assembly details are provided.

E. Karobath and W. Pulitzer; U.S. Patent 3,701,689; October 31, 1972; assigned to Firma Telephon- und Telegraphen- Fabriks-Aktiengesellschaft Kapsch & Sohne, Austria describe a primary cell of the kind having a leak-proof casing made of a thermoplastic resin which has a contact member that is embedded in the bottom of the casing to provide an external electrical connection to the negative electrode of the cell. The cell comprises a contact plate and a tubular conductor piece which is preferably cup-shaped with an inverted frusto-conical wall.

The tubular conductor piece is embedded in the casing bottom with the material of the casing extending through a series of perforations spaced around the wall of the conductor piece, one end of the conductor piece being in electrical contact with the negative electrode and the other end being bonded to the contact plate. The contact plate has an exposed face externally of the casing bottom with a diameter greater than the diameters at both ends of the conductor piece.

Heat-Shrinkable Packaging

A process of F. Przybyla; U.S. Patent 3,630,783; December 28, 1971; assigned to Mallory Battery Company of Canada Limited, Canada relates to dry battery cells and particularly to a dry cell in which an insulating jacket is used, to provide an economical structure for throwaway cells. In addition, a vent passage is provided for the cell to permit excess bubbles to exit from the cell and thereby avoid creating an undesirable pressure within the cell, that would cause bulging of the usual container that serves as one of the electrodes, and that would also disturb a predesigned and preestablished physical relationship and disposition between the components of the cell, appropriate for optimum operation.

In prior conventional-type dry battery cells, a can has been used as one electric terminal of the cell, and has served also as a container for the other elements of the cell, which generally included a centrally or coaxially disposed anode assembly with suitable electrolyte material separated by an absorbent and inert membrane material from a depolarizer material, whose mutual operation would oxidize hydrogen bubbles formed during operation of the cell.

With recent and current improvements in the structure of the anode, the depolarizer material fails near the end of discharge, and as a result, substantial amounts of hydrogen are evolved. Due to increased corrosion rates of the improved anodes, a larger amount of hydrogen is evolved on storage. Hydrogen evolved increases pressure in cells considerably and tends to deform cell containers towards the end of discharge. This process provides a modified cell construction in which a vent passage is provided in a simple way to permit excess bubbles to exit from the cell and thus avoid generating an excess pressure within the cell, that might otherwise disturb the physical relation between the elements, as predesigned and disposed in the cell during manufacture, for optimum operation.

In the process, the cathode can container for the elements of the dry cell is substantially closed by a separator top insulating cover disc, and either the top edge of the can or the bottom edge of the disc is provided with a vent or exit port for bubbles from the cell. That vent or exit port is arranged to communicate with an elongated exit passage leading from that vent or exit port to the ambient external air. The exit passage from the vent or exit port is located and defined between the outer surface of the cathode can and the inner surface of an outer plastic jacket of insulating material that is heat-pressed onto the can for tight physical peripheral contact and anchorage, except for a narrow elongated exit air passage defined along a filamentary-shaped element disposed between the can and the insulating jacket, at least along the length of the can from the exit port, or vent, at the top of the can, to a region near the lower end of the can. Such filamentary element serves as a spacer between the cathode can and the surrounding insulating jacket, and serves to define and locate two elongated regions immediately adjacent and parallel to the filament and running alongside the full length of that filament. Such a passage need not be very large transversely, in section, to permit gas venting of the cell, to permit exit passage of the bubbles formed in the cell.

In addition, however, this passage serves another function. It is sufficiently small in section, transversely, and also long, either along the length of the cell cathode can, or along a helical path around the can, to serve as a reservoir to hold enough of the exiting bubbles to permit those bubbles to shift in reciprocation, and to act as a moving seal according to the temperature and pressure within the cell, and thus serve as a shifting seal to hold the cell substantially closed to the ambient air, while at the same time permitting exiting of the excess bubbles from the cell, and at the same time preventing breathing that would otherwise permit any external air to enter the cell and interfere with its normal intended chemical operation.

An important feature of this process is that efficient venting is accomplished independently of the position of the cell. Pressure of hydrogen bubbles generated anywhere in the cell will increase above the sum of pressure in the venting passage and the small hydrostatic pressure of electrolyte; therefore, a pressure differential will drive gas bubbles towards the venting passage even at the "upside-down" position of the cell. In conventional practice, the cathode can is encircled with a tube of insulating material, such as paper, and an outer metallic tube is then applied to encircle the paper tube, and both are suitably crimped at top and bottom ends, to suitably crimp a closure element on the can, and to crimp top and bottom electrode elements to the can, to enable the battery cell to be readily applied to an external circuit.

The external metal tube and the two metal-crimping operations are expensive in the manufacture of a cell. An important object of this process is to provide a structure that eliminates the need for the outer metal jacket, as part of the closure elements of the cell, and thus also eliminate the concomitant handling operations for applying and crimping the metal jacket, by substituting a hear-pressed plastic jacket of suitable material to harden upon cooling and to anchor itself to the cathode can, with suitable heat-deformed end crimps, easily and readily formed and more economic than similar metal deformations.

As shown in Figure 1.7a, a cell (10) comprises a central axial negative electrode (12) shown as made up of several sectional elements of electrode material surrounded by a cylindrical body of absorbent or cellulosic or felt material (14) soaked with a suitable electrolyte consistent with the central anode material to establish desired appropriate chemical action for the battery cell operation.

A cylindrical barrier (16) surrounds and confines the paste material (14) and serves as a separator between the cathode and the anode regions of the cell. Separator (16) is made of relatively porous paper or other suitable insulating material such as a plastic with sufficient porosity to permit ionic transfer but not chemical reaction between the materials in the two regions of the cell, while at the same time separating the anode material (12) physically from a cylindrical body of depolarizing material (18) that is disposed concentrically around the separator (16) and in contact with a metal can (20) that serves as a container and cathode for the cell. As shown in Figure 1.7a, the external can (20) is closed at the bottom to provide a bottom floor (22), and is also coaxially depressed to provide a central button contact (24) to serve as an external contact terminal for the cell. For insulation purposes between the can (20), as cathode and the central anode elements (12), an insulating disc (26) is provided which serves both as a spacing separator between the anode (12) and the cathode can (20), and serves also as an insulating support for the electrolyte paste material (14), the separator (16) and the depolarizing material (18).

At the other end of the can (20), an insulating closure disc (28) is disposed and it serves to close the can (20) excpet for a small exit port or vent (30), shown at the upper left-hand side of the can in Figure 1.7a. That exit port (30) may be a small notch in the wall at the top end of the can (20), or it may be provided by a notch in the surface of closure disc (28). The top closure disc (28) serves also as a support for a terminal contact button (35) electrically connected to anode (12), to which the button is connected by a suitable element shown in the form of a compressible contact spring (34). The cell is made up of the foregoing components in which the button (24) at the bottom of the cell serves as one contact terminal of the cell, and the metal button (35) at the top of the cell serves as the other terminal of the cell for connection in an external circuit.

The cell can (20) is covered by a tubular plastic jacket (40), which is illustrated as a heat-shrinkable tube of plastic material. In addition, in order to provide a venting air passage for the cell, a filament (42), which preferentially is a plastic or plastic-coated metal, is supported between the cell can (20) and the external insulating jacket (40), to be held between those two cylindrical bodies when external jacket (40) is heat shrunk onto the metal can (20). When the external insulating jacket (40) is shrunk onto the can (20), the jacket and the can will engage in close surface-to-surface contact except for a narrow region defined along the filamentary spacing element (42), as illustrated and shown in Figures 1.7b and 1.7d. As shown, the filamentary element (42) serves to separate the outer jacket (40) from the can (20) by and along the two elongated vent passages (42A) and (42B) running lengthwise along the two opposite sides of the filamentary element (42).

The opening of the vent port (30), at the top of the can of the cell, communicates with the two elongated passages (42A) and (42B). Any bubbles passing out through the vent port (30) will move down through passages (42A) and (42B) to the outlet port (42C) of those two passages, where the bubbles generated in the cell will then pass out into the ambient air to be dissipated. As shown further in Figure 1.7a, the upper end of the outer plastic jacket (40) is heat-pressed over as a flange (44) to fit over the outer border of the top closure disc (28), and that flange (44) is bonded to the disc (28), by any suitable process, to provide a hermetic seal at the top of the can except for the intended vent (30) and the communicating exit passages (42A) and (42B).

Similarly, the plastic insulating jacket (40) is flanged at its bottom at (45), around the lower end of the can (20) for relatively close fit except for the outlet venting port (42C). Thus the jacket (40) is heat shrunk tightly onto the can (20) along the can body and at the top and bottom heat-pressed flanges of the jacket, except along the filament (42) and its two adjacent air passages. The jacket locks the top closure disc (28) in place to seal the top of can (20), except for the vent, as desired. This construction provides economy and eliminates many handling operations, as well as the

conventional outer metal jacket currently used to crimp and hold the top and bottom
terminal electrode caps for the battery cell.

FIGURE 1.7:   HEAT-SHRINKABLE PACKAGING FOR BATTERIES

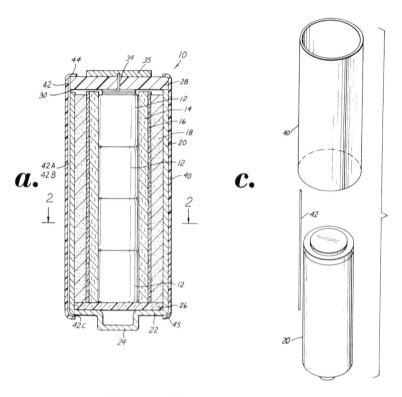

a.   *c.*

Vertical Section of Cell   Filamentary Element Unit Construction

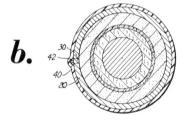   

b.   *d.*

Horizontal Section Along (2—2) of   Enlarged Sectional View of the
Figure 1.7a   Arrangement of the Filamentary Unit

Source: F. Przybyla; U.S. Patent 3,630,783; December 28, 1971

Tubular Electrode

F. Alf and W. Brill; U.S. Patent 3,657,019; April 18, 1972; assigned to Varta GmbH, Germany describe a primary cell which comprises a tubular member constituting a first electrode and having opposite open ends and a circumferential radially inwardly extending annular portion at a first of the open ends. A depolarizing agent is contained within the tubular member and electrically nonconductive sealing means are provided at the opposite open ends of the former to sealably close the depolarizing agent in the tubular member. A second electrode extends axially through the depolarizing agent and through the closure means at the second of the open ends and is attached to an electrically conductive member of this latter open end.

A second electrically conductive member overlies the sealing means at the first of the open ends of the tubular member and firmly engages both the sealing means and the circumferential radially inwardly extending annular portion at the latter open end to thereby fluid-tightly seal this open end. Figure 1.8a illustrates an electric primary element or cell according to the process which comprises a metallic tubular member (1) which constitutes a first or external electrode and has opposite open ends (1a), (1b) and a circumferential radially inwardly extending annular substantially bead-shaped portion (1c) at the open end (1b).

Contained within the interior of the tubular electrode is a depolarizing agent (2) through which a second or inner electrode (3), made essentially of carbon material, axially extends. Fluid-tightly closing the tubular electrode (1) within the confines of the upper end (1a), is an electrically nonconductive flexible sealing member (4), preferably made of a plastic material, and having a central passage (4a) through which the end portion (3a) of the inner electrode (3) projects outwardly.

The projecting end portion (3a) of the electrode (3) is connected with an electrically conductive cap member (5) which is substantially disc-shaped, overlies the sealing member (4), and is retained in position peripherally by a radially inwardly extending circumferential end portion (6a) of a dielectric jacket (6) which surrounds the tubular electrode (1). Sealably closing the tubular electrode (1) within the confines of the lower end (1b) thereof, is a substantially disc-shaped electrically nonconductive flexible sealing member (7) which lies across the end (1b) and has a peripheral flange-shaped enlargement (7a) which in the example shown in Figures 1.8a and 1.8b is substantially V-shaped but which, as shown in Figure 1.8c at (7b), may have a configuration other than V-shaped.

The sealing member (7) is pressed into the open end (1b) into abutting engagement with the end portion (3b) of the inner electrode (3) and has its flange-shaped enlargement (7a) abuttingly engaging the inner peripheral wall (1d) at the lower end (1b). Engaging the sealing member (7) in overlying relation therewith, is an electrically conductive disc-shaped end member (8) having a flanged periphery (8a) which, at the one hand, tightly abuttingly engages a seat (7c) of the peripheral enlargement (7a) of the sealing member (7) and, at the other hand, firmly interlocks with the radially inwardly extending circumferential annular portion (1c) at the lower end (1b) of the tubular electrode (1) to thereby fluid-tightly close the lower end (1b).

Surrounding the dielectric jacket (6) preferably of cardboard is a protective casing (9), which in the preferred embodiment is made of steel and includes radially inwardly extending circumferential end portions (9a), (9b) which respectively overly the end portions (6a) of the dielectric jacket (6). The tubular electrode (1) is constituted by a helically wound metallic band with the adjoining edges of each two adjacent windings of the tubular structure welded to each other. However, as an alternative, the tubular electrode (1) may be constituted by an elongated metallic band having a prerequisite length and which is folded transfersely to its elongation and has its adjoining elongated edges welded to each other.

FIGURE 1.8:   PRIMARY CELL CONSTRUCTION

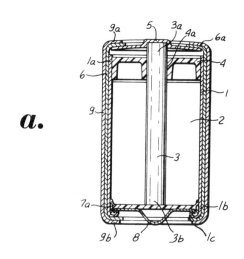

Cross–Sectional View of Cell

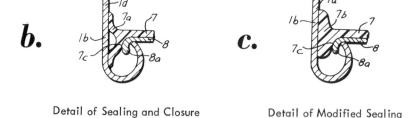

Detail of Sealing and Closure   Detail of Modified Sealing
Arrangement   Arrangement

Source: F. Alf and W. Brill; U.S. Patent 3,657,019; April 18, 1972;

In manufacturing the electric primary element described above, a mixture of a de- polarizer substance and a liquid electrolyte is subjected to extrusion and formed into a strand which, if required, is provided with an additional amount of electrolyte, for example, by spraying or by dipping the strand into an electrolytic bath. The mix- ture strand is sufficiently dried and provided with the tubular electrode (1) by helically winding a metallic band around the mixture strand and welding the adjoining edges of each two adjacent windings to each other in any suitable conventional manner. If a separation layer is required between the depolarizer (2) and the tubular electrode (1), a layer composed either of an ion permeable binder, or similar material, may be interposed between the depolarizer (2) and the tubular electrode (1) or may be applied at the outer surface of the tubular electrode (1).

Upon welding of the windings to each other, the thus formed combination is severed at a prerequisite length and one end of the tubular electrode (1) is closed with the sealing member (4) and the inner electrode (3) introduced into the depolarizer (2) with the end portion (3a) of the electrode (3) projecting outwardly through the opening (4a) of the sealing member (4). At this stage, the upper end (1a) of the tubular electrode (1) is fluid-tightly sealed. Pressure is applied to the depolarizer prior to and following closing of the upper end (1a) and introduction of the inner electrode (3) so as to compress the depolarizer into a compact substance.

Following the latter step, the flexible sealing member (7) and the electrically conductive end member (8) are placed within the confines of the lower end (1b) of the tubular electrode (1) with the flanged periphery (8a) of the end member (8) seated against the seat (7c) of the sealing member (7). Thereupon, the lower most end (1c) of the tubular electrode is circumferentially flanged into a radially inwardly extending annular bead which fluid-tightly interlocks with the flanged periphery (8a) and thereby simultaneously presses the flanged enlargement (7a) into abutting engagement with the inner peripheral wall (1d) of the tubular electrode (1) at the lower end (1b).

The above indicated sealing methods are only applicable for primary elements whose external electrodes are separated from the depolarizing agent by means of an ion permeable electrically nonconductive layer having a sufficiently high stability. For providing primary cells of the Leclanche-type with a closure and sealing arrangement according to the process, first the zinc tubular outer electrode is to be formed, then one end of the tubular electrode to be sealed and closed with the members (7) and (8), in the manner as described, subsequently the electrolyte is to be filled into the tubular electrode, and finally the pre-prepared carbon electrode is to be introduced into the electrolyte and the other end of the tubular electrode to be closed, as hereinbefore described.

With the above construction of the primary element a positively fluid-tight sealing of the bottom portion of the tubular electrode is obtained which resolves the usage of the previously superfluous and electrochemically unnecessary bottom of conventional cup-shaped primary elements.

Annular Projection in Can Bottom

In a process described by T. Tsuchida, K. Hirukawa, Y. Okajima and F. Iguma; U.S. Patent 3,573,990; April 6, 1971; assigned to Fuji Denki Kagaku Kabushiki Kaisha, Japan the structure of a battery dry cell includes a cup-shaped bottom member having an annular projection vertically projecting near its bottom circumference. The bottom member is fitted over a bottom surface of an anode zinc can. An annular gasket is fitted to the bottom member and a cavity in the gasket is fitted to the annular projection. A skirt portion of a plastic cylindrical jacket, which is bent along the gasket, is radially constricted against the gasket by means of a metal member.

The sealing effect of the dry cell structure, is not dependent upon the elasticity of the anode zinc can. The sealing effect always depends on the repelling power of the annular projection which tends to restore to the original state. Accordingly, a reliable constricting and sealing effect is obtained and maintained unchanged for a long period. The annular projection may be relatively short and, therefore, can be manufactured of only a little material.

The shape of the annular projection is variously alterable. Preferably, the shape of the projection is such that a root portion thereof is narrower than a head portion. The annular gasket, unlike the conventional O-ring, allows the annular metal member to be completely insulated from the anode zinc can and the cup-shaped bottom member.

The dry cell (Figure 1.9) comprises an anode zinc can (4) containing cell elements. A plastic cylindrical jacket (10), with top wall portion (14), seals an opening of zinc can (4). Top wall portion (14) is opened at its center to expose a cathode terminal (13) electrically connected with the cell elements. Side wall portion (15) is integral with top wall portion (14) and covers the side wall portion of zinc can (4). Bottom member (7) is fitted over the bottom surface of zinc can (4), the bottom member (7) having an annular projection (11) vertically projecting near its bottom circumference. An annular elastic gasket (8), with cavity (16) is fitted to projection (11). An annular metal member (9) is utilized to constrict the plastic cylindrical jacket.

As shown in Figure 1.9, the cell components include, within the anode zinc can (4) and inside of a paste layer (1), a cathode mixture (3) holding a carbon electrode (2) in the middle. Reference numeral (5) indicates a paper for holding the cathode mixture (3) in a fixed position, and reference numeral (6) indicates a bottom spacer of paper. The cup-shaped bottom member (7), having annular projection (11), has a root portion which is made narrower than the head portion and can be molded by pressing (stamping) a metal sheet. The cup-shaped bottom member (7) also serves as the anode terminal.

Bottom member (7) is fitted over the bottom surface of the bucket-shaped anode zinc can (4), so that the side cylindrical surface of the bottom member (7) may be on the same exterior surface level as the zinc anode can (4). Thus the plastic cylindrical jacket (10), covering the bottom member (7) and the anode zinc can (4), has no shoulder at their abutting portions. The annular projection (11) engages with a cavity (16) of the insulating elastic gasket (8), preventing the gasket (8) from falling off during assembly. The skirt of the plastic cylindrical jacket (10) is bent inwardly along the outer peripheral surface of the elastic gasket (8) and constricted from the outside by the metal member (9).

In Figures 1.9b and 1.9c, the metal member (9) is substantially U-shaped in cross-section and constricts the elastic gasket (8) radially by its opposing walls. The metal member (9) is depressed especially at (12). The metal member (9) is electrically insulated from the bottom member (7) by means of an inner annular tongue (17). In the second example of the process, illustrated in Figure 1.9d, the metal member (9) is substantially L-shaped in cross-section. According to this example, a part of the bottom member (7) is substituted for one of the opposite walls of the U-shaped member in the first example. The bottom member (7) has a vertical wall (18) opposite to a vertical wall of the annular projection (11), and the inner portion of the elastic gasket (8) is filled. The outer portion of the elastic gasket (8) is constricted radially by the L-shaped metal member (9). In both cases, a highly excellent sealing effect is provided without depending on the anode zinc can (4), since the elasticities of the annular projection (11) and the annular elastic gasket (8) cooperate with each other to retain the seal.

FIGURE 1.9:   SEALED TYPE DRY CELL

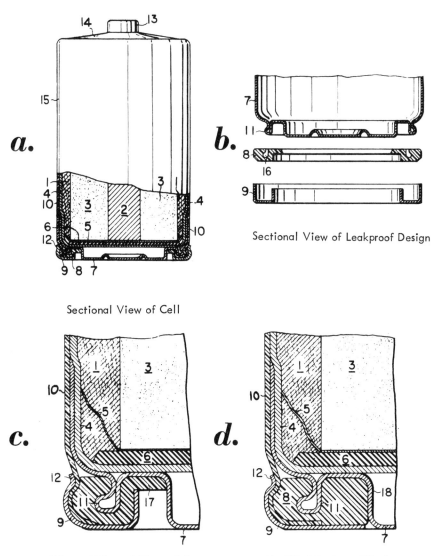

a.

Sectional View of Cell

b.

Sectional View of Leakproof Design

c.

Enlarged View of Figure 1.9a

d.

Modified Leakproof Design

Source: T. Tsuchida, K. Hirukawa, Y. Okajima and F. Iguma; U.S. Patent
3,573,990; April 6, 1971

Aluminum-Electrolyte Reaction Products

<u>S.J. Angelovich; U.S. Patent 3,457,117; July 22, 1969; assigned to P.R. Mallory and Co., Inc.</u> describes a leakproof sealing structure for electrochemical cells having an alkaline electrolyte in which a piece of aluminum is arranged in the leakage path of the electrolyte. The solid reaction products resulting from the contact of aluminum with the electrolyte dissipate the leaking electrolyte so that none of it can reach the exterior.

In alkaline dry cells of the general type described in Ruben Patent 2,422,045, considerable difficulties were experienced with electrolyte creepage. Even though the elastic insulating sealing grommet provided a generally air-tight seal for the cell, a small amount of electrolyte would frequently creep along the inner surface of the top closure disc and progressing around the edge of the disc would arrive to the top surface.

The first step toward solution of the problem of electrolyte leakage was made by introduction of the so-called "double-top" construction described in Williams Patent 2,712,565. In this construction the top closure member combined with the usual metal casing or inner can comprised a pair of metal plates or discs, the center portions of which were interfitting or were nested in each other while their edges were slightly separated. The marginal portions of the inner can and of the top closure discs had an insulating sealing member or collar of elastic material interposed between, with a portion of the sealing collar extending around the edges of the discs and being continued for a short distance.

An outer can or jacket encircled the inner can and was constricted at one end directly above the sealing collar to apply sealing pressure, thereby defining a substantially air-tight enclosure for the cell. In the assembled position, an intermediate portion of the sealing collar was forced to extend into the interspace between the edges of the two top discs and thereby further improved the air-tight character of the seal. For best results, the inner top disc was formed of, or was at least plated on both faces with a metal having a low potential with respect to the anode, such as tin for a zinc anode, whereas the outer top disc was composed, or was at least plated on both faces with a metal having good corrosion resistance to the electrolyte, such as nickel.

While the "double-top" structure represented a definite improvement over prior cell constructions and was used in the quantity production of alkaline dry cells, incrustations would still frequently appear on the outer top disc, particularly after prolonged storage. The latest major forward step in sealing structures was the so-called "bi-plate-molded grommet" double-top cell described in Clune Patent 3,096,217. This structure was based on the recognition that the difficulties experienced with conventional "double-top" cells were due to the interaction of minute amounts of electrolyte trapped in the interspace between the two top discs, with the dissimilar metals in such interspace.

In the usual form of "double-top" cell, the surface of the inner top disc was generally formed of tin, whereas the surface of the outer top disc was generally formed of nickel. Accordingly, two dissimilar metals, tin and nickel, were present in the two faces defining the interspace between the top discs. Thus, even extremely minute quantities of electrolyte leaking into the interspace, or trapped in such interspace during the closing operation at the time of manufacture, formed a galvanic couple between the outer (tin) surface of the inner top disc and the inner (nickel) surface of the outer top disc. As the couple was effectively short-circuited by the contacting center portions of the top discs, gas was produced by partial electrolytic decomposition of the electrolyte that was between the top discs, forcing the remainder of such electrolyte past the closure of the outer top disc to the outer surface of the disc. This alkaline electrolyte would react with the carbon dioxide in the atmospheric air forming alkali metal carbonates. As is known, these incrustations, in addition to their unsightly appearance, would cause corrosion, poor contact and other operating difficulties.

In the structure of the Clune patent, the inner surface of the inner top disc was plated with a metal having low potential with respect to the anode, such as tin, and the outer surface of the inner top disc was plated with the same metal that was plated on both surfaces of the outer top disc and having good corrosion resistance to the electrolyte, such as nickel. As the dissimilar metals formerly present in the interspace between the two top discs, tin and nickel, have thus been replaced with the same metal, nickel, no galvanic couple was formed.

Thus, no gas generation could take place in the interspace, eliminating the principal cause underlying the formation of incrustations. Further important advantages were obtained in the Clune structure by replacing the prefabricated sealing grommet by molding, preferably injection molding, the grommet around the top discs. This expedient prevented the trapping of electrolyte in the interspace of the top discs during the closing operation at the time of manufacture. Also, combination of the "double-top" structure with the grommet integrally molded around the top discs greatly increased the length of any leakage path from the inner surface of the inner disc to the outer surface of the outer disc.

As it appears from the above, substantial progress has been made in recent years in controlling electrolyte creepage, notably through the use of the molded grommet-double-top assembly described in the abovementioned Clune patent. Although many millions of satisfactory cells embodying the structure were manufactured, the problem of creepage still exists though on a statistically reduced scale. The creepage, which occurs in these cells generally manifests itself by the appearance of incrustations, most often on the negative terminals of the cells. These incrustations form when the alkaline electrolyte used in the cells has by-passed the cells' sealing mechanism and reacted with the carbon dioxide in the atmosphere. These incrustations severely mar the appearance of the cells.

In addition, they can corrode components, primarily contact elements, which are either in direct contact with or are close to the cells during their normal operating conditions. Thus, in this process, a suitable material such as aluminum, adapted to react with an alkaline electrolyte, is placed somewhere along the path by which the electrolyte must travel if the cell is to leak. A chemical reaction is then initiated at that point between the electrolyte and the aluminum. Once this happens, the electrolyte becomes dissipated, solid reaction products form and external cell leakage is prevented.

Referring to Figure 1.10a reference numeral (10) denotes a cylindrical metal can made of nickel-plated steel having a body of depolarizer compound (12) compressed in the bottom portion which compound may be composed of a major portion of mercuric oxide and/or manganese dioxide intimately mixed with a minor portion of graphite, or may be composed of silver oxide in its entirety. A depolarizer sleeve (14) having a slightly depressed flange portion (16) is lining the inner lateral surface of the can and may likewise be formed of nickel-plated steel. A microporous barrier layer (18), which may be constituted of a layer of Synpor is located on flange portion (16) of the depolarizer sleeve. An electrolyte absorbent spacer (20) of Webril (a nonwoven fabric made by blending thermoplastic fibers into a cotton web and applying heat and pressure), is resting on the barrier layer and its top is in contact with a porous pressed amalgamated zinc powder anode (22).

The top closure of the cell comprises a pair of nested top discs of which the outer one (24), is nickel-plated on both of its faces, whereas the inner disc (26) is nickel-plated on its outer face and tin-plated on its inner face, which is in pressure contact with the zinc anode. A circular disc (28) of thin aluminum foil, such as one having a thickness of 1 mil, is inserted between inner top disc (26) and outer top disc (24) prior to their being welded together at (30). To this welded assembly, a grommet (32) of a suitable elastomer, such as polyethylene, is injection molded.

The aluminum foil thus comes into intimate contact with the molding material and with the interior surfaces of the metal top discs. In assembling the cell, a metered quantity of electrolyte, such as an aqueous potassium hydroxide solution containing a substantial percentage of potassium zincate, is introduced into the cell. The described top closure structure is then inserted, so that upon crimping down the mouth portions of can (10), as indicated at (34), grommet (32) will be strongly compressed between flange (16) of depolarizer sleeve and crimped portion (34), thereby sealing the cell.

The operation of the cell of the process will be best understood by reference to the greatly enlarged fragmentary view shown in Figure 1.10b. Let it be considered first the manner in which electrolyte would travel were not the aluminum disc (28) present. The electrolyte would creep first along the interior surface of inner top disc (26) (Region A) and, going around the edge of the disc, continue along the outer surface thereof until it entered the space between the two top discs (26) and (24). From here, it would reverse its direction, would continue along the inner surface of the outer top disc (24), around the edge of the disc and along the outer surface, until it reached the atmosphere (Region B). This path is indicated by arrows in Figure 1.10b.

With aluminum disc (28) in place, the electrolyte comes into intimate contact with the aluminum "dam" prior to it ever reaching the atmosphere. The electrolyte used is an alkali metal hydroxide, preferably potassium hydroxide, which contains the zincate ion. The probable reactions which occur are as follows:

$$2Al + 3Zn(OH)_4^{=} \longrightarrow 2Al(OH)_4^{-} + 4OH^{-} + 3Zn$$

The zinc is thus deposited on portions of the aluminum by the above reaction. In addition, the aluminum reacts with the potassium hydroxide according to:

$$2Al + 2KOH + 2H_2O \longrightarrow 2KAlO_2 + 3H_2$$

The net effect is that the corrosive action of the leaking electrolyte takes place within the sealing mechanism, immobilizing the electrolyte in the form of solid reaction products. There is also an additional benefit derived, in that with the aluminum, the negative terminal becomes electro-positive and therefore tends to repel moisture from its surface. Comparative tests carried out with the cells have indicated that after ten weeks at 90% RH and 70°F., 50% of the conventional (control) cells leaked, while 0% of the sealing structures leaked. At 45% RH and 90°F., 67% of the control cells leaked, while 0% of the sealing structures leaked. Chemical indicators were used to determine leakage.

FIGURE 1.10:   LEAKPROOF ALKALINE CELL

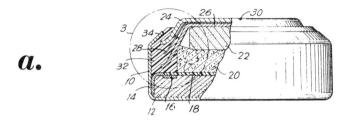

a.

Vertical Sectional View of Cell

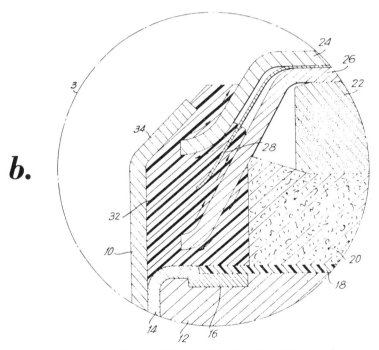

b.

Enlarged View of Section (3) of Figure 1.10a

Source: S.J. Angelovich; U.S. Patent 3,457,117; July 22, 1969

Seal Arrangements

In a process described by T.A. Reilly and W. Bemer; U.S. Patent 3,506,495; April 14, 1970; assigned to Union Carbide Corporation a primary dry cell of the type utilizing a noncorrodible jacket which is impermeable to liquid and having a locked metallic top and bottom closure, is provided with a seal arrangement for retaining liquid exudate formed on discharge while at the same time providing a continuous venting path for venting gas from within the cell. The seal arrangement is characterized in that it occupies a minimum amount of space within the cell.

Referring to Figure 1.11a, a primary dry cell of the process may comprise a cupped electrode (10) of a consumable metal (e.g., zinc) having therein a depolarizer mix (12), an immobilized electrolyte (14) and a porous carbon electrode (16) embedded within the depolarizer mix (12). Both the depolarizer mix (12) and the carbon electrode (16) may be suitably provided in the form of a conventional bobbin. Separating the bobbin from the bottom of the cupped electrode (10) is a conventional bottom insulator washer (18), suitably of cardboard or paper. Atop the washer (18) is a fibrous or paper cup (20) which fits around the bottom edges of the depolarizer mix (12).

Positioned within the upper open end of the cupped electrode (10) and so placed as to define a lower free space (22) above the depolarizer mix (12) and an upper free space (24) is an inner seal composed of a liquid impermeable top collar (26). The top collar (26) is fitted tightly within the upper end of the cupped electrode (10) and around the carbon electrode (16). The top collar (26) is permeable to gas but impermeable to liquid and serves as a liquid barrier, and may be made of paper or other fibrous material which is coated with a liquid-repellent material, for example, a plastic such as polyethylene.

Spaced above the top collar (26) and within the upper free space (24) is a liquid and gas impermeable seal washer (28). The seal washer (28) fits tightly around the carbon electrode (16) and rests on the upper peripheral edges of the cupped electrode (10), which peripheral edges are turned slightly inwardly as indicated at (30). The seal washer (28) is impermeable to both liquid and gas and is electrically nonconductive and may be composed of a suitable plastic material such as polystyrene. Mounted on top of the seal washer (28) is a layer (32) of a highly absorbent or bibulous material such as conventional blotting paper, for example. The seal washer (28) and absorbent layer (32) may be formed from a laminated sheet of polystyrene having adhered to one side a layer of blotting paper or other suitable bibulous material.

The top closure of the cell may comprise a one-piece metal plate (34). As clearly shown in Figure 1.11a, this top closure plate (34) is shaped to fit over the top edges of the carbon electrode (16) and has its outer peripheral edges locked in liquid-tight engagement with the noncorrodible jacket (36). Similarly, the bottom closure may comprise a metal plate (38) positioned beneath the cupped electrode (10) and having its outer peripheral edges locked in liquid-tight engagement with the jacket (36). The bottom closure plate (38) may also be provided with a central indentation as at (40) which makes electrical contact with the bottom of the cupped electrode (10). In the construction of the dry cell shown in Figure 1.11a, the top closure plate (34) is locked in liquid but not gas-tight engagement with the jacket (36).

As shown in Figure 1.11a, the juncture between the top closure plate (34) and the jacket (36) is positioned over the top of the seal washer (28) and absorbent layer (32) and just above the peripheral edges of the cupped electrode (10). By this construction, the top closure plate (34) is electrically insulated from the cupped electrode (10) and the seal washer (28) is firmly held in place. It will be noted that the seal washer (28) is slightly larger in diameter than the upper end of the cupped electrode (10) and that its outer edges abut tightly against the interior side walls of the jacket (36). Underlying the seal washer (28) in the space left by the inwardly turned edges of the cupped electrode (10) is a peripheral bead seal (42) of wax, for example. This bead seal (42) extends around the periphery of the cupped electrode (10) and the jacket (36). It will be seen that any electrolyte or exudate which may escape from within the cupped electrode (10), due to perforation of its side walls during discharge and which passes between the cupped electrode (10) and jacket (36) is barred by the bead seal (42) from contact with the top closure plate (34).

Underlying the seal washer (28) on the opposite interior side of the cupped electrode (10) is a bead seal (44) which extends around the inner peripheral edges of the cupped electrode (10). Similarly, an electrode bead seal (46) underlies the innermost edges of the seal washer (28) surrounding the carbon electrode (16). The combination of these bead seals (44), (46) underlying the seal washer (28) assures that the seal washer (28) effectively seals off the upper open end of the cupped electrode (10), preventing liquid from coming into contact with the top closure plate (34) of the cell.

Figure 1.11b shows in enlarged detail the multiple-ply, laminated tube structure of the jacket used in the dry cell. As shown, the jacket (36) comprises a first or innermost ply (48) of a high strength, liquid impermeable plastic material, e.g., polyethylene terephthalate, which is positioned adjacent to the outer side walls of the cupped electrode (10) and which serves as liquid impermeable barrier, and a second ply (50) of a thermoplastic material, e.g., polyethylene. Both the first and second plies (48), (50) of liquid impermeable and thermoplastic material constitute the first laminate in the multiple-ply, laminated tube. The second laminate of the tube comprises a ply (52) of thermoplastic material and three plies (54), (56) and (58) of a fibrous cellulosic material, such as kraft paper.

During discharge of the cell, both liquid and gas are formed as a by-product of the cell reaction. The gas normally follows a path from within the depolarizer mix (12) directly into the porous carbon electrode (16) or into the lower free space (22) above the depolarizer mix (12). The gas that passes into the free space (22) enters the carbon electrode (16) or portions of the gas may pass through or around the top collar (26) into the upper free space (24) below the seal washer (28) from whence it also enters the carbon electrode (16). Eventually all of the gas is vented from the cell through the carbon electrode (16) by way of the locked junction between the top closure plate (34) and jacket (36).

As indicated above, this locked juncture is made liquid but not gas-tight and is capable of venting gas from the cell. The venting path so provided is maintained continuously open and free from liquid blocking the passage of gas by the seal arrangement of the process. The liquid exudate passes into the lower free space (22) where it collects together with gas that is generated in the cell. Under normal conditions, the liquid exudate will be substantially confined within the free space (22) by virtue of the liquid barrier formed by the liquid impermeable top collar (26), but should the cell be subjected to severe conditions of use more liquid exudate may be formed then can be held within the free space (22).

FIGURE 1.11:   PRIMARY DRY CELL HAVING SPECIAL SEAL ARRANGEMENT

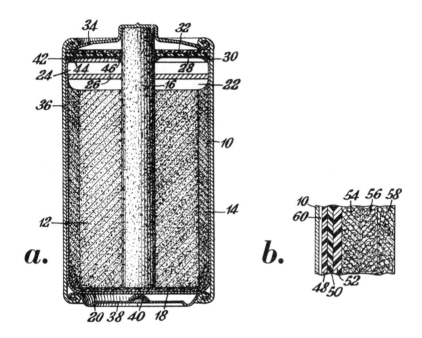

Sectional View of Primary Dry Cell   Enlarged View of Jacket Section

Source: T.A. Reilly and W. Bemer; U.S. Patent 3,506,495; April 14, 1970

The additional liquid exudate which is formed may cause the top collar (26) to become displaced upwardly so as to allow some of the liquid to pass into the upper free
space (24). This liquid which passes into the upper free space (24) is barred from
contact with the top closure plate (34) [or more importantly with the locked juncture
between the top closure plate (34) and jacket (36) through which gas is vented] by
the seal arrangement composed of the combined seal washer (28) and peripheral bead
seals (44), (46) formed respectively around the upper peripheral edges of the cupped
electrode (10) and the carbon electrode (16).

The seal washer (28) being both liquid and gas impermeable and being held firmly in
place by the top closure plate (34) as described above constitutes a permanent or
nondisplaceable liquid barrier which liquid cannot penetrate, while at the same time
the bead seals (44), (46) effectively prohibit liquid from passing around the edges of
the seal washer (28) even under the influence of gas pressure developed with the cell.
It should also be mentioned that in order to prevent the liquid exudate which collects
within the free space (22) from passing into the porous carbon electrode (16), thus
blocking the passage of gas, the carbon electrode (16) is preferably made of a fine
grain carbon and is suitably waterproofed in order to prevent liquid penetration.
A suitable waterproof treatment may be achieved by impregnation with a solution

containing a microcrystalline wax dissolved in a suitable solvent, such as ethylene dichloride. It will thus be seen that the seal arrangement effectively retains liquid exudate which might otherwise result in leakage from the cell but without at the same time taking up a large amount of space within the cell. The seal arrangement also provides a continuous venting path for venting gases so that no substantial gas pressure is allowed to build up within the cell.

In the event perforation of the cupped electrode (10) should occur due to normal consumption of the zinc during discharge, the liquid exudate that escapes from within the cell is prohibited from penetrating and becoming soaked into the jacket (36) by the innermost layer or ply (48) of liquid impermeable material, e.g., polyethylene terephthalate, which constitutes a liquid impermeable barrier. This liquid exudate will normally collect between the jacket (36) and the side walls of the cupped electrode (10).

For the purpose of retaining this liquid but without causing the jacket (36) to bulge it is generally good practice to fit the jacket (36) loosely around the cupped electrode (10) in order to provide a free space or exudate chamber for the liquid as generally indicated at (60) in Figure 1.11b. If the liquid exudate should collect within this space or chamber in any significant amount, the liquid may force its way towards the top closure, thus endangering the free passage of gas through the locked juncture between the top closure plate (34) and the jacket (36). It will be noted however, that that in this dry cell construction the peripheral bead seal (42), which is positioned between the jacket (36) and upper edges of the cupped electrode (10), bars the liquid from reaching the top closure plate (34) or more importantly its locked juncture with the jacket (36).

An important advantage of the dry cell jacket is that leakage of liquid exudate through the locked juncture between the top closure (34) and the jacket (36) due to so-called "wicking effect" is eliminated. In prior dry cell constructions where the jacket has been made of a fibrous cellulosic material, or even in those cells where a liquid impermeable barrier layer has been incorporated but as an intermediate ply in the jacket, the liquid exudate would normally become soaked into the first or innermost fibrous ply of the jacket and would eventually creep into or wick through the locked juncture, resulting in blockage of the venting path and even leakage from the cell.

In the construction of this dry cell, the jacket (36) incorporates a liquid impermeable barrier in the first or innermost ply (48) which the liquid cannot penetrate and additionally this first or innermost ply (48) actually froms a seal between the jacket and the top closure plate (34) as shown in Figure 1.11.

A process described by L.F. Urry; U.S. Patent 3,338,750; August 29, 1967; assigned to Union Carbide Corporation relates to leak-resistant dry cells of the type having a closed container surrounding the cell proper. The process provides an improved gas venting path in a leak-resistant dry cell particularly of the type of construction utilizing a noncorrodible container, which gas venting path is not prone to obstruction during use of the cell. In the construction of the dry cell a partition seal is placed between the closure and depolarizer mix of a leak-resistant dry cell defining a barrier which protects at least one of the electrodes of the cell from contact by liquid cell exudate and also providing a path for venting gas from the cell. Complete details of construction are provided.

Cell Casing

A process described by M.D. Kocherginsky, L.F. Penkova, V.A. Naumenko and S.L. Kalachev; U.S. Patent 3,607,429; September 21, 1971 provides a galvanic battery consisting of alkaline manganese dioxide–zinc cells, where each cell includes a plastic casing with current collectors which accommodates a cathode, two diaphragms disposed on both sides of the cathode, and two zinc powder anodes separated from the cathode by the diaphragms, the cell casing being made in the form of two shells within which the three electrodes are disposed.

As a feature of the process, each shell has provision on its edge periphery for an extension for insulating the negative and the positive current collectors from each other by forming a lap joint from the edge extensions of both shells when assembling the cell. Additionally, a bank of assembled cells are housed in a hermetic plastic bag to be contained within a cellulosic outer casing.

In the galvanic battery the electrodes' surface areas are twice as great as that of the known batteries and the discharge current will be correspondingly higher. When discharged across 15 ohms during 10 minutes daily until the cutoff voltage equals 2.25 v., as stipulated by the Tentative Standard of the International Electrotechnical Commission, the flashlight batteries have an operational life of 1,200 minutes and outperform in this respect sal-ammoniac batteries for flashlights by a factor of five (the standard operational life of flashlight batteries should equal 210 minutes in compliance with the International Electrotechnical Commission's requirements).

An alkaline manganese dioxide–zinc cell (1) (Figures 1.12a and 1.12b) of the galvanic battery is housed in a casing consisting of first and second shells (2) and (3) of dissimilar size, provision being made for an improved hermetic seal for the casing by allowing shell (2) to partially enter or overlap shell (3). Disposed inside the casing are a cathode (4), diaphragms (5) and (6) and zinc powder anode (7) and (8) adjoining to and separated by the diaphragms from both sides of the cathode. A positive current collector or terminal (9) is pressed into cathode (4), while negative current collectors or terminals (12) and (13) connected to anodes (7) and (8) are brought out of the cell casing via openings (10) and (11) in shells (2) and (3).

To seal openings (10) and (11), an insulating composition is used, provision also being made to insulate negative current collectors (12) and (13). Shown in Figure 1.12c is the alkaline manganese dioxide–zinc cell (1), which is contained in a casing in the form of two shells (2') and (3') of identical size. In this instance, the entire galvanic battery is made of identical shells (2') and (3'), which simplifies the process of assembling the galvanic batteries and diminishes the production cost.

To prevent cathode (4) from contacting negative current collectors (12) and (13), in shells (2') and (3') provision is made on the edge periphery for extensions (14) and (15) which extensions form a lap joint when assembling cell (1) and eliminate the need to employ additional casing components for insulating negative current collectors (12) and (13). Figure 1.12d shows one possible form of cathode (4), which is rectangular and contains a wire helix serving as positive current collector (9). Alkaline manganese dioxide–zinc cells (1) are assembled in banks (Figures 1.12e and 1.12f) by placing cells (1) one above another and using a binding band (16) to tie the cells together to panel (17) to which are secured positive and negative current collectors (18) and (19).

FIGURE 1.12:   GALVANIC BATTERY

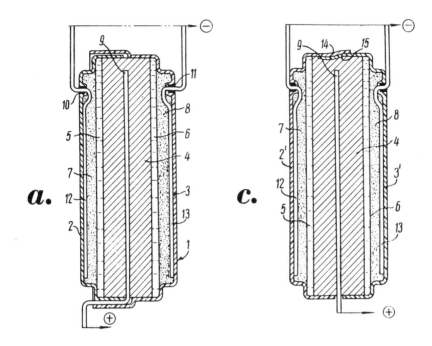

Cross-Sectional View of Cell   Cross-Sectional View of Modified Cell

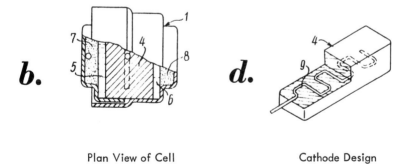

Plan View of Cell   Cathode Design

(continued)

FIGURE 1.12:   (continued)

e.

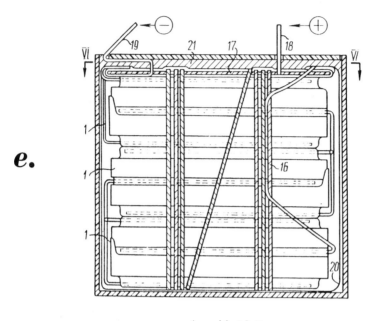

Assembled Battery

f.

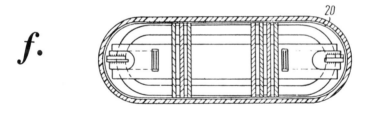

Plan View of Battery of Figure 1.12e

Source: M.D. Kocherginsky, L.F. Penkova, V.A. Naumenko and S.L. Kalachev;
          U.S. Patent 3,607,429; September 21, 1971

In each cell (1) negative current collectors (12) and (13) should be connected in pairs and also joined to positive current collector (9) of an adjacent cell. The connection of current collectors is effected by twisting together or welding appropriate current collectors. The terminal current collectors in a bank of cells are soldered or welded to appropriate current collectors (18) and (19) of the galvanic battery.

In order to eliminate galvanic battery shorting by current collectors leading from cells (1) disposed in the bottom part of a bank, the current collectors pass under binding band (16) and a thin insulating coating is applied thereto. A bank of cells (1) is placed in an insulating plastic bag (20), e.g., a polyethylene film, which is sealed except for apertures through which current collectors (18) and (19) pass. Next the assembled banks of cells (1) are inserted into an outer casing made, for example, from paper and the apertures through which current collectors (18) and (19) emerge from the plastic bag are sealed with an insulating composition (21).

OTHER DEPOLARIZERS (ZINC AND MAGNESIUM)

Ammonium Peroxydisulfate

B. Cohen and P.R. Mucenieks; U.S. Patent 3,532,552; October 6, 1970; assigned to FMC Corporation describe primary electric cells which employ mixtures of peroxydisulfates and phosphates as depolarizers with a zinc or magnesium electrode and an indifferent electrode. The cell may be operated by placing the electrodes directly into a solution of the depolarizer, by adding a conducting liquid to a container which holds the electrodes and the depolarizer in a solid form, or by impregnating a carrier, such as paper or similar nonmetallic material, with the depolarizer, and wetting the impregnated carrier with a conducting liquid either before or after insertion of the impregnated carrier between the proper electrodes.

Peroxydisulfates have been used as depolarizers in primary electric cells (see "Primary Battery Improvements", The Electric Review, 1932, by A.M. Codd). A simple cell was designed and operated; however, many problems were experienced and the work was not pursued. An improvement in this primary cell was described by Blake et al in U.S. Patent 2,534,403, in 1950. Blake et al. describes a primary cell employing a depolarizer composed of a peroxydisulfate and a silver base catalyst for reduction of the persulfate.

One of the major difficulties with primary cells employing peroxydisulfates as a depolarizer is the corrosive action of these compounds on the zinc or magnesium anode material. Ammonium peroxydisulfate is known to attack zinc at the rate of 0.01 mm./min. or 1.1 mmols./cm.^{2}/min. at 25°C. and at approximately twice this rate at 40°C. The attack of magnesium at 25°C. is 0.33 mm./min. or 2.36 mmols./cm.^{2} per minute. The result of this corrosive attack is that primarily heat energy and not electrical energy, is produced by peroxydisulfate depolarizers.

British Patent 1,055,472, issued January 18, 1967, described a method of overcoming the electrode corrosion problem encountered when using peroxydisulfate depolarizers. This patent describes an electric cell, using peroxydisulfate as the depolarizer, in which the anode is separated from the electrolyte by a diaphragm. The diaphragm prevents attack of the anode by the peroxydisulfate so that an electrochemical reaction rather than a chemical reaction occurs. The diaphragm adds materially to the cost and complexity of the cell.

It has been found that primary electric cells containing a peroxydisulfate as the depolarizer, with a zinc, zinc-base alloy, magnesium or magnesium-base alloy electrode and an indifferent electrode are improved by including an ionizable phosphate in the depolarizer. The addition of phosphates allows direct contact of the depolarizer solution with the negative electrode without unduly corroding the electrode, and the cells produce predominantly electrical energy rather than heat energy.

Orthophosphates, pyrophosphates and peroxydiphosphates are effective phosphate additives, in trace quantities to several percent based on the peroxydisulfate with zinc and zinc alloy electrodes. Orthophosphates and pyrophosphates are not as suitable with magnesium and magnesium alloy electrodes, but peroxydiphosphates are very effective in preventing corrosion of magnesium electrodes. The peroxydiphosphate is used from trace amounts to about 50% of the peroxydisulfates but preferably about 2 to 5% is used.

The primary electric cells of this process employing a mixture of a peroxydisulfate and a phosphate as a depolarizer can be operated without a cell diaphragm. There is no lag in current flow in going from an open to a closed circuit using a magnesium electrode. The depolarizer is used as a solid to which an appropriate conducting liquid is added; the polarizer may be impregnated into a suitable carrier which is subsequently wetted with an appropriate liquid. The system is portable, stable and does not deteriorate on storage when dry. The depolarizers are not toxic.

Example 1: A solution containing 4.5 g. of $(NH_4)_2S_2O_8$ and 0.04 g. of $Na_4P_2O_7 \cdot 10H_2O$ in 5.5 g. of water was prepared. A comparison solution was prepared, omitting the sodium pyrophosphate (i.e., 4.5 g. of ammonium peroxydisulfate dissolved in 5.5 g. of water). One ml. of each solution was placed in different cells, each cell containing a zinc electrode and a carbon electrode. Each electrode had an area of 20 cm.^{2}. The current was drawn from each cell at a constant amperage of 25 ma. The cell containing the depolarizer solution of this process yielded 3.5 watt minutes whereas the cell containing the comparison depolarizer solution yielded only 1.3 watt minutes.

Figure 1.13a shows the discharge curve for each solution; Cell (A) is the example of this process. It is evident that without the phosphate in the depolarizer the useful life and energy output of the cell are greatly reduced. This is due to excessive heat energy rather than electrical energy being produced by this comparison cell.

Example 2: Ordinary paper towel was thoroughly wetted in a 50% by weight solution of $(NH_4)_2S_2O_8$, which contained 2% of $K_4P_2O_8$. The impregnated paper was air dried. Pieces of impregnated paper, 1 1/2" x 3", when dry, each contained 3 g. of depolarizer mixture. Two sheets of the impregnated papers were wetted and placed between a magnesium electrode and a carbon electrode which were also 1 1/2" by 3" by about 1/16" thick. The cell was pressed tightly together to insure good contact. The cell yielded 2.0 watt minutes. The experiment was repeated and it was found that the output could be maintained constant for 2 hours, after which the output slowly decreased until the $(NH_4)_2S_2O_8$ was consumed.

Example 3: A solution containing 4.5 g. $(NH_4)_2S_2O_8$, 2 g. $(NH_4)_4P_2O_8 \cdot 2H_2O$ and 7.5 g. H_2O was prepared. One ml. of the solution was placed between a magnesium electrode and a carbon electrode. Each electrode had an area of 10 cm.^{2}. The current output of the cell was at a constant amperage of 30 ma. The solution in the cell yielded 6.9 watt minutes. A comparison solution was prepared using the same amount of ingredients except the ammonium peroxydiphosphate was deleted. Figure 1.13b shows the discharge curve of the cell using a depolarizer solution of this process, Cell (B), and also the comparison discharge curve for the cell that did not contain a peroxydiphosphate in the depolarizer solution.

FIGURE 1.13: PRIMARY CELL WITH PEROXYDISULFATE DEPOLARIZER

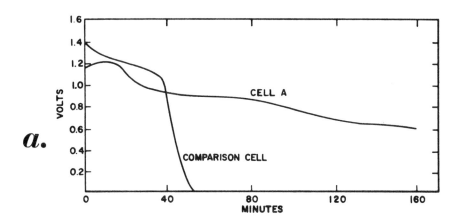

Discharge Curves for Zinc Cells

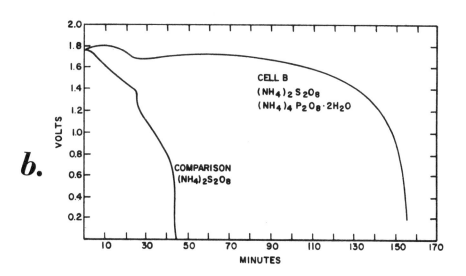

Discharge Curves for Magnesium Cells

Source: B. Cohen and P.R. Mucenieks; U.S. Patent 3,532,552; October 6, 1970

Ionizable Peroxymonosulfates

B. Cohen and P.R. Mucenieks; U.S. Patent 3,532,554; October 6, 1970; assigned to FMC Corporation have found that primary electric cells can be made which employ ionizable peroxymonosulfates as depolarizers, a zinc or zinc-base alloy, magnesium or or magnesium-base alloy negative electrode and a positive electrode. These cells have higher storage capacity than cells made with peroxydisulfate depolarizers and they have an unusually high potential. The peroxymonosulfate attacks both the zinc and the magnesium metals which are used as electrode materials. This attack can be stopped by use of a diaphragm so that an electrochemical rather than a chemical reaction takes place at the negative electrode. The corrosive attack can also be stopped or greatly reduced by adding an ionizable phosphate to the peroxymonosulfate depolarizer.

Thus, the cell may be operated either with or without a cell diaphragm such as a cation exchange membrane. These primary cells employing peroxymonosulfate as depolarizers give unusually high potential for an aqueous system when magnesium is used as the negative electrode; voltages on the order of 2.3 to 2.5 volts at a 2.5 milliampere/cm.^{2} current density are produced. There is no current lag going from an open to a closed circuit when using magnesium as the negative electrode.

Ionizable peroxymonosulfates were found to be suitable depolarizers for use in primary electric cells employing zinc, zinc-base alloy, magnesium or magnesium-base alloy negative electrode materials. Any of the ionizable peroxymonosulfates, such as the peroxymonosulfates of potassium, sodium, ammonium, lithium, may be used as depolarizers. Potassium peroxymonosulfate is preferred as it is a satisfactory depolarizer and is readily available.

Example 1: One milliliter of a solution made of 3.5 g. $KHSO_5$, 1 g. $K_4P_2O_7$ and 6 g. H_2O was placed in a cell between a zinc and a platinum coated titanium electrode. Each electrode had an area of 20 cm.^{2}. The space between the electrode was 1 mm. and contained a 20 cm.^{2} piece of glass cloth so that the added liquid and glass cloth completely filled the cell. The current was drawn at a constant amperage of 25 ma. The cell yielded 2.75 watt minutes.

Example 2: Ordinary paper towel was thoroughly wetted in a 50% solution of $KHSO_5$ which contained 1% $Na_4P_2O_7 \cdot 10H_2O$. The paper was air-dried. Each piece of paper 1 1/2" x 3" contained 1.5 g. of the depolarizer mixture. Two of these dried, impregnated sheets were wetted and placed between a zinc and a platinum coated titanium electrode, each electrode being 1 1/2" x 3" x 1/16" thick. The cell was pressed tightly together to insure good contact. The cell yielded 0.7 watt minute. The output could be easily maintained constant for 1 hour, after which the yield slowly decreased until all of the $KHSO_5$ was consumed.

Example 3: Several peroxygen chemicals and potassium chloride were compared to peroxymonosulfate as depolarizers. Each cell used in making the comparisons consisted of two electrodes each with a surface area of 20 cm.^{2}, separated by a rubber gasket 1 mm. thick. The negative electrode was either pure zinc metal or essentially pure magnesium metal. The inert electrode was platinum coated titanium. The same amount of each depolarizer 0.3 g. in each, was added to each cell. The cells using peroxydisulfate and peroxymonosulfate depolarizer also contained 0.06 g. Na_3PO_4. One cc of water was added to each cell and the open circuit voltage measured.

The following table contains a list of depolarizers and open circuit voltages of cells using the listed materials as depolarizer.

Open Circuit Voltages of Cells Using Peroxygen Chemicals as Depolarizer

	Open Circuit Voltages	
Depolarizer	Zn–Pt	Mg–Pt
$KHSO_5$	1.9	2.6
$KHSO_5$*	1.9	2.6
Comparison Examples:		
(a) KCl	0.5	---
(b) Sodium perborate	0.35	---
(c) $K_4P_2O_8$	0.5	---
(d) $(NH_4)_2S_2O_8$	1.4	1.9
(e) $Na_4P_2O_8$	0.5	---
(f) $Na_2H_2P_2O_8$	---	1.9

*Sample did not contain phosphate

Acidified Peroxydiphosphate

P.R. Mucenieks, B. Cohen and L.R. Darbee; U.S. Patent 3,532,553; October 6, 1970; assigned to FMC Corporation have found that primary electric cells can be made which employ an acidified peroxydiphosphate as a depolarizer, a zinc, zinc-base alloy, magnesium, or magnesium-base alloy, negative electrode and a positive electrode. The cell may be operated by placing the electrodes directly into a solution of the depolarizer, by adding water to a protonated peroxydiphosphate anion, or water and acid to an unprotonated peroxydiphosphate anion in a solid form, or by impregnating a carrier, such as paper or similar nonmetallic material, with the depolarizer and wetting the carrier either before or after insertion between the proper electrodes. Water is not unique in its ability to activate these batteries. Any conducting liquid constitutes a suitable medium for the depolarizer.

Typical useful peroxydiphosphates are the peroxydiphosphates of potassium, sodium, ammonium and lithium. The peroxydiphosphate may be used in solution or loaded in the cell as a solid to which a conducting liquid is added. If the anion of the peroxydiphosphate is unprotonated, i.e., $P_2O_8^{-4}$, an acid species must also be added to the depolarizer. Suitable acidic species, include but are not limited to, acetic acid, $NaHSO_4$, H_2SO_4, HCl, H_3PO_4, KH_2PO_4 and other species containing an ionizable hydrogen ion. A suitable protonated form of the peroxydiphosphate is represented by the formula $X_2H_2P_2O_8$ where $X = NH_4^+$, Na^+, K^+, or Li^+. The cell may be operated at any pH below 6.

The preferred pH range for operating the cell is 3 to 4 as some balance is needed between the acid attack of the anode and the phosphatizing protection offered by the peroxydiphosphate. When the anion of the peroxydiphosphate is unprotonated, that is when the $P_2O_8^{-4}$ is used, an acidic species containing an ionizable hydrogen ion should be added in sufficient quantity as to form one or all of the ions $HP_2O_8^{-3}$, $H_2P_2O_8^{-2}$ or $H_3P_2O_8^{-1}$. Sufficient acid to form $H_4P_2O_8$ should be avoided. The negative electrode can be zinc, a zinc-base alloy, magnesium, or a magnesium-base alloy. The positive electrode may be platinum, carbon or other materials which do not react or react only to a limited extent with the components of the depolarizer.

Example 1:   Two-tenths of a gram of solid $K_2H_2P_2O_8$ was placed in a cell between a magnesium electrode and a carbon electrode. Each electrode had an area of 20 square centimeters. One ml. of water was added to the cell. The current drawn was constant at 30 ma. The cell had an open circuit voltage of 1.93 and a closed circuit voltage of 1.66 volts. The cell yielded 1.08 watt minutes and had an efficiency of 34%.

Example 2:   One gram of a solution which was 20% by weight $Na_2H_2P_2O_8$ was placed in the cell described in Example 1. The cell was operated at a constant current of 30 ma. The cell had an open circuit voltage of 1.97 and a closed circuit voltage of 1.70 volts. The cell yielded 1.27 watt minutes and had an efficiency of 42%.

Example 3:   A cell as described in Example 1 employing a zinc electrode and a carbon electrode was charged with 0.34 g. $K_4P_2O_8$. To this 0.28 g. of perchloric acid (73% aqueous) was added. The cell was operated at a constant current of 30 ma. The cell had an open circuit voltage of 1.33 and a closed circuit voltage of 0.90 volt. The cell yielded 0.62 watt minute and had a cell efficiency of 18%.

GENERAL DESIGN AND CONSTRUCTION

Terminal Protector

A process described by D.O. Hamel; U.S. Patent 3,655,456; April 11, 1972; assigned to ESB Incorporated relates to a dry cell battery of the type having an insulated terminal board with one or more battery terminals and a terminal protector. The terminal protector is characterized by being fastened directly to the terminal board, the terminal protector being fastened to the terminal board prior to the final cell assembly operation thus assuring its permanent and accurate location.

In Figure 1.14a there is shown at (10) an elongated tape of paper or plastic film. Adhesive material is applied to selected locations (12) on one side of this ribbon. Many adhesive types can be used for this application. The principal types will be found among solvent dispersed, pressure sensitive, hot melt and thermoset materials. The choice will depend upon the material to which the tape will be adhered, the method chosen to perform the fastening step and the time available for the setting of the adhesive after application. The other side of the ribbon may be suitably imprinted with removal instructions and brand or other advertising display. It is indicated by dotted lines (14) that the ribbon (10) is cut to length, as will be described below, to form each terminal protector.

Figure 1.14b shows in outline form the operation of fastening the terminal protector tape (10) to a typical battery terminal board (16) to provide a protective covering to the terminals (18) and (20). The positive terminal (18) and negative terminal (20) are mounted to the board (16) by rivets (22). Two moveable heated jaws (24) press the tape (10) onto the board (16) and against fixed jaws (26). The tape (10), fed from coil (28) by rollers (30) is positioned to locate the adhesive spots (12), (12) under the heated jaws (24). A shear arrangement shown as (32) is provided to sever the tape (10) from the feed roll (28). In some cases, it may be desirable to allow the tape (10) to extend beyond the edges of the terminal board (16) as shown at (34).

FIGURE 1.14:   DRY CELL BATTERY

a.

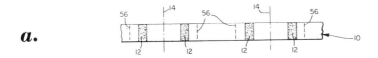

Insulating Material for Use as Terminal Protector

b.

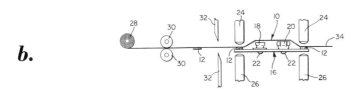

Application of Protector

c.

d.

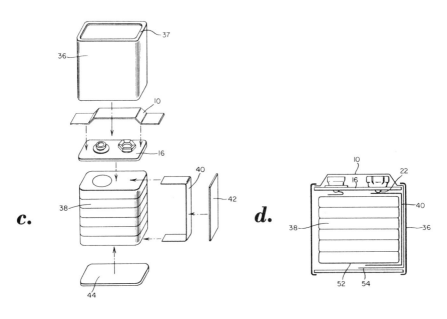

Exploded View of Battery Components   Sectional View of Battery

Source: D.O. Hamel; U.S. Patent 3,655,456; April 11, 1972

Figure 1.14c shows the several parts of the battery ready for assembly into an outer sleeve (36). This sleeve is provided with a flange (37). The terminal board (16) with terminal protector (10) attached thereto is pushed into sleeve (36) until it rests against flange (37). This is followed by a cell pack (38), internal end cell conductor (40) and conductor insulator (42). The bottom board (44) is located and then the whole assembly is placed in a press where the bottom edge of the sleeve (36) is crimped around the bottom board (44) to complete the battery.

In Figure 1.14d, a single insulating terminal protector and end cell conductor insulating member (10) is shown having a first end (54) overlapping a first end of the end cell conductor (40). A portion of the protector and insulating member (10) is positioned between the end cell conductor (40) and the metallic outer container (36), a further portion of the protector and insulating member (10) passes over the terminal board (16) and the one or more terminals mounted thereon [(18) and (20), Figure 1.14b] to a second end of the protector and insulating member, the second end being adhesively attached to the terminal board (16).

In Figure 1.14d, the internal end cell conductor (40) is shown connecting the negative end (52) of the battery pack (38) to the negative terminal rivet (22). Tail (54) to the terminal protective tape (10) is shown brought down parallel to conductor (40) and between conductor (40) and outer battery sleeve (36). By this means, the loose insulator piece (42) shown in Figure 1.14c is eliminated, thereby further simplifying the battery assembly operation.

It should be pointed out that the material from which the protective tape is made should be comparatively weak so that it can be readily removed when the battery is put in use. To this end, it may be desirable to weaken the tape by punching or scoring it near its point of attachment to the terminal board. In the case of the laminated tape, the heat sealing operation may serve to weaken the tape to a desirable degree. In Figure 1.14a, score lines are shown at (56).

Prevention of Short Circuiting

In a process described by F.K. Nabiullin, I.I. Koval, Z.M. Buzova, E.M. Gertsik, B.V. Marfin and V.A. Rabinovich; U.S. Patent 3,510,358; May 5, 1970, a primary alkaline cell comprises a tubular negative electrode with internal current collector disposed within a tubular positive electrode with external current collector, a partition being disposed between the electrodes and functioning as an ion-conducting diaphragm. Washers of electrical insulating material are at the ends of the electrodes and the washers each have at least one annular rib embedded in the ends of the partition to reliably separate the electrodes and prevent short circuit through the diaphragm.

Referring to Figure 1.15, the cell is contained in a metal case (1), which accommodates positive electrode (2) comprised of manganese dioxide, carbonaceous materials and electrolyte, metal case (1) serving simultaneously as the current collector of positive electrode (2). Negative electrode (3), which is made of pasted zinc and pressed onto tubular current collector (4) comprised of a metal sheet bent in the form of a helix, is contained within the cavity of positive electrode (2) and separated therefrom by ion-conducting diaphragm (5) which consists of thickened electrolyte. In the bottom part of metal case (1) provision is made for a ridge against which there abut sealing ring (6) and metal bottom (7). To seal the cell, recourse is had to rolling in the edges of case (1). In order to provide a reliable electric contact, the protrusion of bottom (7) enters the inner cavity of current collector (4). Disposed at the butt ends of electrodes (2) and (3) are washers (8) and (9) made of an insulating material, i.e., polyethylene. In each washer provision is made for an annular rib (10) which protrudes into ion-conducting diaphragm (5) along the entire periphery of the diaphragm butt end. Washer (8) is mounted in case (1) prior to the fabrication of electrodes (2) and (3), whereas washer (9) is inserted after negative electrode (3) and diaphragm (5) have been fabricated. Ribs (10) are instrumental in reliably separating positive and negative electrodes (2) and (3) from each other and preventing short circuits due to the presence of electrode material particles which happen to be contained within the diaphragm.

FIGURE 1.15:   PRIMARY ALKALINE CELL

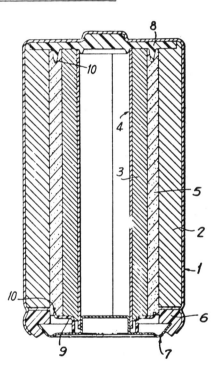

Source: F.K. Nabiullin, I.I. Koval, Z.M. Buzova, E.M. Gertsik, B.V. Marfin
          and V.A. Rabinovich; U.S. Patent 3,510,358; May 5, 1970

Plastic Envelope Design for Flat Cylindrical Cell

M.D. Kocherginsky, S.L. Kalachev, L.F. Penkova, V.A. Naumenko and K.N. Gilmanov; U.S. Patent 3,576,678; April 27, 1971, describe an alkaline flat-type cylindrical galvanic cell of a manganese-zinc system which comprises electrodes arranged in parallel to each other in two plastic envelopes whose portions accomodate negative cylindrical electrodes. The envelopes are drawn together by a bandage and include electrode masses and as a result the galvanic cell has a substantially cylindrical shape. The galvanic cell is made without sealing elements and the galvanic cell outer casing is made of paper.

Referring to Figure 1.16a, the alkaline galvanic cell is composed of two plastic envelopes (1). Placed on the bottom of each envelope (1) are negative powder zinc electrodes (2), with the portions (3) of the envelopes (1) adjoining the negative electrodes (2) being made cylindrical and having annular slots (4) accommodating bandages (5) (Figure 1.16d, 1.16e and 1.16f), the bandages intended for clamping or securing the cell together and imparting to it a cylindrical shape.

Located inside the negative electrodes (2) (Figures 1.16a and 1.16b) are negative current leads (6) to be insulated at the output from the envelopes (1) by means of a compound (7). Placed on the negative powder zinc electrodes (2) are diaphragms (8), between which there are provided the positive electrode (9) complete with a positive current lead (10). The clamping effected by the aid of the bandage (5) (Figures 1.16d and 1.16f) ensures a dependable contact between the negative electrodes (2), diaphragms (8) and positive electrode (9).

The cell can be clamped or drawn together by the aid of a casing (11) (Figure 1.16d) for which purpose the total thickness of the negative electrodes (2), diaphragms (8) and the positive electrode (9) will have to be adopted by as much as 0.2 to 0.5 mm. greater than the internal dimension of the casing (11). In this case, owing to the clamping of the diaphragms (8) when inserting the cell into the casing (11) there is obtained a dependable contact between the portions of the cell.

However, the use of the cell is ensured by providing annular slots (4) with the bandage (5) disposed therein, as the projections formed by the slots (4) allow the negative current leads (6) to be situated in the middle of the active mass of the negative electrode (2) (Figure 1.16a), thus improving its operation when manufactured to the maximum overall dimensions. The envelopes (1) (Figures 1.16d and 1.16e) complete with the assembled electrodes are placed into a plastic casing (12) made, for example, of a polyethylene film, which is hermetically welded together with the exception of points (13) of output of the current leads (6) and (10). Welding or brazing is employed to connect these current leads at point (13) to the upper and lower covers (14) and (15).

Thereafter, the cell enclosed in the plastic casing (12) is to be put into a paper sheath (11) which is rolled up at points (16) so as to be secured to the covers (14) and (15). As it appears from Figure 1.16e, formed between the envelopes (1) and casing (11) of the cell are air chambers (17) providing for expansion of the envelopes (1) upon the swelling of the electrode mass.

The operating efficiency of the proposed cell is by as much as 3 to 3.5 times greater than that of the conventional galvanic cells made to the R-20 type overall dimensions, which is confirmed by graphs (Figures 1.16g and 1.16h) illustrating the dependance of voltage on the time of discharge when operating at resistances of 5 and 10 ohms. The discharge at a resistance of 5 ohms was carried into effect for 30 minutes a day, whereas the discharge at a resistance of 10 ohms required 4 hours a day. On the drawings, lines (18) and (19) indicate the discharge of the proposed cell, whereas lines (20) and (21), indicate the discharge of the existing cells.

It is also possible to provide cells featuring a positive electrode for air depolarization. In this case, the capacitance of the cells manufactured to the R-20 type overall dimensions will be equal to at least 20 amp.-hours.

FIGURE 1.16:   ALKALINE CELL

a.
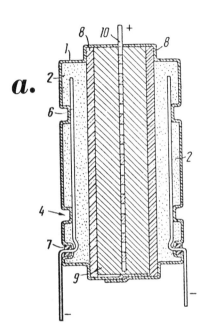

Longitudinal Cross-Sectional View
of Cell

b.
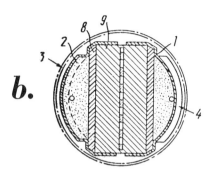

Transverse Cross-Sectional View of Cell

d.
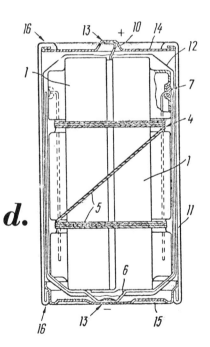

Battery Assembly

c.
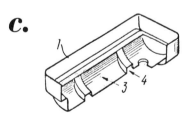

Plastic Body Element

(continued)

FIGURE 1.16:   (continued)

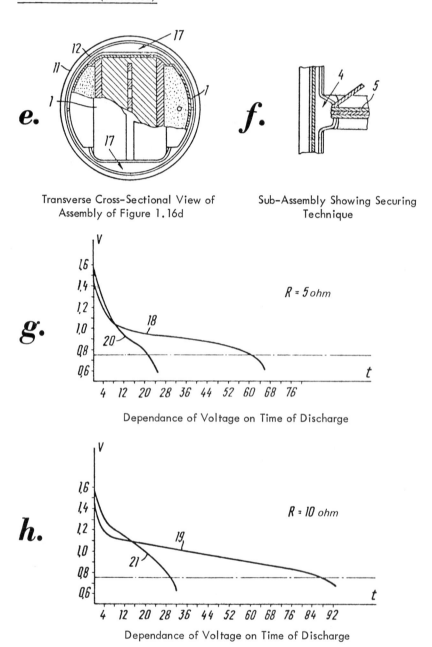

e.

Transverse Cross-Sectional View of
Assembly of Figure 1.16d

f.

Sub-Assembly Showing Securing
Technique

g.

$R = 5\,ohm$

Dependance of Voltage on Time of Discharge

h.

$R = 10\ ohm$

Dependance of Voltage on Time of Discharge

Source: M.D. Kocherginsky, S.L. Kalachev, L.F. Penkova, V.A. Naumenko
and K.N. Gilmanov; U.S. Patent 3,576,678; April 27, 1971

Flexible Tape

A process described by C.A. Grulke and R.A. Powers; U.S. Patent 3,494,796; February 10, 1970; assigned to Union Carbide Corporation involves a galvanic primary cell construction in the form of a thin, flexible tape adapted to feed past current collectors through which an external voltage may be derived, with all the active elements enclosed from atmospheric contact by carrier sheets having sealed edges, and with elements of the structure, including the anode and cathode materials, arranged as separated, discrete layers to form a plurality of individual cells spaced along the tape length.

Figure 1.17a schematically illustrates a primary dry-tape cell system in the form of an elongated, hermetically sealed, thin flexible tape (10) conveniently stored in the form of a roll (12), although a source of supply in a different form could be used. From the roll (12) or other source of supply, the elongated tape (10) is continuously fed to current collector means, suitably a pair of current collector plates (14), (16). These plates (14), (16) may be composed of metal or any other conductive material.

For the purpose of activating the reserve-type dry-tape cells, a pair of pressure plates (18), (20) may be provided just ahead of the current collector plates (14), (16). Although not shown in Figure 1.17a, the tape (10) is provided with a conductive surface on each of its sides for the purpose of maintaining electrical contact with the current collector plates (14), (16). Upon passage between the plates (14), (16), the cells of the tape (10) are discharged and an electrical current is produced that can be supplied by means of suitable leads to any desired load circuit.

In Figures 1.17b and 1.17c, there is illustrated one example of the dry-tape cell construction. As shown, the tape (10) is a multiple layer or laminate structure comprising a series of layers of an anode material (24), arranged as a series of discrete forms, e.g., squares, supported on a suitable carrier sheet (26), and a series of layers of cathode material (28), each coterminous with an anode layer (24), supported on the carrier sheet (30). Between the layers of anode material (24) and cathode material (28) is an intermediate layer of bibulous separator material (32) impregnated with an electrolyte.

The carrier sheets (26), (30) may be composed of a nonconductive plastic film material in the form of a thin, elongated ribbon or the like. Additionally, the carrier sheets (26), (30) comprise conductive surfaces which, as indicated above, are provided on both sides of the dry-tape cell construction for maintaining electrical contact with the pair of current collector plates (14), (16). The conductive surface may be achieved upon the carrier sheets in a number of ways in accordance with the process. For example, in Figures 1.17b and 1.17c, elongated strips of metal foil (42), (44) are provided, each of which are suitably adhered, respectively, to carrier sheets (26) and (30), by means of an adhesive or the like.

The foil strips (42), (44) are then turned over the peripheral edges of the carrier sheets (26), (30), respectively, and lie flat against the opposite sides thereof as generally indicated at (46) and (47). Suitably, the metal foil may be composed of tin, aluminum, zinc or other suitable conventional metal compatible with the cell system. It may be mentioned that in the instance of the foil strip (44) which makes electrical contact with the layer of cathode material (28), a conductive coating of a carbon-in-resin composition should be applied over the foil strip (44) in order to prevent direct contact and shorting between the metal foil and the depolarizer. The foil strips (42), (44) may be provided as discontinuous discrete forms, e.g., squares, coterminous with anode and cathode layers (24), (28). The anode material (24) may be provided onto one side of the carrier sheet (26) over the metal foil (42) in the form of finely divided particles or powders of one of the conventional anode metals such as zinc, cadmium, aluminum or magnesium. Metal evaporation or other suitable techniques for forming the thin anode layer are well-known in the art.

The cathode material (28) is provided in the form of a thin porous layer of finely divided particles or powders of, for example, one of the conventional oxidic depolarizers such as manganese dioxide or the oxides of silver, copper and lead. Suitably, a mixture of the oxidic depolarizer and a conductive material such as graphite or acetylene black may be employed. The cathode material (28) may be provided on the carrier sheet (30) by means of so-called printing techniques known in the art. Such printing techniques involve the use of a paste containing the cathode material, a binder, and a volatile solvent which is evaporated, leaving behind a residue or thin deposit of the cathode material which is firmly bonded to the carrier sheet (30).

As more clearly shown in Figure 1.17b, the laminates containing the anode material (24) and the cathode material (28) are sealed directly in contact with the intermediate layer of separator material (32) by means of thin narrow marginal seal layers as indicated at (34) and (36). These marginal seal layers are adhered to the peripheral edges of the carrier sheets (26), (30) and preferably penetrate through the edges of the separator (32) in a manner whereby the layers of anode material (24) and cathode material (28) are hermetically enclosed or enveloped from atmospheric contact. Furthermore, as a result of the action of the marginal seal layers (34) and (36) to penetrate the separator material, a tendency develops to prevent electrolyte bridging thereby maintaining the electrolyte for one cell separate from the electrolyte for an adjacent cell. Suitable adhesive materials include a wide variety of nonconductive adhesive cements such as a vinyl resin composition.

The basic contribution of the process is that there is provided the specific structure including the anode and cathode materials, in the form of individual segments disposed along the tape construction to provide a series of discrete cells. As depicted in Figure 1.17b, this may be accomplished by providing discrete layers of anode material (24) and cathode material (28) each, respectively, having its periphery bounded by the marginal seal layers (34) or (36). The seal layers (34) and (36) provide the sealing effect well as separation of electrolyte from one cell to an adjacent cell, and there is provided by the construction herein described a series of galvanic primary cells with all the active elements enclosed from atmospheric contact, the entire assembly being in the form of a thin, flexible, sealed tape of an extended length capable of providing discrete sources of electrical output therealong.

The carrier sheets may be composed of any plastic film material such as phenoxy, styrene and polyethylene. The electrolyte may be any one of the conventional electrolyte solutions such as potassium hydroxide and may be provided in gel form if desired. However, the preferred electrolyte is one of the class of "polyelectrolytes" or those polymeric structures capable of carrying ionic charges. Suitable polyelectrolyte materials include polyacrylyl urea coreacted with oxamide, maleimide, malonimide or ethylenediamine. In the operation of the device the tape (10) may be fed past the current collectors (14) and (16) in an intermittent manner, thereby permitting each individual cell defined by a square of anode material (24) and cathode material (28) to pass between current collectors (14), (16) and remain so positioned with sufficient dwell time to allow discharge thereof prior to placement of the next cell in discharging position. It will be apparent that for each discharging position an appropriate segment of foil strips (42), (44) will be in contact with an appropriate current collector (14), (16). In this mode of operation, the output from the tape (10) achieved through current collectors (14), (16) is of a pulse-like nature thereby enabling expanded versatility in the utilization and application of the process.

FIGURE 1.17:   DRY TAPE CELL CONSTRUCTION

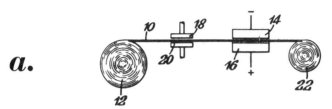

a.

Schematic of Dry Tape Cell System

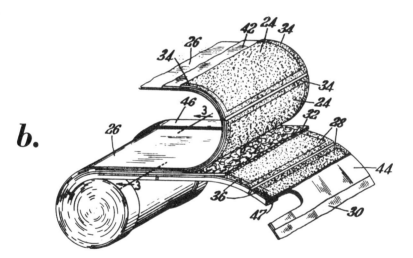

b.

Detail of Dry Tape Cell Construction

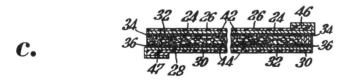

c.

Sectional View Along Line (3—3) of Figure 1.17b

Source: C.A. Grulke and R.A. Powers; U.S. Patent 3,494,796; Feb. 10, 1970

Due to the fact that the tape (10) is constructed with separated, discrete cells, certain capabilities will be produced in the system of the process enabling a variety of modifications and uses within the scope of the process. For example, a series of current collectors such as (14) and (16) could be arranged along the tape and connected in a parallel circuit arrangement to permit a plurality of cells to be discharged simultaneously thereby giving increased output. Further, the foil contact strips (42), (44) may be interconnected to provide either series or parallel interconnection of cells. For example several strips (42) on one side of the tape (10) could be interconnected with each other, with an equal number of strips (44) on the opposite side of tape (10) being similarly interconnected thereby providing a parallel circuit connection between the cells involved.

Of course, such an interconnection could be accomplished merely by making foil strips (42), (44) continuous for a specified length. In a similar fashion, cells could be joined in a series circuit arrangement by connecting together a strip (42) with a next adjacent strip (44), with any specified number of cells being joined by this mode of series connection. Thus, the anode of one cell would be joined to the cathode of a next adjacent cell for as many cells as may be desired.

BASIC MAGNESIUM DRY CELLS

CELL DESIGN

Hermetically Sealed Closure Disc

M.E. Wilke and H.J. Strauss; U.S. Patent 3,615,866; October 26, 1971; assigned to Clevite Corporation describe a method for making a magnesium cell whereby gases formed during the initial period after assembly are permitted to escape in order to prevent the buildup of excessive gas pressure after which the cell is properly sealed so that leakage will not subsequently take place.

Referring to Figures 2.1a and 2.1b, one type of cell of the process is shown comprising a tubular magnesium can having a sealing disk (2) formed of a suitable stiff but somewhat resilient plastic composition inserted in closure position in the open top of the can. The sealing disk (2) is provided with a hole (3) having a carbon rod (4). The carbon rod is provided at its end with a metal cap (5) of a material such as brass. The sealing disk (2) is additionally provided with a restricted aperture in the form of a notch (6) extending across the thickness of the disk.

The aperture is of sufficient size to ensure a minimum passageway for a gas such as hydrogen to pass therethrough when the disk is placed in the top of can (1), but sufficiently small to prevent a substantial amount of water vapor from escaping. The cell additionally contains a paper separator (7) having a suitable cell electrolyte absorbed therein. Disposed within the paper separator is a depolarizer core (8) of a composition comprised of about nine parts of a depolarizer material such as manganese dioxide and one part of an acetylene black, and additionally including a suitable electrolyte such as magnesium chloride. A paper washer (9) covers one end of the depolarizer core.

The cell is assembled by inserting the separator (7) in the open can (1). The depolarizer cathode (8) is then inserted inside the separator and suitably tamped. The washer (9) may then be placed over the cathode (8). The brass-capped carbon rod (4) having the sealing disk (2) is inserted through a hole provided in the washer (9) and into the cathode material.

FIGURE 2.1:   SEALING METHOD FOR MAGNESIUM CELLS

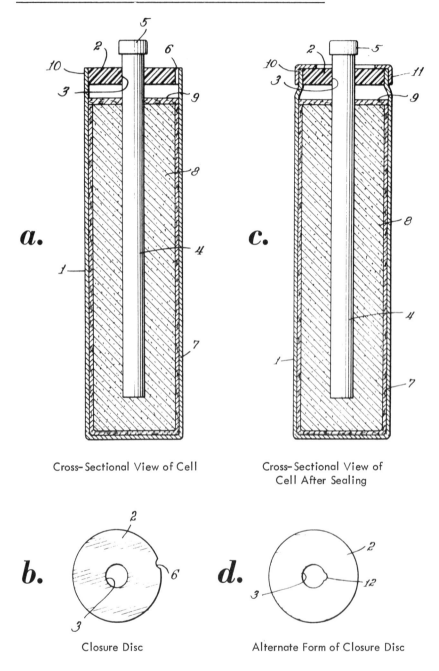

Cross-Sectional View of Cell

Cross-Sectional View of
Cell After Sealing

Closure Disc

Alternate Form of Closure Disc

Source: M.E. Wilke and H.J. Strauss; U.S. Patent 3,615,866; October 26, 1971

After the cell has been thus assembled, it is permitted to stand for a suitable length of time, as for example 4 or 5 days, to permit gases which initially form when the various component parts and materials of the cell are first brought together to escape from the cell through the aperture (6) prior to the final sealing. At the end of the period, the cell is hermetically sealed. To this end the top portion (10) of the can is constricted or reduced in diameter (see Figure 2.1c) by means of a suitable tool which applies radial pressure thereto to make it smaller and compress the disk and place it under radial strain and thus provide sealing pressure at the interfaces between the disk (2) and the carbon rod (4) and top portion (10) of the can, respectively.

The composition and thickness of the disk (2) is such that upon removal of the forming tool, the end portion (10) of the can retains its deformed shape with reduced diameter. In this process, the radius of the closure disk is reduced somewhat more than the depth of the notch (6) so that the gas escape aperture which it provided is completely closed. To complete the closure and ensure a permanent seal, the ring (11) of steel or other suitable material is forced onto the reduced portion (10) of the magnesium can. To facilitate the application of the ring (11) to the end of the can, the end portion (10) of the can and/or the engaging portion of the ring (11) may be slightly tapered. The cell, thus sealed, is ready for service.

An alternative form of closure disk is shown in Figure 2.1d. In this example, the aperture, instead of being a notch in the outer periphery of disk (2), is a notch or scratch provided at the periphery of the hole (3) in which the carbon rod is to be inserted; i.e., a notch in the inner periphery of the disk. The cell is assembled in the same manner as previously described with reference to Figure 2.1a. When, at the end of the gassing period, radial constriction or clamping force is applied to the end of the can and a holding ring (11) applied, as above described, the aperture formed by notch (12) is closed and the disk–carbon rod interface securely sealed under pressure.

Resilient Disc with Shaped Cover

A process described by M.E. Wilke; U.S. Patent 3,427,202; February 11, 1969; assigned to Clevite Corporation involves the improvement of the seals of cells having magnesium cans the production of cylindrical magnesium negative terminals from magnesium tubing as an improvement over the forming of cans by extrusion or drawing of the metal, and the provision of metal surfaces electrically connected with the magnesium terminals to which intercell conductors may readily be welded or soldered.

The can of Figure 2.2b is made from a length of magnesium tubing (1) which is cut to the desired length from standard magnesium tubing. A disk (2) of a suitable stiff but somewhat resilient plastic composition is inserted into the bottom end of the tubing after which the bottom end portion of the tubing which is engaged by the periphery of disk (2) is reduced in diameter by means of a suitable tool which applies radial pressure to this end portion of tubing (1) and to the body of disk (2) to compress the latter and place it under radial strain. The composition and thickness of disk (2) is such that upon removal of the forming tool, the end portion of magnesium tubing (1) will retain its deformed shape with reduced diameter and the interface between disk (2) and portion (3) of the tubing will be subject to sealing pressure due to the compressed resilience of disk (2).

FIGURE 2.2: MAGNESIUM BATTERIES

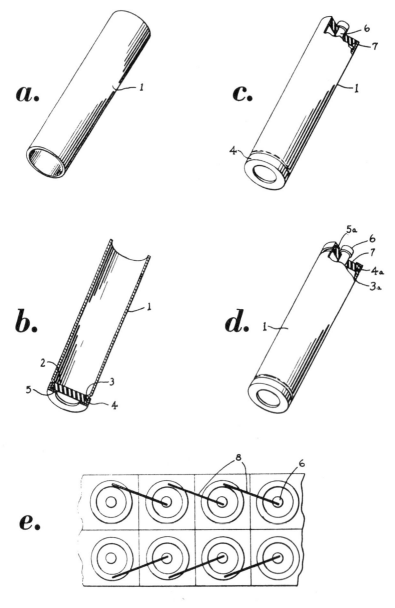

(a)-(d)   Assembly Process Using Magnesium Tubing
(e)   Multicell Battery with Intercell Connections

Source: M.E. Wilke; U.S. Patent 3,427,202; February 11, 1969

To complete the closure, a ring (4) of steel or other suitable material is forced onto the reduced portion (3) of magnesium tubing (1). To facilitate the application of ring (4) to the bottom end of tubing (1), the end portion (3) and/or the engaging portion of ring (4) may be slightly tapered. A flange (5) may be provided as an integral part of ring (4) to engage the end of tubing (1) and thus serve as a stop and also to provide support for disk (2) to prevent outward movement or bulging of the disk.

When the bottom closure is completed, ring (4) is firmly fixed in position. It prevents any enlargement of the reduced end portion (3) of the tubing due to relaxation of the magnesium and ensures a permanently sealed closure for the bottom end of the tubing. Thus, a "can" is formed for the reception of the contents of the dry cell. The materials and parts of the cell are then inserted into the can. Included is the usual brass-capped positive terminal (6) which emerges from the cell through a hole in washer (7). This washer may be similar in composition to disk (2), suitably a stiff and somewhat resilient plastic material.

Although, as will be seen by reference to Figure 2.2d, the completely assembled cell includes a top seal construction similar to that described for the bottom of the tubing, Figure 2.2b, the uncompleted cell, as shown in Figure 2.2c, is , in accordance with one aspect of the process, permitted to stand for a suitable length of time, say four or five days, to permit gases which initially form upon the bringing together of the several component parts and materials of the cell to escape from the cell prior to the final sealing.

To this end, the hole in the center of disk (7) to receive terminal rod (6) is dimensioned to provide a snug but not a sealing pressure fit. Similarly, the outside diameter of disk (7) is such that the disk, when inserted into the top end portion of tube (1) will hold its position, but is not under sealing pressure at the interface of the disk and the tube. Under these conditions, gases under only slight pressure within the cell may escape past disk (7) to the atmosphere.

After aging, the top seals of the cells are completed in a manner similar to that described with reference to the bottom closures of the tubing. The top portion of the tubing, along with the body of disk (7), is subjected to inward radial pressure which permanently deforms the metal as the diameter of this portion is reduced. A ring (4a), preferably having a flange (5a), is then forced upon the reduced portion (3a) of the top end of tube (1). The result is a permanent pressure seal at the engaging surfaces of disk (7) with terminal (6) and the inside surface of reduced portion (3a) of the magnesium tube (1).

Electrical connection of the several cells of a multi-cell battery, such as that shown more or less diagrammatically in Figure 2.2e, is greatly facilitated by the use of top closure rings (4a) as a base for soldering or welding the intercell conductors (8). The vexing problem of securely and electrically connecting such intercell conductors to the magnesium negative electrodes of the cells is eliminated. Rings (4a) intimately engage the top portions of the magnesium tubes, providing assured electrical connection between the rings and the magnesium cell terminals, and the intercell conductors (8), welded or soldered at one of their ends to the positive terminals (6), may be welded or soldered either to the cylindrical portions of rings (4a) or, as is shown in Figure 2.2e, to the flanges (5a) at the tops of the cells, whichever is more convenient. The end result is a more dependable, yet less costly, battery.

Flat Cells Enclosed in Fiber Glass Reinforced Resin

A process described by R.R. Balaguer; U.S. Patent 3,490,952; January 20, 1970; assigned to Battery Corporation of America provides a flat dry cell battery in which the metal anode is a flat plate located at the center of the cell which is surrounded by a graphite or carbon cloth cathode. Battery mix is disposed between the anode and cathode on both sides of the anode. The entire cell is enclosed in an envelope which may be a fiber glass reinforced resin. Swelling of the cell or separation of the elements is inhibited by having the elements sewn together with fiber glass thread. External contact to the cloth electrode is made through a graphite rope or metal contact elements.

Referring to Figures 2.3a and 2.3b, the dry cell battery comprises a metallic battery anode (10) which is preferably magnesium and preferably in the form of a flat sheet. The anode (10) is covered on both flat surfaces and on all edges with a bibulous covering (11), e.g., a kraft paper separator, the function of which is to prevent direct contact between the anode metal and the battery mix.

Battery mix, i.e., depolarizing mix and electrolyte, is coated in layers (12) and (13) on respective sides of the anode (10). The battery mix is enclosed by sheets of carbon or graphite cloth (14) and (15) disposed on opposite sides of the anode (10) and forming a first envelope. Layers (16) and (17) of vinyl film overlie the outer surfaces of the cloth layers (14) and (15), respectively.

The entire assembly is enclosed within a heat sealed plastic envelope (18) which may be of any suitable type which will resist penetration by moisture or by the corrosive products of cell discharge. One such material is the heat sealable, flexible, rubber hydrochloride sheet sold under the trademark Pliofilm. The ends of the carbon or graphite cloth layers (14) and (15) at opposite edges are clamped between tinned steel channel shaped contact elements (19) and (20).

External metal contacts (21) and (22) extend through the envelope (18) at one end of the cell and are clamped by the channels (19) and (20), respectively, to afford the external contact to the battery cathode formed by cloth layers (14) and (15). The contact elements may be joined by an insulated wire (23) and external connections to the cathode may be provided by a wire (24). A contact element (25) formed at one end of anode (10) projects outwardly through the envelope (18) and may be connected to the external circuit through a wire (26).

The meal anode (10) is preferably magnesium and might be a flat sheet of magnesium 0.035" or 0.020" thick and of any desired length and width, e.g., 7 1/2" x 3 1/4". The bibulous covering (11) may be of any suitable type as is well known in the art. A salt free kraft paper has been found well suited for the purpose. Metals other than magnesium can be used, e.g., zinc.

The layers of battery mix (12) and (13) should be formed to correspond to the battery electrochemical system and preferably will be about 10% wetter than customary when used in a conventional battery. A typical battery mix composition for use with a magnesium anode would be the following, percentages being by weight: 88% type M manganese dioxide (synthetic) chemical ore; 1% $Mg(OH)_2$; 3% $Ba(CrO_4)$; and 8% acetylene black wet with 660 ml. of 250 grams/liter of $MgBr_2$ and 0.25 gram per liter of Na_2CrO_4 per 1,000 grams of dry cathode mix.

For a cell having roughly twice the capacity of a standard size "D" cell about 100 g. of the battery mix would be used. The carbon electrodes (14) and (15) are flexible, electrically conductive carbonized or graphitized fabric which is woven or otherwise formed so as to have openings between the yarns. The carbon fabric may be formed by carbonizing or graphitizing a prewoven or formed cellulosic fabric or may be woven or otherwise formed from carbonized or graphitized filaments or yarns. The fabric may be prepared in any suitable way, for example as described in any of U.S. Patents 3,011,981; 3,107,152 or 3,116,975.

It is desirable that battery mix fill the interstices between the carbon yarns so as to maximize electrical contact between the carbon fabric and the battery mix. The reason the mix is made more moist than usual is to facilitate filling the interstices of the fabric with battery mix. As is best shown in Figure 2.3b, the carbon cloth electrodes (14) and (15) approach each other at each side of the cell and are clamped together by the clamping members (19) and (20).

The carbon surfaces extending from clamping elements (19) and (20) to the respective adjacent edges of the battery mix layers are preferably not coated with battery mix and preferably are waxed or otherwise treated to inhibit penetration of battery mix or corrosive products which would tend to corrode the members (19) and (20). The inner surfaces of the vinyl film layers (16) and (17) are coated with a pressure sensitive adhesive which will adhere to the underlying carbon surface. In this way the vinyl films reinforce the carbon cloth and assist the latter in resisting expansive forces within the cell.

The flat battery construction described above and illustrated in Figures 2.3a and 2.3b, is satisfactory for many purposes. However, a superior flat battery construction is illustrated in Figures 2.3c, 2.3d and 2.3e. Referring to Figures 2.3c to 2.3e, the dry cell battery is similar in construction to the one of Figures 2.3a and 2.3b, and like elements are given like but primed reference numerals. The cell of Figures 2.3c to 2.3e comprises a metallic anode sheet (10') which is preferably magnesium, a bibulous covering (11') covering all surfaces of the anode, layers of battery mix (12') and (13') and carbon cloth electrode layers (14') and (15'). The latter may be formed or sewn in the shape of a continuous envelope or sleeve, as shown in Figure 2.3d, with arcuate portions (27) and (28) joining the flat portions (14') and (15').

Vinyl plastic films (16') and (17') having their inside surfaces coated with a pressure sensitive adhesive overlie the outer surfaces of the carbon cloth electrodes (14') and (15') respectively. A polyester resin impregnated fiber glass cloth sheath or envelope (29) surrounds the assembly and forms a moisture proof sealing layer at all sides, the top and the bottom. The fiber glass fabric and the resin may be of the type commonly used in fiber glass reinforced resin structures, e.g., boat hulls.

In making the cell, the fiber glass fabric is preferably wrapped around the carbon cloth and vinyl as a tube with a substantial overlap, e.g., about one inch. Before the polyester resin and catalyst are applied to the fiber glass, a fiber glass thread (30) is sewn into the assembly through numerous holes (31) provided in the anode sheet (10'). The fiber glass thread (30) may be sewn in any desired pattern, a typical pattern being illustrated in Figure 2.3c. It is desirable that the fiber glass threads (30) extend also around the cell edges to assist in binding the entire cell into a unitary structure. It is not necessary that a single thread (30) be used since multiple threads may be employed, as desired.

FIGURE 2.3:   FLAT DRY CELL BATTERY

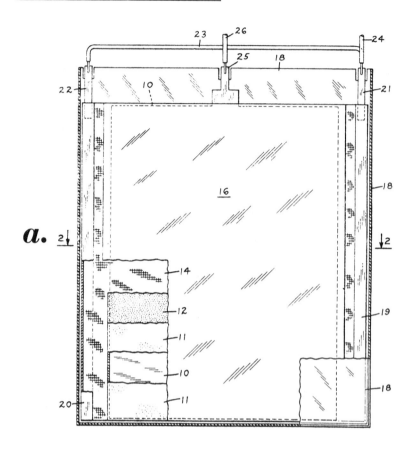

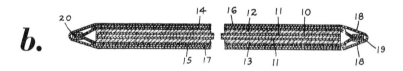

(a)   Top View Showing Internal Construction
(b)   Cross Sectional View Along Line (2—2) of Figure 2.3a

(continued)

FIGURE 2.3: (continued)

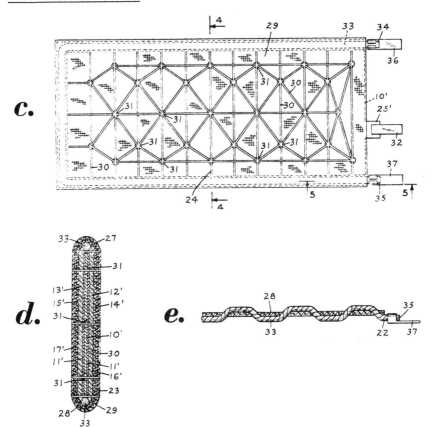

(c) Top View of Fiber Glass Polyester Covered Unit
(d) Cross-Sectional View Along Line (4—4) of Figure 2.3c
(e) Cross-Sectional View Along Line (5—5) of Figure 2.3c

Source: R.R. Balaguer; U.S. Patent 3,490,952; January 20, 1970

As many passes of the thread (30) through the cell as is desired may be made. In the limiting case sufficient thread passes may be used that the thread itself will serve as the fiber glass reinforcing fabric thereby eliminating the need for the separate fabric (29). After the thread (30) is sewn in place, the polyester resin and catalyst are applied as a layer impregnating the fiber glass fabric and thread with which it comes into contact. The resin is allowed to harden, yielding a strong, moisture and gas proof cell closure envelope. The battery external negative terminal is provided by a contact element (25') projecting from anode (10') and extending through the fiber glass and resin covering at one end of the cell. A suitable clip (32) may be provided for making electrical connection.

A carbon rope (33) extends down one side of the cell, along the bottom and up the other side. The rope (33) emerges from the fiber glass on both sides to form connecting tails (34) and (35) to which are clamped contact elements (36) and (37), respectively. The carbon rope (33) provides electrical contact between the carbon electrode and the outside of the cell. To enhance the contact between rope (33) and the carbon cloth it has been found desirable to sew the rope (33) through the carbon cloth in long stitches, as shown in Figure 2.3e.

An important advantage of the dry cell battery of Figures 2.3c to 2.3e is that no metal at all, except for anode (10'), is located within the fiber glass reinforced plastic covering (29). By extending bibulous covering (11') to the plastic sealing area around projecting contact element (25'), all portions of anode (10') which may come into contact with the battery mix or corrosive fluids or gases produced during cell discharge are covered with the bibulous covering and hence are protected from corrosion other than as desired for chemical reaction during cell discharge.

Another important advantage of the cell construction is that the fiber glass thread (30) and the fiber glass fabric (29) serve as reinforcements to prevent cell elements from being separated, as by the action of hydrogen gas generated during cell discharge. The thread and the glass fabric are essentially bonded into a unitary assembly by the plastic. And since fiber glass is essentially unstretchable, the threads extending through the thickness of the cell bond the two sides together to resist strongly any expansion forces. The importance of resistance to separation of cell elements will be realized when it is understood that internal gas pressures in the cell tend to lift the mix away from the anode thereby increasing cell internal resistance and decreasing cell capacity.

Multiple Anode and Cathode Arrangement

A process described by P.F. George; U.S. Patent 3,490,951; January 20, 1970; assigned to The Dow Chemical Company relates to a primary cell having multiple anodes and cathodes arranged in series within a tubular shaped envelope. The cell assembly is such that both positive and negative terminals are at one end while a negative terminal is at the opposite end.

Referring to Figure 2.4, there is shown a primary cell assembly, indicated generally by the numeral (10), having an axially disposed central cathode electrode, usually a carbon rod (12), which is surrounded by an inner or first anode (14) and an outer or second anode (16). The bottom (18) of the cathode electrode (12) is separated from the inner anode (14) by a paper spacer which, as illustrated, is the bottom of the paper bag-like separator member (20) which abuts against the surface of the anode (14) which faces the cathode electrode (12). The member (20) is folded inwardly towards the electrode (12) at the top of the anode (14). The space between the cathode electrode (12) and the paper separator (20) adjacent the surface of the anode (14) is occupied by a suitable cathode mix-electrolyte composition (22) which is compatible with the anode (14).

The outer surface which faces away from the cathode electrode (12) is coated, for example, with a tightly adhering, fluid impervious, electrically conductive carbon loaded epoxy coating (24) of the type described in U.S. Patent 3,343,995. The bottom of the anode (14) is separated from the bottom of the outer anode (16) by a waxed paper washer or other suitable insulating element. The inner surface of the

FIGURE 2.4:   PRIMARY CELL ASSEMBLY

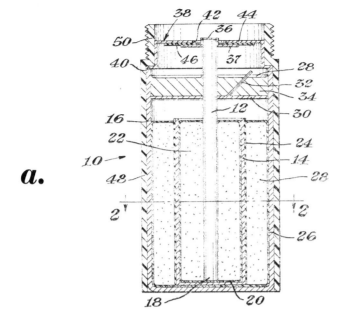

a.

Side, Sectional View of Cell

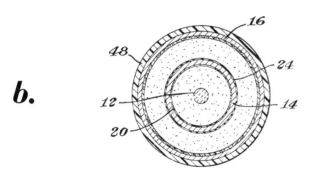

b.

Sectional View Along Line (2—2) of Figure 2.4a

Source: P.F. George; U.S. Patent 3,490,951; January 20, 1970

anode (16), which is illustrated as a cup-shaped anode, has a layer (26) of paper
separating material covering that part of its inwardly facing surface against which a
second quantity (28) of cathode mix-electrolyte abuts. The cathode mix-electrolyte
(28) is disposed between the coating (24) of the inner anode (14) and the paper separator (26) abutting against the outer anode (16). The anode (16) is substantially
longer than the anode (14) but is usually shorter than the carbon cathode rod (12).
A sealing-venting assembly composed of top wafer (28), bottom wafer (30), including

a vent tab (32), and a wax seal (34) sandwiched between the wafers extends between the cathode electrode (12) and the anode (16) at or near the upper end of the anode (16), effectively preventing moisture loss from the cell assembly but permitting hydrogen (released during operation of the cell assembly) to be vented from the cell. The top end of the cathode electrode (12) has the usual metal end cap (36), having outwardly extending flange (37), fitted over it. A somewhat hat-shaped metal connector cap, indicated generally by the numeral (38) has its outwardly extending flange at its lower end or "brim" part spot welded, as at (40), to the upper end of the anode. The connector cap (38) has an aperture (42), substantially larger than the diameter of the electrode (12), in its top (44), with the end cap (36) of the cathode (12) extending through the aperture (42).

A waxed paper or other suitably insulating washer (46) is sandwiched between the flanged part of the end cap (36) and the top (44) of the connector cap (38). A tubular outer case (48), made of any suitable, but usually electrically insulating, material (commonly plastic sheet material) is tubular in configuration and fits closely around the outer anode (16) from the lower end thereof to a threaded inset collar (50) disposed above the upper end of the anode (16). The case (48) fits against the anode (16), with its inset collar (50) retaining the metal connector cap (38) in position atop the upper end of the anode (16).

In operation, the coating (24) on the side of the anode (14) which is remote from the cathode electrode (12), functions in effect as the cathode for the outer cell whose anode is the anode (16), thus placing the inner and outer cells in series with each other to provide a higher output voltage. In one cell assembly made in accordance with this process, the outside anode (16) is made of AZ31 magnesium alloy and is an impact extruded cup. The inside anode (14) is made of machined AZ21 magnesium alloy. The cathode mix-electrolyte composition is 88% manganese dioxide (chem ore), 3% barium chromate, 8% acetylene black and 1% magnesium hydroxide, wet with 550 cubic centimeters of 250 grams/liter of magnesium bromide and 0.25 gram per liter of sodium chromate per 1,000 grams of dry cathode mix.

The separators (20), (26) are made of kraft paper, and the coating (24) is about 2.5 to 3.0 mils thick and is made of carbon loaded epoxy. The open circuit voltage of the cell assembly showed 3.80 volts.

OTHER ASPECTS

Protective Anode Coating

In a process described by L.W. Eaton; U.S. Patent 3,634,142; January 11, 1972; assigned to Clevite Corporation the external, nonreacting surface of the magnesium anode of a dry cell is covered by a protective coating to prevent impairment or loss of electrical contact with the adjacent cell of a battery which, it has been discovered, frequently results from limited but effective chemical action that takes place at this exposed surface and proceeds under presumably sealing barriers to undermine electrical contact with the adjoining cell. The protective coating is applied to one entire surface of the magnesium anode, excepting only the central portion which is left exposed for making intercell electrical contact. The coating comprises a thin, dense layer of inert material which substantially excludes water molecules from penetrating to the covered surface of the magnesium anode. A suitable material for this purpose is an epoxy primer paint. A suitable material is that commercially known as Epoxy-Cote primer 15-404 with curing agent 15-014. Other primer-type paints and hot melt adhesive materials have been found to be satisfactory. It has been found that use of the protective coating for the external surfaces of the magnesium anodes effectively prevents deterioration of the normally excellent electrical contact at the intercell connector-anode interface.

Potassium Permanganate-Magnesium Chloride Electrolyte

A cell described by S. Ruben; U.S. Patent 3,539,398; November 10, 1970 utilizes an anode of magnesium, a depolarizing cathode composed substantially of a water soluble oxygen yielding permanganate such as potassium permanganate mixed with an electronic conductor, such as finely divided carbon, for example, Shawinigan black or micronized graphite, and an aqueous halogen metal salt which is stable with magnesium and in which the permanganate is soluble. The alkaline earth halides are preferred, specifically an alkaline earth chloride, such as magnesium chloride ($MgCl_2 \cdot 6H_2O$) containing permanganate in solution. The anode and cathode are desirably separated by an oxidation resistant submicroporous spacer, such as Acropor, a nylon cloth impregnated with a copolymer of vinyl chloride and acrylonitrile, Synpor, a microporous polyvinyl chloride, or Permion, a microporous polyethylene.

The depolarizer may be made by grinding potassium permanganate to a powder then mixing it with Shawinigan black in the ratio of 7 parts of $KMnO_4$ and 3 parts by weight of Shawinigan black. To each 10 grams of this mixture is added 10 ml. of a solution of 50 g. of $MgCl_2 \cdot 6H_2O$ per 100 ml. of H_2O saturated with $KMnO_4$. The electrolyte is no longer only a halogen solution, but specifically a combination of a solution of potassium permanganate and magnesium chloride. Potassium permanganate dissolves in the aqueous magnesium chloride solution raising the specific gravity from 1.14 to 1.16. It is desirable to initially saturate the electrolyte with the permanganate so as to avoid dissolution of the permanganate from the depolarizer bobbin. The proportion of micronized graphite or other finely divided conductor added to the permanganate may generally vary between 10% and 30% of the mix.

In magnesium cells of the prior art utilizing a halogen electrolyte, the depolarizer, such as manganese dioxide, is substantially insoluble in the electrolyte. In the cell of this process, the water soluble permanganate is not readily soluble in the permanganate saturated electrolyte, and this gives the electrolyte a double function: as an ionic conductor and, in part, as a depolarizer at the cathode surface. If the electrolyte is not initially saturated with $KMnO_4$, saturation will occur over a period of time, due to dissolution of the $KMnO_4$ in the electrolyte.

ZINC AND MERCURY OR SILVER OXIDE CELLS

ZINC-MERCURY OXIDE

Resilient Nickel-Bronze Laminated Cover for Button Cell

P. Ruetschi; U.S. Patent 3,657,018; April 18, 1972; assigned to Leclanche SA, Switzerland describes a galvanic cell which is characterized in that the cover which is in contact with the zinc is in the form of an elastically resilient laminate which has an exterior side consisting of a layer of nickel or a layer of a rustproof nickel alloy and having a thickness of 0.05 to 0.5 mm. and an interior side consisting of a layer of copper, zinc, tin, lead or their alloys and having a thickness of 0.05 to 0.5 mm. whereby at least one layer of the laminate consists of a spring alloy. The interior side of the cover thus consists of a metal which in its amalgamated condition has a high hydrogen overvoltage.

Referring to Figure 3.1, the cover is made of a laminate the upper part (1) of which consists of pure nickel or of a rustproof nickel alloy, whereas the lower part (2) consists, for example, of a spring bronze that contains 5 to 15% tin, preferably 8%, the balance being copper and has a hardness of more than 100 Vickers degrees of hardness.

The interior layer may also consist of spring brass containing 30 to 40% zinc, preferably 37%, and 60 to 70% copper, preferably 63%. It has a thickness of 0.05 to 0.5 mm., preferably about 0.1 to 0.2 mm. The thickness of the metallic layer (1) amounts to 0.05 to 0.5 mm., preferably 0.01 to 0.2 mm. The two layers need not be of the same thickness. The advantage of this composition of the cover is the fact that it provides exceedingly high tightness or imperviousness.

The edge (3) of the resilient cover is preferably slightly bent in an upward direction. Upon closing the galvanic cell by bending in the edge of the metallic housing (4), the packing ring (8) is normally clamped between the cover and the housing. The cover in this case acts like a spring. The elastic deformation holds the packing ring under constant pressure so that even after remaining stored for several years, and even if the packing material gives away a little due to cold-flowing, the spring

FIGURE 3.1: ALKALINE CELL WITH RESILIENT NICKEL-BRONZE LAMINATED COVER

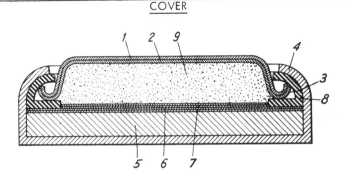

Source: P. Ruetschi; U.S. Patent 3,657,018; April 18, 1972

action of the cover makes it possible to maintain a steady undiminished pressure upon the packing material. For example, it has been found that a laminated cover having an exterior side of nickel and an interior side of a resilient spring bronze (92% copper, 8% tin) having a hardness of more than 100 Vickers degrees of hardness, or a resilient spring brass (63% copper, 37% zinc), provides a better spring action with respect to the packing ring and thus insures greater imperviousness than a corresponding laminate having an interior side consisting of nonresilient pure copper, lead, zinc, tin and nonresilient alloys of such metals.

The packing ring is made of nylon which is less subject to cold-flowing. The separators (6) and (7) are provided between the negative electrode of amalgamated zinc powder (9) and the positive electrode (5) which, for example, contains mercury oxide or silver oxide.

It is also possible to have the exterior layer of the laminate cover, instead of the interior layer, made in a resilient manner from nickel or rustproof nickel alloys. However, in that case, based upon practical experience, this outer layer must have a hardness of more than 250 Vickers units when it consists of a nickel alloy. When the outer layer consists of pure nickel, it must have a hardness of more than 150 Vickers degrees of hardness and the interior layer should then consist of spring bronze or spring brass.

Since the closing of the cell by inwardly bending the edge of the housing (4) is to effect an elastic deformation of the laminate cover so that upon removal of the tool used for bending the edge the packing ring (8) is held under resilient pressure by the laminate cover, it is necessary that the metalliz housing (4) be made to have an elasticity that is smaller than that of the laminate cover.

The galvanic cell according to the process is to remain impervious even after it is kept in storage over a period of years and at elevated temperatures. The increased imperviousness is achieved by the spring action of the laminate cover.

Example: A button galvanic cell of 11.4 mm. diameter and 5.2 mm. height according to Figure 3.1, contains in a housing (4) made of nickel-plated sheet steel 1.0 to 1.4 g. of a mixture consisting of 86% mercury oxide, 5% graphite and 9% manganese dioxide. Above this positive electrode (5) lies the separator (6). The packing ring (8) made of nylon encloses the edge (3) of the cover. The negative electrode (9) consists of 0.3 to 0.4 g. coarsely granular amalgamated zinc powder that contains 10 to 15% mercury. An element having the abovementioned outer dimensions displays a capacity up to 240 mah, whereas conventional galvanic cell having the same dimensions only go as high as 165 to 200 mah.

In related work, P. Ruetschi; U.S. Patent 3,673,000; June 27, 1972; assigned to Leclanche SA, Switzerland describes a cylindrical button element of similar construction which is antimagnetic. These devices are especially useful in electronic timepieces.

Stacked Battery Construction

W.H. Bach; U.S. Patent 3,676,221; July 11, 1972 describes a battery comprising a plurality of stacked, disk-like sealed cells secured together by cups fitted over one cell of each pair and having bottoms spot welded to the next cell and sidewalls spot welded to the interfitting cell. A heat shrunk sheath encloses the battery and has caps forming the poles. Between each pair of cells is a circular disk of insulating material against which the cup bottoms bulge upon expansion of the contents of the cells, thereby breaking the welds and electrically disconnecting the cells.

Referring to Figure 3.2, a battery (10) formed by a stack of series connected mercury cells (11) of a type having chemical contents producing a voltage differential between the opposite end surfaces of each cell. Each cell is generally disk shaped and has a cylindrical sidewall (12), a generally flat circular wall (13) (Figure 3.2d) forming the lower end of the disk as viewed in the drawings, and a circular upper end wall (14) fitted tightly into the sidewall and sealed there by a suitable compound shown at (15) in Figures 3.2b and 3.2e.

At least two, and usually more than two, such cells (11) are stacked coaxially to form the battery (10), four cells being shown. To secure the cells together in the stack, each pair of cells is joined together in a suitable manner, such as spot-welding of the adjacent contact surfaces together and here the stack is encased in a plastic sheath (17) fitted snugly around the stack with a shrink fit. Metal caps (18) at the ends of the stack close the ends of the sheath and are electrically connected to the adjacent contact surfaces of the end cells to form the poles of the battery.

To facilitate the welding operation, the lower portion of each cell (11) above the lower end cell is fitted in a cup having a sidewall (19) secured to the cylindrical sidewall (12) of the cell and a bottom (20) secured to the adjacent end of the next cell within the stack.

The bottom (20) of each cup thus can be spot welded to the adjacent lower cell, the next cell fitted in the cup and the sidewalls (12) and (19) spot welded together. In this manner the cup bottoms (20) become extensions of the lower contact walls (13) of the cells. The cells thus are joined electrically by the spot welds, which are indicated at (21) and are held securely in stacked relation by the sheath (17). The spot welds between the cup walls (19) and the disk walls (12) are indicated at (22).

FIGURE 3.2: STACKED BATTERY CONSTRUCTION

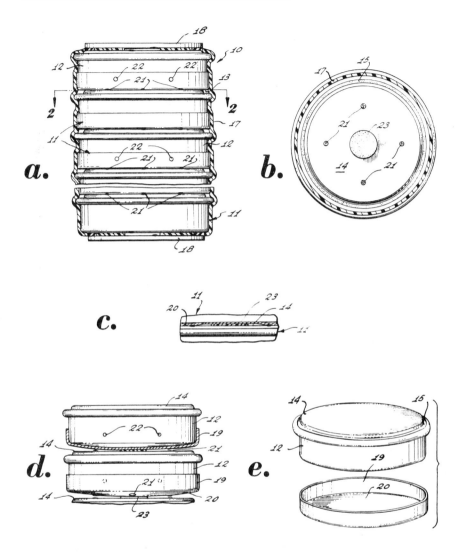

(a) Side View of Battery

(b) Cross-Sectional View Along Line (2—2) of Figure 3.2a

(c) Enlarged Side View Along Line (2—2) of Figure 3.2a

(d) Part of Cell Stack After Expansion of Contact Walls

(e) Exploded View of One Cell

Source: W.H. Bach; U.S. Patent 3,676,221; July 11, 1972

The battery (10) is protected against excessive expansion of the individual cells (11) and thus against rupturing and explosion of the cells, by insulating spacers (23) centrally disposed between the contact walls (14) and (20) of the pairs of cells, thereby to disconnect the spaced portions of the walls electrically from each other.

The joint between the contact walls herein formed by the spot welds (21) is disposed around the spacer and is frangible or breakable in response to an expansion force within a cell great enough to bulge one end wall thereof outwardly against the spacer but insufficient to explode the cell. Thus, the bulging of the central portion of the cell wall against the spacer axially spreads the cells to shift the spot welded portions away from each other, leaving the bulged central portion in engagement with the insulating spacer and the peripheral portions axially spaced apart.

In this instance, each spacer (23) is a relatively thin, circular disk of suitable material such as Mylar, and has a diameter equal to about one-fourth the diameter of the contact walls (14) and (20) of the cells and a thickness on the order of 0.010 of an inch. For convenience in assembly and to insure retention of the spacer in the selected central position, it preferably is cemented to one of the end walls, either on the underside of the cup bottom (20) or on top of the upper contact wall (14).

As shown most clearly in Figures 3.2b and 3.2d, the representative frangible joint comprises four spot welds (21) distributed around the spacer (23) approximately midway between the latter and the periphery of the upper contact wall (14) of the lower cell in the pair. In the illustrative cells, the upper wall is relatively rigid while the lower wall (13) and the cup bottom (20) are more flexible so that the spot-welding operation draws each cup bottom down around the spacer in an inverted, dished, configuration (Figure 3.2c) for conducting contact along a substantially uninterrupted and generally circular line passing through the four welds. The central portions of the contacting end walls, however, are separated by the spacers.

The expanded condition of one of the cells (11) is illustrated in Figure 3.2d where it will be seen that the lower end wall (13) and the cup bottom (20) have bulged downwardly from the normal condition, the central portion having bulged to a greater extent than the peripheral portion, as would be expected.

Thus, the central portion is the primary pressure point through which the expansion acts against the spacer in tending to spread the two cells of the pair, and the result is a breaking and separation of the four spot welds (21). This leaves the cells in the condition shown, in which the cup bottom (20) is insulated from the adjacent upper contact wall (14) by the spacer in the central portion and by the air gap outside the periphery of the spacer.

Accordingly, there is no conducting contact between the two cells and the electrical connection is effectively broken to prevent further conduction that could lead to additional expansion and possible rupture of the expanded cell.

From the foregoing, it will be seen that the simple insulating spacers (23) between adjacent cells (11) may be incorporated in a battery (10) without significant expense in labor or materials, and without interfering with normal operation of the battery, but are operable automatically as an incident to the beginning of excessive expansion to disable the battery and avoid possibly serious damage.

Addition of Mn_2O_3 to Mercury Depolarizer Mix

R.J. Dawson; U.S. Patent 3,600,231; August 17, 1971; assigned to ESB Inc. describes a mercury cell having additives in the mercuric depolarizer mix which have the effect of stabilizing the open circuit voltage and inhibiting agglomeration of mercury in the depolarizer mix. The preferred additive is Mn_2O_3 although Mn_3O_4 has been found to be satisfactory. The additive is mixed with the depolarizer mix in amounts ranging from at least about 1% by weight of the mix to a maximum percentage consistent with capacity and volume requirements.

The basic elements of a mercury cell include a cathode cup containing a mercuric oxide depolarizer mix, a zinc anode pellet with a barrier material placed between the pellet and the depolarizer mix, an electrolyte absorbent material between the zinc pellet and the barrier, an alkaline electrolyte, an anode cup which surrounds the zinc pellet and mates with the cathode cup and a means to seal the cups which is normally provided by an insulating grommet or washer. Usually a small amount of graphite is included in the depolarizer mix to improve its conductivity.

A cell in accordance with this process has a metallic cathode cup which preferably is made of nickel-plated steel. Pressed into this cup is the mercuric oxide depolarized mix which comprises mercuric oxide with minor amounts of graphite and a polystyrene binder. The depolarizer mix also contains the Mn_2O_3 or Mn_3O_4 additive to stabilize the open circuit potential. A woven nylon material with an ion exchange resin therein acts as a barrier between the depolarizer mix and the zinc pellet, and in addition, an electrolyte absorbent material is placed between the zinc pellet and the barrier.

This absorbent material preferably is a nonwoven matted cellulosic such as cotton fibers. The anode cup then covers the zinc pellet and is sealed to the cathode cup by means of an insulating grommet molded to the edge of the anode cup, and the cathode cup is crimped upon the insulating grommet to form a sealed cell.

Prior to placing the mercuric oxide depolarizer mix in the cell, the mix may be prewetted with alkaline electrolyte. This prewetting of the depolarizer mix aids in inhibiting excessive electrolyte absorption by the mix. The use of a hydrophobic plastic binder such as polystyrene is also quite helpful in giving structural support to the mix as well as preventing excessive absorption of electrolyte.

Example: A series of cells were made according to this process using various amounts of Mn_2O_3 in the depolarizer mix. In all series, the test cells were negative limited and contained the same amount of negative active material. The additive was mixed with the mercuric oxide depolarizer mix having a small amount of graphite. In some instances a polystyrene binder was added to the mix while other cells were made without the binder.

The test cells were constructed without the nylon barrier between the electrodes to demonstrate the effect of the additives. The final mixture of depolarizer mix and additive was then pressed into the proper shape and pressed into the cathode cup. In Table 1 are listed series of 625 size cells having the components shown in the depolarizer mix. These test cells were subjected to open circuit voltage and capacity tests. The quantities of the components of the depolarizer mix are given in terms of parts by weight of the mix.

TABLE 1

Cell series:	Open circuit voltage (days)			Capacity hours, 0.9 v.
	0	7	14	
A	1.320	1.320	1.320	290
B	1.304	1.298	1.296	320
C	1.298	1.285	1.284	252
D	1.335	1.335	1.335	482
E	1.334	1.335	1.332	500
F	1.328	1.330	1.330	420
G	1.335	1.335	1.335	502
H	1.335	1.336	1.340	505
I	1.333	1.335	1.335	500

The cells were tested on a 1,000 ohms/v. voltmeter over a period of time as shown in the table. Capacity measurements were taken after three weeks storage at room temperature. The capacity was measured in terms of hours elapsed before the cell voltage reached 0.9 v.

TABLE 2

Series:	Parts by weight			
	HgO	Mn₂O₃	Graphite	Binder
A	95	0	5	0
B	95	0	5	½
C	95	0	5	1
D	85	10	5	0
E	85	10	5	½
F	85	10	5	1
G	75	20	5	0
H	75	20	5	½
I	75	20	5	1

Table 2 clearly illustrates the improved open circuit voltage stability provided by the Mn_2O_3 in cells having a binder, and the improved capacity in all cells provided by incorporating Mn_2O_3 in the mercuric oxide depolarizer. Other part by weight variations of the components in the depolarizer mix were tested with corresponding improvements in the open circuit voltages and capacities. The quantity of Mn_2O_3 or Mn_3O_4 added to the depolarizer mix affects the cell capacity and consequently, the capacity requirement of the cell is an important factor in determining the amount of additive to be included in the depolarizer mix.

ZINC–SILVER OXIDE

Zinc Powder–Sodium Hydroxide Electrode

E.C. Jerabek and R.P. Hamlen; U.S. Patent 3,533,843; October 13, 1970; assigned to General Electric Company describe a zinc electrode which comprises a support and a mixture of amalgamated zinc powder and metallic or alkyl substituted ammonium hydroxide particles on opposite surfaces of the support.

The method of forming a zinc electrode comprises mixing together amalgamated zinc powder and metallic or alkyl substituted ammonium hydroxide particles, applying the mixture to opposite surfaces of a support and pressing the mixture against the support.

Example 1: A zinc electrode was prepared by mixing zinc powder in a mercury salt electrolyte of mercuric chloride so that the zinc powder became amalgamated to 5 weight percent mercury. 10 g. of this amalgamated zinc powder and 2 g. of crushed sodium hydroxide were mixed together. A support in the form of a silver screen was used.

A die was filled with the above mixture after which the screen support with an electrical lead extending therefrom was positioned in the center of the powder. In this manner, the powder mixture was applied to opposite surfaces of the support. The powder was pressed against the support at a pressure of 5,000 psi to thereby form a zinc electrode.

Example 2: The zinc electrode of Example 1 was positioned as the anode in an electrically insulated cell casing. This electrode was 7.86 amp.-hours in capacity. A pressed silver oxide electrode, which was 10 amp.-hours in capacity was positioned in the cell as the cathode. The electrodes were separated by a cellophane membrane separator. No loss in capacity due to self discharge results since no aqueous electrolyte is present.

Subsequently, distilled water was added in the amount of 3.6 cc to make an aqueous electrolyte of 35.7 weight percent sodium hydroxide. The cell was operated through 148 ohms load for 670 hours giving 6 amp.-hours capacity. The cell was operated then through 682 ohms load providing an additional 0.5 amp.-hour capacity.

Potassium Hydroxide-Magnesium Oxide Activator

A process described by R.P. Hamlen and E.C. Jerabek; U.S. Patent 3,578,504; May 11, 1971; assigned to General Electric Company relates to silver oxide-zinc primary cells and to such cells containing magnesium oxide powder with a high viscosity electrolytic solution added to the cell prior to use.

Normally, a primary silver oxide-zinc cell has a metallic casing with central zinc anode, normally a cellophane separator wrapped around the anode, a divalent silver oxide cathode on each opposite side of the anode and spaced therefrom by the separator and an aqueous alkaline electrolyte in contact with the electrodes. Such batteries are manufactured generally in a charged state thereby available to produce electrical energy upon discharge.

One serious problem associated with silver oxide-zinc primary cells is that the silver oxide is slightly soluble in most alkaline electrolytes. This allows silver to migrate to the zinc electrode and to the various cell components. The deposition of silver on the zinc electrode increases the direct reaction of the zinc with the electrolyte, causing a greater pressure buildup in the cell as a result of hydrogen evolution. Electrolyte leakage is also caused during gas venting. Deposition on cell components, such as on the ceramic insulator, may also result in short circuits in the cell.

Frequently, such cells employ cellophane or other microporous membrane-type separators adjacent the electrodes to minimize silver migration. This introduces another serious problem of electro-osmotic pumping of electrolyte toward the silver oxide cathode, which decreases both the volume and the concentration of electrolyte adjacent the anode. This process involves a method of forming a silver oxide-zinc primary cell comprising providing a casing, providing a zinc anode positioned in the casing, providing at least one silver oxide cathode positioned in the casing and spaced from the anode, adding magnesium oxide powder within the casing, closing the casing and filling the assembled cell prior to use with an alkaline electrolyte.

In Figure 3.3a there is shown generally at (10) a cell which has a metallic casing (11) including a metallic body portion (12) and a metallic cover portion (13). A central zinc anode (14) is shown positioned within body portion (12) and surrounded except at the top with one or more chemically inert, cellophane separators (15). An electrical lead (16) extends from anode (14) through an aperture (17) in cover portion (13). An electrically insulating, ceramic insulator (18) is positioned in aperture (17) and lead (16) passes through the insulator to the exterior of the cell.

A divalent silver oxide cathode (19) is shown positioned on each side of anode (14) and spaced therefrom by separators (15). An electrical lead (20) is attached to cover portion (13) for cathodes (19) which are in electrical contact with both body portion (12) and cover portion (13). A vent opening (21) is provided in cover portion (13), which opening is closed by a removable vent cover (22). The electrolytic solution of magnesium oxide in the electrolyte is introduced prior to cell use through vent opening (21) into a chamber (23) defined by the space above the electrode and separators and below the cover portion.

In Figure 3.3b there is shown a partial sectional view of a cell similar to Figure 3.3a which is modified by chamber (23) being filled with magnesium oxide powder (24) prior to metallic cover (13) being sealed to body portion (12). The alkaline electrolyte is introduced prior to cell use through the vent opening.

It has been found that an assembled silver oxide–zinc primary cell could be activated by mixing together 5 to 40 weight percent of magnesium oxide powder in an aqueous alkaline electrolyte thereby forming a pourable electrolytic solution, filling the assembled cell with the electrolytic solution and allowing the viscosity of the solution to increase prior to discharging the cell. Examples of silver oxide–zinc primary cells are set forth below. A cell, which was not made in accordance with this process, is set forth below in Example 1. Cells, which were made in accordance with this process, are set forth in Examples 2 and 3.

Example 1:   A silver oxide–zinc primary cell was made as shown generally in Figure 3.3a. A zinc anode was wrapped on all sides except the top with a chemically inert cellophane separator material and positioned centrally within the stainless steel body portion of a cell casing. A divalent silver oxide cathode was positioned on each side of the anode and spaced therefrom by the separator material. Each cathode was in electrical contact with the metallic body portion of the cell casing while the anode was electrically connected to an electrical lead extending upwardly from the anode.

An electrolyte of 30% potassium hydroxide was added to the cell after which a metallic cover portion was sealed to the upper open end of the body portion. The electrical lead from the anode extended through an aperture in the cover portion, which aperture included a ceramic insulator to electrically insulate the lead from the metallic cover. An electrical lead for the cathode was attached to the metal cover. A vent opening was also provided which included a vent closure for the opening. Each of the silver oxide electrodes had a capacity of 5.3 amp.-hours while the anode had a capacity of 10.5 amp.-hours.

FIGURE 3.3:   SILVER OXIDE-ZINC PRIMARY CELL

a.

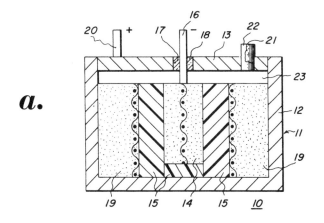

Sectional View of Cell

b.

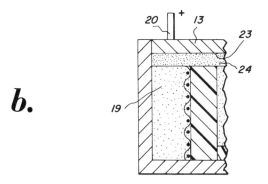

Sectional View of Modified Cell

Source: R.P. Hamlen and E.C. Jerabek; U.S. Patent 3,578,504; May 11, 1971

This cell had an initial voltage of 1.609 v. when tested after assembly. It was noted that this voltage fell to 0.712 v. under no load condition within 800 hours after assembly. A load was then connected across the electrodes of the cell and the cell would not discharge above 1 v. The cell was disassembled and examined visually. It was noted that the ceramic insulator was coated with a dark deposit of silver which apparently caused an initial short circuit across the ceramic insulator resulting in the initial drop in voltage under no load conditions.

Example 2:   A silver oxide-zinc cell was made as described above in Example 1 with the exception that no electrolyte was added to the cell prior to assembly of the cell.

A mixture of 10.0% magnesium oxide in 30% sodium hydroxide was formed into a pourable electrolytic solution which was then added immediately to the interior of the cell through the vent opening, which was then closed. The cell was allowed to stand for a period of 24 hours to allow the electrolytic solution to become highly viscous. There was no decrease in initial cell voltage under no load conditions as occurred in Example 1. The cell was then discharged at a current of 10.0 ma. for a period of 940 hours. The efficiency of the cell was 94%.

Example 3:   A silver oxide–zinc cell was made as set forth above in Example 1 with the exception that 0.5 g. of magnesium oxide powder was added to the cell as shown in Figure 3.3b prior to completing the assembly of the cell. 30 weight percent sodium hydroxide was added to the cell to provide an electrolytic solution having 15% magnesium oxide powder in the electrolyte. The cell was allowed to stand for a period of 24 hours so that the electrolytic solution became highly viscous. This cell did not exhibit a decrease in initial cell voltage under no load as occurred in Example 1.

The cell was then discharged at a current of 23.5 ma. for a period of 350 hours. The cell was then discharged further at a current of 3.5 ma. for 330 hours giving 9.38 amp.-hours or 92% efficiency. Upon subsequent disassembly and visual observation, there was no evidence of silver migration on the insulator or anode.

Mono- and Divalent Silver Oxide

L.A. Soto-Krebs; U.S. Patent 3,655,450; April 11, 1972; assigned to ESB Inc. describes an electrode means for achieving a single potential discharge from a multivalent oxide, such as divalent silver oxide, that discharges at two or more potentials. The principle of the process can be illustrated by means of a divalent silver oxide electrode which discharges at two different potentials. During the initial discharge of such an electrode, the divalent silver oxide is reduced to monovalent silver oxide.

Theoretically, this reaction proceeds until all of the divalent silver oxide has been reacted. Next, the monovalent silver oxide is further reduced to metallic silver. The first reaction, while divalent silver oxide is present, provides an open circuit potential, in an alkaline electrolyte of approximately 1.8 v. versus zinc. The second reaction while monovalent silver oxide is present, provides an open circuit potential in an alkaline electrolyte of approximately 1.6 v. versus zinc.

In practice, due to such effects as polarization and the masking of the divalent oxide by the formation of monovalent oxide, before a divalent silver oxide cell is halfway through its useful life, its output voltage will drop 0.2 v. Many types of battery-operated electronic equipment cannot tolerate a voltage change of this magnitude. A divalent silver oxide electrode in accordance with the process, however, will deliver all of its capacity at the monovalent potential, thus providing a substantially constant output voltage.

To achieve a lower potential discharge from an active material in accordance with this process, three conditions must be met. First, the discharge product of the lower potential active material must be readily oxidizable in the battery electrolyte by the higher potential active material. Second the higher potential active material must be electronic contact with the lower potential material. Third, the discharge circuit must have electronic connection only with the lower potential active material. Divalent silver oxide and monovalent silver oxide are examples of active materials which satisfy the first condition stated above. The discharge product of the lower potential active material, monovalent silver oxide, is metallic silver. Metallic silver is readily oxidizable by the higher potential active material, divalent silver oxide, to monovalent silver oxide.

The second and third conditions are achieved by an electrode structure utilizing as the principal active material a body of divalent silver oxide having as the secondary active material a layer of monovalent silver oxide. The electrode contact means, that is, the electrical path for the discharge circuit is in contact only with the monovalent silver oxide layer.

By way of example of the applicability of the process to other active material, manganese dioxide can be discharged at the lower potential of copper oxide. Similarly, divalent silver oxide can be discharged at the potential of copper oxide. Potassium permanganate can be discharged at a lower potential such as that of monovalent silver oxide or copper oxide. These examples are illustrative of only a few electrode combinations which are possible by means of this process.

The process can be illustrated in connection with a positive silver electrode. The capacity in milliampere hours per gram and the specific gravity of both divalent and monovalent silver oxide are given in Table 1.

TABLE 1

	Milliampere Hours per Gram	Specific Gravity
Monovalent Silver Oxide (Ag$_2$O)	232	7.14
Divalent Silver Oxide (AgO)	433	7.44

As shown, divalent silver oxide has 1.87 times more capacity per gram than the monovalent oxide and has 1.95 times more capacity per unit volume than the monovalent oxide. The significance of this becomes apparent with reference to Figures 3.4a and 3.4b. Figure 3.4a shows a theoretical discharge curve of a conventional silver-zinc primary cell using monovalent silver oxide as the positive active material. Figure 3.4b shows the theoretical discharge curve of a similar cell utilizing an equal volume of divalent silver oxide as the positive active material.

Both of the discharge characteristics shown represent continuous discharges through a 300 ohm load at 73°F. Obviously, the cell of the Figure 3.4b, the cell utilizing divalent silver oxide as the positive active material, has substantially more usable capacity than the cell in Figure 3.4a.

However, while the divalent silver oxide cell has significantly more capacity than the monovalent silver oxide cell, its discharge is characterized by two distinct voltage plateaus. One, at approximately 1.72 v. and the other at approximately 1.5 v. Many battery applications, particularly transistorized devices such as hearing aids, cannot tolerate a voltage drop such as is exhibited by the cell of Figure 3.4b.

FIGURE 3.4:   DIVALENT SILVER ELECTRODE

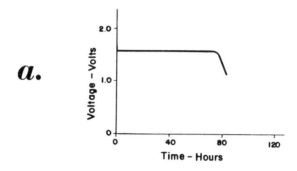

Monovalent Silver Oxide-Zinc Cell

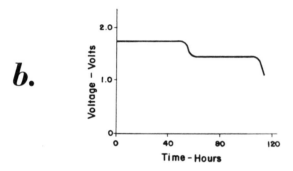

Divalent Silver Oxide-Zinc Cell

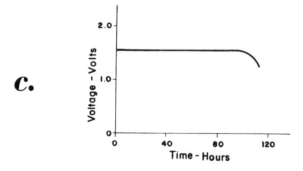

Mono- and Divalent Silver Oxide-Zinc Cell

(continued)

FIGURE 3.4:   (continued)

d.

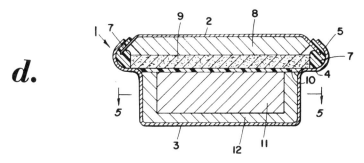

Cross–Sectional View of Primary Cell

e.

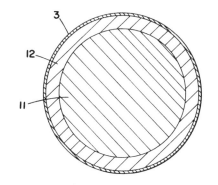

Top View Along Line (5—5) of Figure 3.4d

f.

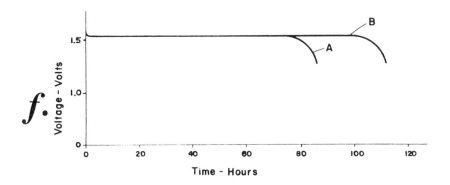

Performance Curve for Special Silver–Zinc Cell

Source: L.A. Soto-Krebs; U.S. Patent 3,655,450; April 11, 1972

The electrode structure is designed to provide means for utilizing the capacity of an inherently high capacity material such as divalent silver oxide at the potential of a second material having a lower potential such as monovalent silver oxide. Referring to Figure 3.4c, there is shown the theoretical discharge curve of a silver–zinc cell identical to those shown in Figures 3.4a and 3.4b, except that it utilizes an electrode of the process. The curve of Figure 3.4c is plotted for discharge conditions with Figures 3.4a and 3.4b. As shown by this curve, the cell exhibits a single potential discharge at the monovalent oxide potential level while utilizing the capacity of divalent silver oxide at that lower potential.

Referring to Figure 3.4d, there is shown a sectional elevation of a silver–zinc primary cell, designated by the numeral (1), having a positive electrode in accordance with the process. The cell (1) is conventional in all respects with the exception of the construction of the positive electrode. The cell (1) has a two-part container comprising an upper section or cap (2) which houses the negative electrode, and a lower section or cup (3) which houses the positive electrode.

As shown, the bottom cup (3) is formed with an annular shoulder (4) having a flange (5) which is crimped inward during assembly to seal the cell. The bottom cup (3) may be made of nickel-plated steel and the cap (2) may be made of tin-plated steel. The cap (2) is insulated from the cup (3) and the flange (5) by means of a grommet (7) which is compressed between the cap (2) and the flange (5) during the crimping operation of cell assembly to provide a compression seal between these parts. The grommet (7) may be made of a suitable resilient electrolyte-resistant material such as neoprene.

The negative electrode of the cell (1) comprises a lightly compacted pellet (8) of finely divided amalgamated zinc. The zinc electrode (8) is separated from the positive electrode by means of an electrolyte absorbent layer (9) and a membrane barrier (10). The electrolyte absorbent layer (9) may be made of electrolyte-resistant, highly absorbent substance such as matted cotton fibers (Webril). The barrier layer (10) may be a suitable semipermeable material such as cellophane, or comprise a suitable organic carrier such as polyethylene or polyvinylchloride having a polyelectrolyte homogeneously dispersed therethrough. Such a material is described in U.S. Patent 2,965,697.

The positive electrode of the cell (1) comprises, in accordance with this process, a first pellet (11) of the divalent silver oxide which is surrounded on the bottom and side surfaces by a layer of monovalent silver oxide (12). The pellet (11) of divalent silver oxide is the principal active material and comprises the majority of the active material in the electrode available for discharge. The layer of monovalent silver oxide (12) is the secondary active material. As shown in Figures 3.4d and 3.4e, the layer (12) of monovalent silver oxide isolates the pellet of divalent silver oxide from all electronic contact with the bottom cup (3) which is the positive terminal of the cell and the electrode contact.

This electrode may be formed in a number of ways. For example, the pellet (11) may be formed by first pelletizing finely divided divalent silver oxide powder in a suitable die. This pellet may then be centered in a bigger pellet die, and finely divided monovalent silver oxide powder compressed around it to form the composite pellet of the type shown in Figures 3.4d and 3.4e. It is possible also to form the electrode by pelletizing a suitable quantity of divalent silver oxide powder and then chemically reducing its surface to the monovalent oxide. In addition, where desired, the surface layer of the divalent silver oxide pellet can be reduced to metallic silver and that layer subsequently reoxidized to monovalent oxide. Electrochemical reduction can also be utilized to reduce the surface layer of a divalent silver oxide pellet to the monovalent oxide.

It should be noted that the thinner the layer of monovalent silver oxide, the more divalent oxide can thus be included in the electrode increasing its capacity. However, under no circumstances should there be a discontinuity in the monovalent oxide layer which would provide direct electronic contact between the divalent oxide body and the electrode to discharge in the conventional manner with two voltage plateaus.

The curves of Figure 3.4f demonstrate the increase in capacity actually available from a cell in accordance with the process. In this figure, curve (A) is the discharge curve of a conventional silver-zinc cell, and curve (B) is the discharge curve for a cell having a positive electrode in accordance with the process. The cell construction utilized in both cells was that shown in Figure 3.4d. The negative electrodes of both cells comprised lightly compacted battery grade metallic zinc amalgamated with 14% mercury.

The cells were sealed with neoprene rubber grommets. In both cells, the separation between the electrodes comprised a layer of Webril absorbent and a 3 mil layer of membrane made in accordance with descriptions in U.S. Patent 2,965,697. The electrolyte absorbent layer was saturated with an electrolyte formulated by dissolving 100 g. of potassium hydroxide and 16 g. of zinc oxide in 100 cc of water.

The positive electrode of cell (A) comprises 1.73 g. of commercially available battery grade monovalent silver oxide compressed into a pellet 0.1 inch thick and 0.485 inch in diameter. The positive electrode of cell (B), the cell in accordance with the process, comprised a central pellet containing 1.2 g. of divalent silver oxide 0.085 inch thick and 0.387 inch in diameter.

This central pellet was surrounded by 0.6 g. of monovalent silver oxide which was compressed to provide a composite pellet 0.1 inch thick and 0.485 inch in diameter. The divalent silver oxide mix from which the central pellet was compressed comprised 96 g. of a specially prepared low-gassing divalent silver oxide and 4 g. of lead dioxide to which there was added 3.5 cc of the electrolyte described hereinbefore for each 100 g. of the dry mix.

Both cells showed an open circuit voltage of between 1.58 and 1.59 v. and each had an impedance lower than 6 ohms. The curves of Figure 3.4f are for a continuous discharge of the cells through a 300 ohms resistance at a temperature of 73°F. As shown from these curves, cell (B), the cell in accordance with this process, exhibited approximately 30% more useful capacity than an identical size cell of the conventional monovalent silver oxide type. Of equal importance is the fact that this increase in capacity is achieved without the usual two voltage plateau discharge usually associated with divalent silver oxide.

Electrodes have been constructed using electrode materials other than divalent and monovalent silver oxides. For example, divalent silver oxide has been discharged at the potential of 0.9 v. versus zinc through a 300 ohms load at 73°F. by using

cupric oxide as the second electrode material. The cupric oxide composite electrode was constructed by mechanically pressing finely divided cupric oxide around
a pellet of divalent silver oxide. Some examples of electrode combinations which
may be constructed in accordance with the process together with open circuit voltages for such electrodes in alkaline electrolyte are listed in Table 2.

TABLE 2

Primary Active Material		Secondary Active Material	
Material	Voltage vs. Zinc (Open Circuit)	Material	Voltage vs. Zinc (Open Circuit)
MnO_2	1.5	CuO	1.1
AgO	1.8	Ag_2O	1.6
AgO	1.8	CuO	1.1
$KMnO_4$	1.8	Ag_2O	1.6
$KMnO_4$	1.8	CuO	1.1

In each case the primary active material must be able to oxidize the discharge product of the secondary active material. If the materials chosen meet this requirement
and the primary active material is in electronic contact with the secondary active
material while the discharge circuit has electronic connection only with the secondary active material as described, the electrode will discharge only at the lower potential of the secondary active material.

Additional work with the divalent silver oxide and monovalent silver oxide type
electrodes is described by L. Soto-Krebs and R.J. Dawson; U.S. Patent 3,476,610;
November 4, 1969; assigned to ESB Incorporated and L. Soto-Krebs; U.S. Patent
3,615,858; October 26, 1971; assigned to ESB Incorporated.

Barrier Layers for Mono- and Divalent Silver Oxide Electrodes

R.J. Dawson; U.S. Patent 3,484,295; December 16, 1969; assigned to ESB Inc. describes a battery having a positive electrode comprising a principal active material,
(e.g., divalent silver oxide, potassium permanganate, cupric oxide, sulfur) and a
secondary active material, (e.g., monovalent silver oxide, mercuric oxide, manganese dioxide) which is stable in the battery electrolyte.

The secondary active material is employed as a substantially electrolyte impermeable
layer, such as being dispersed in a suitable plastic material, which is interposed between the principal active material and the battery components containing the electrolyte so as to isolate the principal active material from contact with the electrolyte
until the secondary active material is discharged. The battery can be discharged at
a single potential if the discharge product of the secondary active material is oxidized by the principal active material in the presence of the battery electrolyte.

The positive electrode for a typical cell comprises a first pellet of divalent silver
oxide consolidated in situ in the cathode cup at 15,000 psi of pressure. The masking layer comprises monovalent silver oxide and polystyrene. The particles of monovalent silver oxide are first coated with polystyrene by mixing the particles in a solution of carbon tetrachloride containing polystyrene in the amount of 1% by weight

of the monovalent silver oxide. The coated particles are then consolidated by means of pressure to produce a structurized layer united with the preformed pellet. In this operation the pressure will compress the monovalent silver oxide into a continuous phase throughout the polystyrene to form an electrolyte impermeable masking layer.

It is also possible to form the masking layer by dispersing the secondary active material throughout a suitable plastic such as polyethylene, polypropylene and polytetrafluoroethylene or the like by means of heat and pressure in a mill. In this process the plastic is first plasticized and the secondary active material added while the binder is in the plasticized state.

The material thus produced can be sheeted to an appropriate thickness by calendaring rolls to produce sheet material from which masking layers can be cut by means of a stamping operation. Masking layers produced in this manner can be united with the remainder of the electrode structure by means of pressure.

The discharge characteristics of a cell incorporating an electrode in accordance with the process will depend upon the active material utilized. Generally, it would be expected that the discharge characteristic of the electrode will show a voltage rise or drop from the potential characteristic of the secondary active material when electrolyte is made available for the discharge of the principal active material.

The expected change in voltage will be to the voltage characteristic of the principal active material. However, with certain active materials, electrodes in accordance with this process produce single potential discharges at voltages characteristic of the secondary active material. However, the single potential discharge is obtained only where the discharge product of the secondary active material is oxidizable by the principal active material.

Secondary active materials such as monovalent silver oxide, mercuric oxide and manganese dioxide which have discharged products which are oxidized by high potential materials such as divalent silver oxide and potassium permanganate may be used in electrodes with either of these materials as the principal active material to provide a single potential discharge.

It is believed that where the single potential discharge is achieved, the geometry of electrode discharge is such that the masking layer functions to effectively isolate the principal active material from ionic contact with the electrode of opposite polarity throughout the discharge of the electrode. During discharge, the secondary active material in the barrier layer is first reduced until the interface with the principal active material is reached.

At this time the electrolyte comes into contact with the principal active material and an internal cell within the positive electrode is formed. This internal cell comprises the discharge product of the secondary active material as the negative electrode and the principal active material as the positive electrode.

Due to the intimate electronic and ionic contact between these materials, an oxidation and reduction reaction occurs between them which takes precedence over the reaction between the principal active material and the negative electrode of cell which is isolated from it electronically. The effect is manifested in the cell voltages which are at the potential characteristic of the couple formed by the secondary active material and negative electrode under load or discharge and at the potential characteristic of the couple formed by the principal active material and negative electrode on open circuit. Examples of the performance of cells having different secondary active materials in a layer masking divalent silver oxide are shown in the table below.

Cell Number	Secondary Active Material	Average Voltage to 0.8 v.
1	HgO	1.18
2	MnO_2	1.31
3	Ag_2O	1.35
4	CuO	0.98

All of the cells tested were button cells and were identical in construction except for the material in the masking layer. The negative electrode in these cells was amalgamated zinc. In cell number 2, the secondary active material of the masking layer comprised manganese dioxide and carbon in the proportions of 9 to 1. Cell number 4 incorporated cupric oxide and carbon in the proportion of 4 to 1 in the masking layer. While cupric oxide is not particularly stable in alkaline electrolyte and hence, not ideally suited for utilization in the masking layer, this cell was constructed to illustrate the unipotential principles of the process. Polystyrene was used in all of the cells as the binder in the masking layer.

All of the cells tabulated in the table discharged at a single voltage level and gave no voltage rise which would show the difference in potential level of the material of the masking layer and the divalent silver oxide. The discharges were continuous and the cells were negative limited. In all cases, inspection of the cells after discharge showed that the divalent silver oxide was used completely while some undischarged material remained in the masking layer. Similar cells made with mercuric oxide in the masking layer and with cupric oxide as the principal active material showed discharge curves at two distinct voltage levels.

This was to be expected since cupric oxide cannot oxidize mercury. These cells discharged at an average closed circuit voltage of 1.19 v. for 40 mah and then the voltage dropped to 0.76 v. and the remainder of the cell discharge was at this level. Similar electrodes utilizing potassium permanganate as the principal active material have been discharged against zinc at the potential characteristic of the secondary active material in the barrier layer.

In this respect, barrier layers in accordance with the process have been particularly effective in preventing electrolyte contact with the potassium permanganate principal active material on prolonged stands. Inspection after several months of stand of the cells so constructed showed none of the electrolyte coloration associated with the soluble potassium permanganate.

From the foregoing, it can be seen that by means of the process there has been provided electrode means which minimizes the effect of the instability of the principal active material. This permits the utilization of highly active, high capacity but unstable active materials in cells and still provides cells with good stand characteristics.

In addition, with certain active materials the electrode provides a discharge at a lower potential than that which normally characterizes the principal active material. In this respect, it should be noted that in the case of divalent silver oxide principal active material, an electrode of the process provides a means for discharging this material at a single potential, whereas this material customarily discharges at two voltage levels.

Addition of Mercury to Divalent Silver Oxide

A. Tvarusko; U.S. Patent 3,650,832; March 21, 1972; assigned to ESB Incorporated has found that certain metallic additives, when incorporated into divalent silver oxide battery active material, will reduce the electrical resistivity of the divalent silver oxide active material, i.e., improve its conductivity, or improve its stability in aqueous alkaline electrolyte without substantially adversely affecting either property.

It is preferred to incorporate the additives into the divalent silver oxide active material by chemical coprecipitation during its preparation, but they can also be physically admixed with the divalent silver oxide active material. It has been found that certain of the additives are also effective when incorporated into the alkaline electrolyte of batteries employing divalent silver oxide positive electrodes, in which case the additives can be omitted from the divalent silver oxide active material.

After extensive tests, it has been determined that mercury improves the electrical conductivity of divalent silver oxide active material and selenium and tellurium improve the stability of divalent silver oxide active material in aqueous alkaline electrolyte. In addition to improving the electrical conductivity, the mercury additive does not substantially adversely affect the stability of the divalent silver oxide active material and in like manner, the selenium and tellurium additives do not substantially adversely affect the conductivity.

It has also been found that combinations of the mercury additive with lead or tin improve both electrical conductivity and the stability of divalent silver oxide active material in aqueous alkaline electrolyte, with particular improvement in the electrical conductivity. This was quite unexpected, for the mercury alone slightly impaired the stability of the divalent silver oxide active material, and the improvement in the electrical conductivity was greater than when the components of the co-additives were used separately.

The divalent silver oxide active material is preferably prepared by the persulfate oxidation of a silver salt, such as silver nitrate, in alkaline solution. The equation of the reaction is written as follows.

$$4AgNO_3 + 2K_2S_2O_8 + 8NaOH \longrightarrow$$

$$4AgO + K_2SO_4 + 3Na_2SO_4 + 2NaNO_3 + 2KNO_3 + 4H_2O$$

This method for preparing divalent silver oxide is reported by Robert N. Hammer and Jacob Kleinberg in "Inorganic Synthesis", Vol. 4, pp. 12 to 14, McGraw-Hill Book Co., New York (1953). The procedure is reported as follows. 72 g. of NaOH in pellet form is added in portions, with constant stirring, 1 l. of water, which is

maintained at approximately 85°C. 75 g. of $K_2S_2O_8$ in the form of an aqueous
slurry is added to the hot alkaline solution; this is followed by the addition of 51 g.
of silver nitrate dissolved in a minimum amount of water. The temperature of the
resulting mixture is raised to 90°C. and stirring is continued for approximately 15
minutes. The precipitate of black divalent silver oxide is filtered on a large Buchner
funnel and sulfate ion is removed by washing with water which is made slightly alkaline with sodium hydroxide. The product is air dried with a yield of 35 g. (94%).

It has been found to be advantageous to incorporate the metallic additives into the
divalent silver oxide active material during its chemical preparation, i.e., oxidation of aqueous silver nitrate solution by alkaline persulfate. The preferred method
for adding a metallic element to the AgO is to add its nitrate salt to the aqueous
silver nitrate solution before chemical oxidation. If the nitrate salt of the metallic
additive is not available or cannot be used because of hydrolysis, the metallic element can be added as an anion in the form of the potassium or sodium salt if one is
available.

If there is a reaction between the silver ion and the anion, the metallic element in
anion form can be added to the aqueous alkaline persulfate solution directly, or dissolved first in the NaOH solution. The metallic additives included within the scope
of this process can be added as follows.

Additive Salt	Added to
H_2SeO_3	$K_2S_2O_8$-NaOH solution
$Hg(NO_3)_2$	$AgNO_3$ solution
Na_2SnO_3	$K_2S_2O_8$-NaOH solution
Na_2TeO_3	$K_2S_2O_8$-NaOH solution
$Pb(NO_3)_2$	$AgNO_3$ solution

This method for incorporating the additives into the divalent silver oxide active material provides a uniform distribution of the additive throughout the AgO. In addition to incorporating the additive by chemical coprecipitation, it may also be added
to the divalent silver oxide by physically admixing it with AgO. Those additives
which improve the stability of the divalent silver oxide (selenium, tellurium, mer-
cury-lead and mercury-tin) are also effective when added to the battery electrolyte
instead of the divalent silver oxide active material. The following examples illustrate the preparation of divalent silver oxide containing metallic additives and the
improved conductivity and stability that is achieved.

Example 1:   102 g. of $AgNO_3$ was dissolved in 85 ml. to form a $AgNO_3$ solution,
and 150 g. of $K_2S_2O_8$ was added to 60 ml. of water to form a $K_2S_2O_8$ slurry.
$Hg(NO_3)_2$ was dissolved in the $AgNO_3$ solution in amount sufficient to provide 1%
by weight of Hg ion per weight of Ag ion. 2,000 ml. of water was placed in a container and it was heated and agitated by a mechanical stirrer. When the water temperature reached about 60°C., 144 g. of NaOH pellets was added in several small
portions to form an alkaline solution. After the alkaline solution reached a temperature of 85°C., it was maintained at this temperature with continued agitation.
The $K_2S_2O_8$ slurry was added to the alkaline solution in small portions, and thereafter, the $AgNO_3$ solution containing Hg ions was very slowly added to the alkaline

solution containing $K_2S_2O_8$. The temperature of the reaction solution was permitted to rise to 90°C. The reaction solution was agitated at 90°C. for 15 minutes to complete the oxidation of the silver ions to divalent silver oxide. After the reaction, the supernatant liquid and then the precipitate was filtered on a Buchner funnel. The filtrate was thoroughly washed with slightly alkaline wash water (3 g. of NaOH per gallon). The AgO filter cake was removed from the funnel and placed in a vacuum oven and was dried for 16 hours at about 60°C. The dried AgO weighed 72 g. equivalent to a yield of 95%.

The AgO active material was chemically analyzed to determine its silver, Hg^{++} and AgO content. The total silver content was 82.6% by weight and the Hg^{115} content was 0.8%. The active oxygen content was determined iodimetrically and from this value the AgO content was calculated to be 96.2% by weight of the active material.

The AgO active material was tested for its electric resistivity and was compared to AgO active material which contained no Hg^{115}. The electric resistivity was pressure dependent, but as the pressure increased, the variance of the resistivity decreased. Therefore, the resistivity was determined at 30,700 psi with a model 503 Keithley Instrument milliohmmeter. The results were as follows.

Sample	Electric Resistivity (ohm-cm.)
AgO	64.0
AgO + Hg^{++}	3.84

The mercury additive provided an improvement in the conductivity of the divalent silver oxide active material.

Example 2: AgO containing tellurium was prepared in accordance with the procedure set forth in Example 1 except for the method of incorporating the additive. The following chart indicates the metallic salt which was added to the AgO during its preparation and the solution in which the salt was added to the reaction solution.

Additive Salt	Added to
Na_2TeO_3	$K_2S_2O_8$-NaOH solution

The additive salt was added in amount sufficient to provide 1% by weight of the tellurium per weight of silver ion in the AgO. The active oxygen content of the AgO active material was determined iodimetrically, and from this value it was calculated that the AgO content was 93.3% by weight of the active material.

The AgO active material containing the tellurium was compared to AgO containing no additive and tested for both stability and electric resistivity. The resistivity was again measured at 30,700 psi. The stability was determined at 100°F. in the following electrolyte.

	Parts by Weight
NaOH	48
ZnO	10
H_2O	100

These tests yielded the following results.

Additive	Electric Resistivity (ohm-cm.)	Gassing Rate (μl./g. AgO/day)
None	64.0	280
Te	69.3	100

The Te greatly improved the stability and had only a slight adverse effect on resistivity.

Example 3:   HgO was ball milled with AgO to determine whether physical mixtures of a metallic element and AgO have improved electric resistivity. The HgO and AgO were ball milled for 5 minutes with 3 balls in a Minimill. The results were as follows.

Additive	Additive (weight percent)	Electric Resistivity (ohm-cm.)
None	—	64.0
HgO	1	58.4
HgO	5	60.3

There was some improvement but not as great as when the Hg is coprecipitated with AgO. One explanation for the poorer results is that the ball milling adversely affected the AgO.

Addition of Silicate Ions to Electrolyte

P.J. Spellman and J.A. Youngquist; U.S. Patent 3,466,195; September 9, 1969; assigned to ESB Incorporated describes an alkaline cell containing silicate ions in the electrolyte which is characterized by an improved initial capacity and improved stability during storage and a limited amount of alkaline electrolyte sufficient only to provide wetting of the cell components without establishing a liquid level of electrolyte in the cell.

The alkaline cells of this process generally comprise a positive active material, a zinc negative active material, a separator placed between the positive active material and the zinc negative active material, and a limited amount of alkaline electrolyte sufficient only to provide wetting of the cell components without establishing a liquid level of electrolyte in the cell, the electrolyte containing a substantial amount of zinc in the form of zincate ions and at least about 0.05% by weight of silicate ions.

An important feature of the process is the provision of a soluble silicon additive which is sufficiently soluble in the alkaline electrolyte to provide silicate ions in an amount of at least about 0.05% by weight of the electrolyte at the time the cell components are thoroughly wetted by the electrolyte.

The cells are made in conventional manner and cells as they come off the assembly line generally require additional time, such as a day or more, for the electrolyte to thoroughly wet the assembly of cell components and to receive the silicate ions where soluble silicon additives have been incorporated in one of the cell components other than the alkaline electrolyte.

Since the silicon additives which are effective in providing alkaline cells having improved initial capacity and improved stability during storage are those which are soluble in the alkaline electrolyte and provide at least the minimum concentration of silicate ions therein, they may be referred to as soluble silicon additives or as silicate ion precursors.

Soluble silicon additives which have been found to be particularly effective are silica (SiO_2), potassium silicate (K_2SiO_3), sodium silicate (Na_2SiO_3), and elemental silicon (Si) may also be used. In addition, silicon compounds such as silicon tetrachloride ($SiCl_4$) and silanes can be used to treat one of the cell components, for example, the separator material, and in this manner they are incorporated into the cell where they form silicate ions when contacted by the alkaline electrolyte.

The improvement in stability during storage and initial capacity for alkaline primary cells containing a soluble silicon additive in accordance with this process is demonstrated in the following examples in which the percentages by weight of the soluble silicon additive are based on the weight of the cell component to which the additive is added unless otherwise specified. The size of the cells which were tested are given as National Bureau of Standards (NBS) sizes if such a size has been established.

Example 1:   The addition of a soluble silicon additive to alkaline electrolyte was tested for a NBS size S-5 silver oxide (Ag_2O)-zinc cell, which is used to supply power for hearing aids. 0.07 g. of alkaline electrolyte solution containing about 46% by weight of KOH and about 10% by weight of ZnO was used in each cell.

About 0.11 g. of zinc negative active material and about 0.45 g. of silver oxide positive active material were used in each cell. The cells were subjected to a 1,500 ohm discharge for 16 hours per day and the initial capacity and the capacity after storage for 12 weeks at 113°F. were determined. The following results are reported in hours to a specified voltage endpoint and each result represents a 3 cell average.

Additive	Amount (g./l.)	Initial		12 Weeks at 113°F.	
		1.40 v.	0.90 v.	1.40 v.	0.90v.
None	—	57.0	72.4	37.1	49.6
$Na_2SiO_3 \cdot H_2O$	10	63.5	77.4	51.3	65.0
K_2SiO_3	10	69.2	73.9	56.6	67.1
SiO_2	1.7	59.4	75.1	43.1	65.6

These results clearly indicate that the presence of silicate ions in the electrolyte improved the initial capacity of the alkaline cells, and in addition, improved the stability of the cells during storage for 12 weeks at 113°F.

Example 2:   In another series of tests, silica (SiO_2) was added to the positive active material in a NBS size S-5 silver oxide (Ag_2O)-zinc alkaline cell. In these cells, the positive active material (Ag_2O) also contained graphite in addition to 0.08% by weight of SiO_2.

Two layers of cellophane were used as the separator barrier material and 0.07 g. of electrolyte which was a 40% solution of KOH containing about 10% ZnO. About 0.11 g. of zinc negative active material and about 0.45 g. of silver oxide positive active material were used in each cell. The same discharge as in Example 1 (1,500 ohm, 16 hour per day) was used and each result represents a 3 cell average. The results were as follows.

	Initial		6 Weeks at 113°F.		6 Months at Room Temperature	
Additive	1.40 v.	0.90 v.	1.40 v.	0.90 v.	1.40 v.	0.90 v.
None	57.8	69.6	42.7	51.7	48.3	56.9
SiO_2	70.1	73.7	61.7	69.8	63.6	70.3

It is apparent that the silica improved initial capacity, and it was especially effective in stabilizing the cell during both room temperature storage and elevated temperature storage.

Example 3:   The effect of adding small amounts of silica to the positive active material of a NBS size S-15 silver oxide (Ag_2O)-zinc alkaline cell was determined for varying amounts of silica. 0.17 g. of 40% KOH electrolyte, about 1 g. of silver oxide positive active material and about 0.27 g. of zinc negative active material were used in each cell.

The cells were tested for initial capacity and stability during storage at room temperature for 6 months. Because the cell size used was different than in the previous examples, the discharge load was reduced to 625 ohms. Each result represents a 3 cell average indicating hours to the voltage endpoint.

	Initial		6 Months at Room Temperature	
Percent SiO_2	1.40 v.	0.90v.	1.40 v.	0.90 v.
0.00	53.3	75.8	40.0	65.0
0.05	54.0	75.0	52.0	75.0
0.10	52.0	80.8	56.0	77.2
0.15	53.3	81.3	52.4	74.6

Whereas there was only a slight improvement in initial cell capacity, the improvement in stability during storage at room temperature for 6 months was outstanding. In fact, there was practically no deterioration of capacity in those cells containing silica.

GENERAL

Button Cell

A process described by E.R. Cich; U.S. Patent 3,655,452; April 11, 1972; assigned to ESB Incorporated relates to multicell batteries made up of a number of dry cells of the button variety. The battery is characterized by having the cells in a planar array so that the entire battery is little thicker than the individual button cells from which it is built.

In the dry cell industry, coloquial names have been given to many of the various shapes and sizes of battery. The name "button cell" has been attached to a series of small, more or less disc shaped cells having a diameter usually greater than the height and having capacities ranging in the main from perhaps 2 to as much as 2,000 milliampere hours. Several electrochemical couples have been used in the button assemblies among the more prominent of which are the mercuric oxide-zinc couple and the nickel cadmium couple. Button cells have wide use in hearing aids, miniature radios, instruments, light meters, etc.

Button cell batteries are normally built as a pile with one cell placed on top of the preceding one and with a single outside clamping member serving to form and maintain the electrical contact between the several cells. This construction is extremely compact and is the normally desirable way to build a button cell battery. The battery is of necessity a cylindrical or, if boxed, a square, prismatic shape. There are cases where this shape is not the most convenient for the battery user and where a rather thin rectangular shape would be more desirable.

Further, the pile arrangement has its contacts at the two ends, requiring some form of spring clamp for making electrical contact. Because of this, the pile type battery is usually mounted within the battery operated device. If, by chance an internally located battery should leak, there would be considerable likelihood that parts of the internal structure of the device might become corroded. Therefore, there is a need for a button cell battery of a more or less rectangular shape having one dimension less than the other two dimensions mounted in a weatherproof pack and suitable for external attachment to the appliance on which it is to be used.

Referring to Figure 3.5a, (6), (7), (8) and (9) represent four button cells from which a battery is to be assembled. It is to be noted that cells (6) and (8) have the positive terminal facing up and that cells (7) and (9) have their negative terminal facing up. This is to provide a series connected battery for maximum voltage.

A center insulating piece (14) of electrically insulating material has four holes (10), (11), (12) and (13) shaped to take the four cells. This center insulating piece is approximately the same thickness as the cells. A top insulating piece (15) carries positive terminal (16), negative terminal (17) and intercell connector (18). The intercell connector (18) is located on the underside of the upper battery insulating

FIGURE 3.5:   BUTTON CELL BATTERY

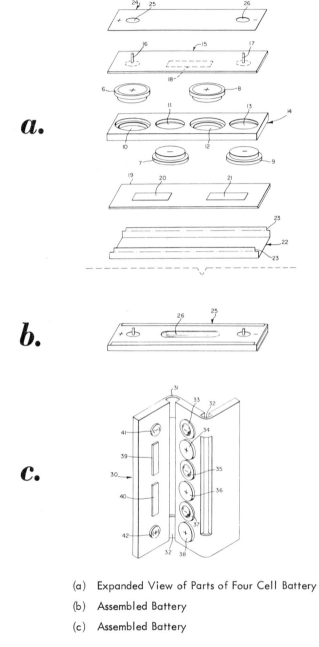

(a) Expanded View of Parts of Four Cell Battery

(b) Assembled Battery

(c) Assembled Battery

Source: E.R. Cich; U.S. Patent 3,655,452; April 11, 1972

means (14). A bottom insulating piece (19) has intercell connections (20) and (21). Terminal pieces (16) and (17), which will contact cells (6) and (9) respectively on assembly, penetrate through the top insulating piece (15) to provide electrical terminal contacts outside of the battery proper. A metal case (22) having one or more deformable ears (23), and a metal cover (24) complete the parts of the battery.

The battery is assembled as a sandwich of the three insulating pieces (19), (14) and (15) with the cells (6), (7), (8) and (9) mounted in the holes of the center insulating piece (14). This assembly is placed in container (22), cover (24) is placed on top and the entire assembly is placed in a press and clamped tightly. The ears (23) are thereby bent down onto cover (24) serving to preserve the clamping pressure after the assembly has been removed from the press.

It has been found that the construction shown in Figure 3.5a can be improved by forming an upset or depressed groove in the metal of cover (24) and the can (22) along the centerline over the intercell connectors. This serves to concentrate the clamping pressure on the contact areas of cell and intercell connector. Figure 3.5b shows such a construction. In Figure 3.5b, (25) represents the battery and (26) an upset area or recessed groove along the centerline of its top, over the top intercell connector (18) (not shown in this view). A similar upset area or recessed groove in the bottom piece (22), Figure 3.5a, is provided.

Figure 3.5c shows an alternate design in which the three insulating pieces are now a single molding (30) having hinge sections (31) and (32) to permit folding into the final shape required by the battery. The hinge sections comprise a web having a thickness less than that of the cell positioning means. Six cells (33), (34), (35), (36), (37) and (38) are shown in this battery. Two metallic intercell connectors are shown at (39) and (40). The terminal cell contacts are shown at (41) and (42).

The material from which the insulating members used in this process are made is not critical. For low cost batteries, materials such as cardboard, impregnated cardboard, vulcanized fiber, etc. will be suitable. For better grade batteries, the entire gamut of thermoplastic and thermoset materials is available to the battery designer, limited only by the resistance to cold flow of the material. In the case of the example shown in Figure 3.5c, a flexible material, suitable for manufacture by extrusion or injection molding, such as polyethylene or polypropylene would be desirable.

It is to be noted that in all the batteries described, the individual cells are arranged so that cell tops and bottoms lie approximately on one or the other of two plane surfaces so that the battery will have a minimum height dimension. For purposes of definition, this is called a planar array of the cells and serves to distinguish the cell assembly from the pile type where cells are located one above the other.

Hermetically Sealed Button Cells

A process described by P. Ruetschi; U.S. Patent 3,556,848; January 19, 1971; assigned to Leclanche SA, Switzerland pertains to a hermetically sealed enclosure, for primary and secondary battery cells. The tiniest alkaline cells are commonly called button cells. The enclosure of these cells consists essentially of two metal cans, open ends opposed, the two cans being electrically insulated from each other by a ring shaped gasket that also forms the hermetic seal. When the enclosure is sealed shut, the rim of the larger, or outer, case is crimped inwards so as to squeeze the gasket between the outer can and the inner can. During this operation, the resulting axial force within the can must be supported. To this end, three basic schemes are presently being used. In one, the chemically active components that are housed in the larger can (usually the active components of the positive electrode) are surrounded by a metal ring, the outer diameter of which is equal to the inner diameter of the larger can, into which it exactly fits. The gasket is seated in the metal ring, which supports the axial force when the cell is sealed shut.

In another method, the two cans are completely telescoped into each other. Two sealing rings provide electrical insulation. The first ring insulates the rim of the inner can from the bottom of the outer can, and the second constitutes the sealing closure between the rim of the outer can and the external wall surface of the inner can.

Thirdly, the axial force can be supported by forming a shoulder in the outer can. The diameter of the lower part of the can, in which the chemically active components are compressed, is smaller than the upper part of the can. The sealing ring, or gasket, which in the simplest form has a cross-section in the shape of an L, of a U on its side, is placed onto the shoulder, which bears the axial force during sealing of the battery enclosure. This approach saves space, because it avoids having a large part of the enclosure double walled, since the two cans overlap only above the sealing ring.

The disadvantage, however, is that the cell is no longer a single cylinder, inasmuch as the lower part of the cell has a smaller diameter. But the space required by the cell is often determined by its maximum diameter and height, so that these cells sometimes do not permit the most efficient use of space, which is particularly disadvantageous with miniaturized electric circuits.

This process provides a battery cell enclosure that incorporates at least one internal shoulder machined into the cell can, so as to make an upwardly inclined shoulder with a sharp edge, pointing inwards, the metal edge forming an acute angle, a synthetic plastic ring member supported by the shoulder and embodying an opening and a metal member for sealing shut the opening.

Referring to Figure 3.6, the wall thickness of the upper end (1) of the metal can (5) is reduced by machining, so as to make an upwardly inclined shoulder with a sharp edge (2), pointing upwards, the metal edge forming an acute angle. The reduction in the thickness of the wall is between 0.1 and 0.4 mm. When the end of the cell is closed by a sealing ring (3), Figure 3.6a, made of a suitable synthetic plastic that is not attacked by the electrolyte of the cell, such as nylon, polyethylene, or that known under the registered trademark Teflon, the sharp edge (2) engages into the ring, assuring a hermetic seal.

With the ring (3) in place, the can edge (4) is crimped over in a known manner, so that the ring is permanently pressed against the edge (2). The result is an extremely good seal, even with smallest cells. In Figure 3.6a, a lead through metal contact (8), in the form of a rivet, is mounted in the ring (3). The contact incorporates an upwardly inclined shoulder with a sharp edge (9) points upwards, the metal edge forming an acute angle. This edge is forced to penetrate into the ring (3). The edge (10) of the contact is bent outwards so as to rivet the latter to the ring. The lead through contact (8) serves as the contact for the negative electrode (12), and the

FIGURE 3.6:   HERMETICALLY SEALED ENCLOSURE FOR BUTTON CELLS

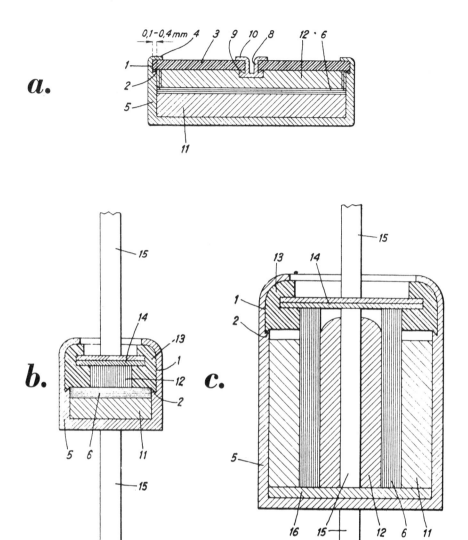

(a)   Cross-Section of Cell Housing

(b)(c)   Additional Housing Designs

Source: P. Ruetschi; U.S. Patent 3,556,848; January 19, 1971

case (5) the contact for the positive electrode (11). These arrangements are particularly suitable for very small nickel-cadmium, mercuric oxide-cadmium, and mercuric oxide-zinc cells. In the latter two systems the positive electrode (11), held in the can (5), is preferably composed of a mixture of about 85% mercuric oxide, 5% graphite and 10% manganese dioxide.

The negative electrode (12) is composed of finely divided metallic cadmium, or finely divided zinc amalgam. The separator (6) consists of microporous felt and membranes that are permeable to the electrolyte. The can (5) can be made of stainless steel, nickel, Kovar or similar alloys. If one deals with the mercuric oxide-zinc system, the part of the enclosure which contacts the negative, zinc, electrode (12), must at least on its inner side be composed of amalgamated copper or silver, or alloys of the metals, in order to avoid the evolution of hydrogen.

The corresponding parts of the three examples, illustrated in Figures 3.6a, 3.6b and 3.6c, respectively, are given the same reference number in the drawing. With reference to Figures 3.6b and 3.6c, the battery cell enclosure includes a synthetic plastic sealing ring (13), in which a metal top (14) is sealingly squeezed. For cells with negatives electrodes of zinc, the cap comprises two layers, an outer layer composed of a nickel or iron alloy and an inner composed of a copper or silver alloy. Because of the cells' extremely small size, the leads (15) of the cells shown in Figures 3.6b and 3.6c are welded on.

The positive electrode (11) of the example shown in Figure 3.6b consists of mercuric oxide or silver oxide, or of sintered or pressed nickel and nickel hydroxide. The negative electrode (12) comprises zinc powder or spongy cadmium. The leads (15) are attached by resistance welding, the welding time being about one millisecond.

The example illustrated in Figure 3.6c shows a concentric cylindrical electrode arrangement. The plate (16) resting on the bottom of the case (5) electrically insulates the negative electrode (12) from the case, which latter serves as the positive pole. The plate (16) is made of a suitable insulator, such as nylon, polyethylene, polypropylene, or that known under the registered trademark Teflon.

The examples shown in Figure 3.6b and 3.6c are particularly suitable for the very tiniest cells, such as those having a diameter of two millimeters or smaller, the example shown in Figure 3.6b being used for short, and that in Figure 3.6c for long, cylindrical cells.

With cells having extremely small diameters, it is difficult or impossible to use the construction, shown in Figure 3.6a, since a sufficiently small rivet cannot be made. For very small cells, the leads are welded to the positive and negative poles of the cells illustrated in Figures 3.6b and 3.6c. Consequently, these cells have the same appearance as other miniature electronic components, e.g., resistors and capacitors, and can therefore be mounted in circuits in the same way as these other components. The miniature cells described are completely leakproof and gas tight.

Lithium Ions Added to Electrolyte

M.H. Johnson and P.J. Spellman; U.S. Patent 3,433,679; March 18, 1969; assigned to F & B Incorporated describe a primary alkaline cell having improved initial capacity and stability during storage or shelf life provided by the presence of lithium ions in the alkaline electrolyte. These alkaline cells comprise a positive active material, a zinc negative active material, a separator between the positive and negative active materials and a sodium or potassium hydroxide electrolyte solution contained substantially wholly within the separator and the positive and negative electrodes.

The higher initial capacity and improved stability during storage are provided by the presence of a lithium additive in amounts ranging from at least about 0.01% by weight up to saturation of lithium ions in the alkaline electrolyte. The lithium additive can be added to any one or more of the cell components so long as it forms an effective concentration of lithium ions in the alkaline electrolyte.

The lithium additive which is incorporated into the cell to improve stability during storage and initial capacity must be sufficiently soluble in the alkaline electrolyte to provide an effective concentration of lithium ions in the electrolyte. Examples of lithium additives which have been found to improve initial capacity include lithium (Li), lithium hydroxide (LiOH), hydrated lithium hydroxide (LiOH·H_2O), hydrated lithium sulfate (Li_2SO_4·H_2O), hydrated lithium acetate ($LiC_2H_3O_2$·$2H_2O$), lithium carbonate (Li_2CO_3), hydrated lithium chromate (Li_2CrO_4·$2H_2O$), lithium chloride (LiCl), lithium phosphate (Li_3PO_4), hydrated lithium citrate ($Li_3C_6H_5O_7$·$4H_2O$), hydrated lithium formate ($LiCHO_2$·H_2O) and lithium nitrate ($LiNO_3$).

The improvement in initial capacity and stability during storage provided by the lithium additives of this process is illustrated in the following examples in which each result represents a 3 cell average unless otherwise indicated. The size of the cells which were tested are given as National Bureau of Standards (NBS) sizes if such a size has been established.

Example 1:   Alkaline electrolyte containing hydrated lithium hydroxide was also tested in an NBS size M-15 mercuric oxide (HgO)-zinc primary alkaline system. The electrolyte concentration was 45% KOH and the electrolyte solution was 53% saturated with zinc oxide. The cells were discharged 16 hours per day through a 625 ohm load, with the total hours to 1.10 and 0.90 volt endpoints used to measure the cell initial capacity. In addition, cells were stored for 12 weeks at 113°F. prior to testing their capacity in order to determine the effect of the lithium additive on stability during storage or shelf life.

Additive	Electrolyte Concentration	Initial Capacity		12 Weeks at 113°F.	
		1.10 v.	0.90 v.	1.10 v.	0.90 v.
None	45% KOH	78.9	80.9	73.9	80.2
0.1% Li	45% KOH	84.2	90.2	86.7	90.2
0.2% Li	45% KOH	92.1	96.7	80.0	86.9

These tests results demonstrate that the hydrated lithium hydroxide improved both the initial capacity and the stability during storage at 113°F. It should be noted that the primary alkaline cells deteriorated more rapidly at elevated temperatures than at normal room temperature.

Example 2:   NBS size S-5 silver oxide (Ag_2O)-zinc primary alkaline cells were used
to test the effectiveness of hydrated lithium hydroxide ($LiOH \cdot H_2O$) and hydrated
lithium sulfate ($Li_2SO_4 \cdot H_2O$) in 45% potassium hydroxide solution.

The hydrated lithium hydroxide was added to the electrolyte in amount of 5 g./l.
and 10 g./l. which corresponds to a concentration of lithium in the electrolyte of
0.05 and 0.1% respectively.   10 g./l. of hydrated lithium sulfate was added to the
electrolyte but this was in excess of saturation and some of the lithium sulfate remained undissolved in the electrolyte. The cells were discharged 16 hours per day
through a 1,500 ohm load. The cells were tested for initial capacity and stability
during storage at 113°F. by recording the total hours to 1.40 and 0.90 volt endpoints.

	Initial Capacity		12 Weeks at 113°F.	
Additive	1.40 v.	0.90 v.	1.40 v.	0.90 v.
None	51.0	72.4	37.1	49.6
$LiOH \cdot H_2O$ (5 g./l.)	65.8	79.2	54.0	67.2
$LiOH \cdot H_2O$ (10 g./l.)	70.7	80.2	63.6	72.5
$Li_2SO_4 \cdot H_2O$ (10 g./l.)	72.0	78.3	63.9	74.8

The results demonstrate the effectiveness of hydrated lithium hydroxide and hydrated
lithium sulfate for improving the initial capacity and stability during storage.

Example 3:   The effectiveness of hydrated lithium hydroxide in an NBS size S-15
silver oxide-zinc alkaline cell was also tested. 1% by weight of $LiOH \cdot H_2O$ was
added to the electrolyte which corresponds to a lithium concentration of about 0.15%.
These cells were tested for initial capacity and stability during storage using 300,
625, and 1,500 ohm 16 hour per day discharges.

	Discharge	Initial Capacity		12 Weeks at 113°F.	
Additive	Load	1.40 v.	0.90 v.	1.40 v.	0.90 v.
None	300 ohm	20.9	35.4	12.7	19.6
1% $LiOH \cdot H_2O$	300 ohm	26.4	35.9	25.1	29.7
None	625 ohm	50.0	69.4	58.3	63.1
1% $LiOH \cdot H_2O$	625 ohm	64.0	79.2	70.8	73.8
None	1,500 ohm	44.7	69.4	46.7	53.0
1% $LiOH \cdot H_2O$	1,500 ohm	59.7	78.9	65.8	70.8

The results indicate that the cells containing hydrated lithium hydroxide had both
improved initial capacity and improved stability during storage at 113°F.

Electrodeposited Metal Hydroxides as Separator

In a process described by J.M. McQuade; U.S. Patent 3,477,875; Nov. 11, 1969;
assigned to General Electric Company a porous, insoluble inorganic hydroxide or

hydrous oxide coating of a metal such as calcium, magnesium, nickel, cadmium, aluminum, or zirconium is deposited on at least one electrode of an alkaline electrolyte cell to act as a separator. Coatings onto such electrodes as nickel, cadmium, silver oxide and zinc electrode plates are specifically described. The coating may be formed by electrolysis of a water-soluble compound of the metal sought to be included in the hydroxide.

Example 1: A nickel plate was provided with a calcium hydroxide deposit to a thickness of approximately 0.001". This plate was sandwiched between two uncoated plates of cadmium. All the plates were of standard size, i.e., 2 3/4" x 2 1/4" and an aqueous solution of 31% potassium hydroxide was utilized as a battery electrolyte in sufficient quantity to completely cover the plates. The battery was charged for 36 minutes and then discharged for 24 minutes. This charge-discharge cycle was continuous. The battery remained operational for approximately 1,000 cycles before internal shorting occurred.

Example 2: A similar test was conducted with a battery provided with a nickel plate coated with magnesium hydroxide. Internal shorting did not occur until after 300 cycles. Silver oxide-zinc batteries may also be provided with at least one plate coated in the manner described above.

It has been found that a water activated battery can be provided by the process. Thus, it has been found that by electrolytically depositing calcium hydroxide on both the silver oxide and zinc plates of AgO-Zn battery and using a saturated solution of calcium hydroxide as electrolyte, a marked improvement in energy density and life at low drain rates can be obtained. Although the parasite consumption of zinc by an aqueous solution of calcium hydroxide is low, the use of the electrolyte by itself does not result in a long-life cell at low drain rates.

Example 3: The battery tested had a silver oxide plate, a zinc plate physically separated by a 1/32" spacing and a saturated solution of calcium hydroxide as electrolyte. This battery was loaded at 1.0 ma. The following results were obtained.

Time in Hours	Battery Voltage in Volts
Start	1.84
10	1.84
82	1.80
126	1.68
142	1.58
157	1.38
162	1.18
169	0.86
178	0.76
190	0.62

After approximately 126 hours it was observed that the zinc plate had turned black. Shortly thereafter, black dendrites extended from the zinc electrode to the silver oxide electrode.

Example 4:   A silver oxide plate and a zinc plate were both coated with a deposit
of calcium hydroxide to a weight of 0.24 and 0.43 g., respectively. The plates
were separated by a standard nylon separator and a saturated solution of calcium hydroxide was utilized as the electrolyte. A resistive load placed on the battery drew
a current of 0.85 ma. The following results were obtained.

Time in Hours	Battery Voltage in Volts
Start	1.80
512	1.22
504	1.21
648	1.12
840	1.12*
1,032	1.40**
1,200	1.32
1,416	1.22

*Load dropped to 0.5 ma. 8.5 ohm
**Load dropped to 0.1 ma. 5.0 ohm

At the end of approximately 1,400 hours the battery was 90% discharged. This was
based on a measurement of the silver capacity of the battery.

Large Surface Area Electrode

R.W. Fletcher; U.S. Patent 3,427,203; February 11, 1969; assigned to ESB Inc.
has found that a large surface area electrode can be prepared from amalgamated
metal powder by heating the powder in the presence of a metal oxide or a silicate
bonding agent and a bonding solution which bonds the metal particles so as to struc-
turize them.

This type of electrode will be referred to as a bonded electrode. It has been found
that the bonded electrodes have greater porosity (surface area) and lower density
than compressed metal powder electrodes and they have sufficient structural strength
to withstand the crimping force which is applied to the cell to close it.

One of the advantages of the bonded electrodes is that they are not compacted upon
closure of the cell and therefore, they maintain their large surface area. As a result of the increased surface area, the bonded active material is more efficiently
utilized as demonstrated by improved high rate capability and low temperature performance. The improved performance of cells having bonded electrodes prepared
in accordance with the process is illustrated in the following examples.

Example 1:   Two primary alkaline cells of the 675 size were prepared, utilizing
mercuric oxide as the cathode material and amalgamated zinc powder as the anode
material. One of the cells (control) was manufactured in accordance with standard
procedures including a compressed zinc powder anode. The other cell was identical
except for the anode which was bonded zinc prepared in situ using zinc oxide as
the bonding agent and a 40% potassium hydroxide (KOH) bonding solution in accordance with this process. These cells were tested for their low temperature

performance. The following results were obtained for a 300 ohm continuous discharge to a 0.90 v. endpoint at 25°F. The low temperature superiority of the bonded zinc anode is apparent.

	Average Voltage (v.)	Duration (hrs.)	Capacity (ma. hrs.)
Bonded zinc anode	1.068	15.2	54.0
Compressed zinc anode	1.032	4.08	14.01

Example 2: Two primary cells of the 675 size were prepared, utilizing silver oxide as the cathode material and amalgamated zinc powder as the anode. One of the cells (control) had a loose zinc powder anode which was not compressed. The other cell was identical except for the anode which was bonded zinc prepared in situ in the anode container. These cells were tested for their performance at room temperature (about 70°F.). The following results were obtained for a 300 ohm continuous drain to a 1.40 v. endpoint followed by a discharge to 0.90 v. The bonded zinc anode performed better when taken to the 0.90 v. endpoint.

	1.40 v. Endpoint			0.90 v. Endpoint			Total Mah
	Volts	Hours	Mah	Volts	Hours	Mah	Mah
Bonded zinc	1.45	28.5	137	1.26	8.02	33.6	170.6
Loose zinc	1.46	28.5	137	1.15	3.55	13.6	150.6

Example 3: Eight primary alkaline cells of the AA size were prepared, utilizing manganese dioxide as the depolarizer and amalgamated zinc powder as the anode material. Four of the cells were manufactured in accordance with standard procedures, including a compressed zinc powder anode. The other cells were identical except for the zinc bonded anodes which were prepared in a cylindrical shape and were then inserted into the anode container.

In addition, four competitive cells of the same type and size were purchased and also tested. All twelve of these cells were substantially identical except that the anodes of the four competitive cells contained a carboxymethylcellulose binder for the zinc particles. These twelve cells were tested for their low temperature performance using two different tests. One test (A) was made with a 20 ohm discharge for 4 minutes per hour and 10 hours per day to a 0.93 endpoint voltage.

The other test (B) had a 20 minute cycle period. The cells were discharged through a 26.89 ohm resistor for 2 minutes and through a 1,110 ohm resistor for 18 minutes. This cycle was continued until the cell voltage dropped to 1.11 v. The following results were obtained. The cells with bonded zinc anodes were superior to both of the other cells for all tests except for test A at -20°F.

Test	Temperature (°F.)	Bonded Zinc		Compressed Zinc		Zinc + CMC Binder	
		1.11 v. (hrs.)	0.93 v. (min.)	1.11 v. (hrs.)	0.93 v. (min.)	1.11 v. (hrs.)	0.93 v. (min.)
A	-20	—	110	—	31	—	115
A	-40	—	96	—	9	—	84
B	-20	28.33	—	8	—	18.26	—
B	-40	14.00	—	3.2	—	11.33	—

CADMIUM CELLS

NICKEL-CADMIUM

Impregnation of Porous Electrode with Nickel Hydroxide

A process described by R.L. Beauchamp; U.S. Patent 3,653,967; April 4, 1972; assigned to Bell Telephone Laboratories, Incorporated involves a method for forming a product which is useful as the positive electrode of an alkaline nickel-cadmium cell, essentially by the electrolytic deposition of nickel hydroxide in a porous structure.

Thus, it has been found that carrying out the electrolytic deposition of nickel hydroxide in a porous electrode structure within the critical temperature range of from about 85°C. to the boiling point of the electrolyte results in a finely divided deposit having a large active surface area, leading to increased loading, increased percent utilization of active material, particularly at high loading levels, and resistance of the electrode to flaking and shedding of the deposit during formation and cell use.

Such filled structures may subsequently be treated electrolytically with an alkaline solution so as to activate the nickel hydroxide prior to their incorporation into electrolytic cells such as alkaline nickel-cadmium cells as positive electrodes. The following examples illustrate the process.

Example 1: A series of sintered nickel plaques of similar size and porosity were prepared. One set of 12 plaques was impregnated at room temperature in a 4M solution of $Ni(NO_3)_2$ at a current of 3 amps for about 8.5 minutes. A second set of 12 plaques was impregnated according to the process in a boiling solution of 2M $Ni(NO_3)_2$ and 0.3M $NaNO_2$ at a current of 1 amp. for about 30 minutes. Each set of plaques was activated by cycling in a 30 percent solution of potassium hydroxide at about the same charge and discharge rates. The results are shown in Table 1, and are expressed as the average values of theoretical capacity (ampere-hour) after forming and percent utilization of active material (measured capacity divided by theoretical capacity).

TABLE 1

	Theoretical Capacity (amp.-hr.)	Percent Utilization
Set 1	0.220	105
Set 2	0.240	122

These results indicate that under conditions of impregnation which result in about the same loading levels, the process results in a significantly increased value of percent utilization over that of the room temperature process.

Example 2: Three sets of sintered nickel plaques of similar size and porosity were prepared. Each set was impregnated by a different method and under a variety of conditions in order to discover the optimum loading obtainable for each method. The first set was loaded by the vacuum impregnation method as follows. The plaques were immersed in an aqueous solution of 4.0M $Ni(NO_3)_2$ under vacuum for a time of 5 minutes, removed and treated electrolytically as a cathode in a hot 25% solution of potassium hydroxide for a time of 20 minutes at a current of 5 amps in order to convert the $Ni(NO_3)_2$ to $Ni(OH)_2$.

Next, the plaques were removed from the electrolytic solution and allowed to dry overnight. It was found that optimum loading was obtained when these steps were repeated from 10 to 11 times. The second set was loaded by the electrolytic method. Optimum loading was obtained as follows. The plaques were treated cathodically in 4M $Ni(NO_3)_2$ solution at room temperature for about 10 minutes and at a current of 5 amps.

The third set was also loaded by the electrolytic method, but in accordance with this process. Optimum loading was obtained for a boiling electrolyte containing 2M $Ni(NO_3)_2$ and 0.3M $NaNO_2$ at a current of 1 amp. for about 90 minutes. Results were obtained as amount of active material loaded (grams per cubic centimeter), theoretical capacity and measured capacity after forming (ampere-hours) and percent utilization of active material. Forming in each case consisted of subjecting the loaded plaques to at least one charge-discharge cycle in an electrolyte solution containing about 30% potassium hydroxide, at charge and discharge currents of 0.150 amp. and 0.075 amp., respectively. Results are shown in Table 2.

TABLE 2

	Loading (g./cc)	Theoretical Capacity (amp.-hr.)	Measured Capacity (amp.-hr.)	Percent Utilization
Vacuum impregnation	1.36	0.322	0.290	90
Electrolytic (room temp.)	1.17	0.275	0.313	114
Electrolytic (elevated temp.)	1.93	0.416	0.499	120

It will be noted that significantly increased loading and capacity and higher energy density cells are obtainable by use of this method, as compared to either of the other methods. In addition, improved percent utilization of active material is also obtainable by the method.

Welding of Nickel Lead Wire

In a process described by K.N. Johnson; U.S. Patent 3,607,432; September 21, 1971; assigned to Texas Instruments Incorporated a porous battery plate comprising nickel metal powder sintered to nickel metal mesh and having an electrochemically active material such as cadmium hydroxide, is shown to be secured to a nickel lead member for use in electrically connecting the plate as an electrode in an alkaline electrolyte battery by inserting a nickel-plated nickel wire between portions of the lead member and plate and by welding the lead member to the plate by electrical resistance welding.

The nickel wire plating contains a phosphor constituent such as is inherently deposited during formation of the plating by electroless nickel-plating techniques for lowering the melting temperature of the nickel plating below the melting temperature of the nickel lead member. The plated nickel wire then serves as a weld projection for securing the lead member to the plate. Complete details of the process are provided.

Sealed Type Oxygen Cell

M. Fukuda and T. Iijma; U.S. Patent 3,658,591; April 25, 1972; assigned to Matsushita Electric Industrial Co., Ltd., Japan describe a sealed type cell having an electrochemical element which enables an oxygen ionizing reaction and an oxygen generating reaction, caused by overcharging, to take place cyclically and the internal resistance of which depends on the pressure of oxygen in the cell.

The electrochemical element is composed of an oxygen ionizing electrode and an oxygen generating electrode which are connected to the negative electrode and the positive electrode of the cell respectively. In Figure 4.1a, a completely sealed container is composed of a jar (1) and a lid (2), which are made from a durable, thick synthetic resin plate or steel plate with an internal cell pressure of up to 1 atmosphere. In an actual product, a safety valve, so designed as to be opened under a pressure higher than a predetermined value, may be provided to ensure safety of the cell.

Reference numeral (3) designates a positive electrode of nickel oxide which may be either the so-called sintered-type electrode or pocket-type electrode of the open type cell, and (4) designates a negative electrode of cadmium which may similarly be either the sintered-type, pocket-type or paste-type electrode. An electrolyte (5) consists mainly of aqueous solution of KOH and a suitable amount of LiOH, etc. may be added as required. Reference numeral (6) generally designates an electrochemical element composed of an oxygen generating electrode (7), an oxygen ionizing electrode (8) and a separator (9).

The oxygen generating electrode (7) of the electrochemical element (6) is made from a thin plate, porous plate or screen of nickel or stainless steel, etc., while the oxygen ionizing electrode (8) is made from a porous plaque consisting primarily

FIGURE 4.1: SEALED TYPE OXYGEN CELL

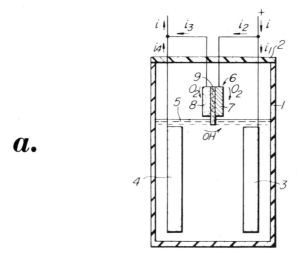

a.

Cross–Sectional View of Cell

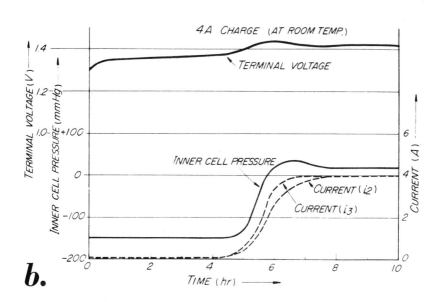

b.

Performance Characteristics of Battery During Charging

(continued)

FIGURE 4.1: (continued)

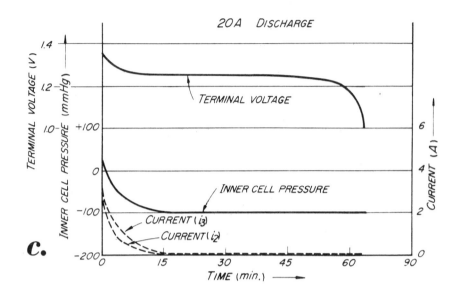

Performance Characteristics of Battery During Discharge

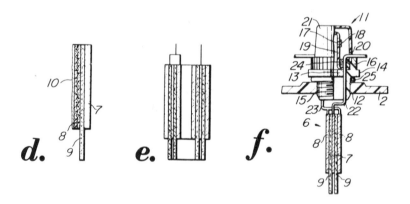

Cross-Sectional View of Different   Electrochemical Element Connected
      Electrochemical Elements   to a Vent Plug

Source: M. Fukuda and T. Iijma; U.S. Patent 3,658,591; April 25, 1972

of nickel or carbon which is subjected to a water-proofing treatment with the addition of such an oxygen ionizing catalyst as silver. The separator (9) through which hydroxyl ion moves, consists of an electrolyte-resistant nonwoven fabric of synthetic resin or Japanese paper, which is impregnated with the electrolyte, and has its lower end dipped in the electrolyte (5).

Example 1: As a sealed type cell, a 20 ah sintered-type nickel-cadmium alkaline battery is used. A method of producing an electrochemical element which enables the cell to be charged with a current of 4A (5 hour rate) and the charging and discharging characteristics of the sealed type cell having the electrochemical element, will be explained.

A mixture of carbonyl nickel powder and carbon powder in a proportion of 9 to 1 (by weight) is attached to both sides of a nickel net in the size of 30 x 40 x 0.7 mm. and sintered at 900° C. for ten minutes in a hydrogen stream, to obtain a porous plaque, to which silver is added in the proportion of about 5 mg. per cm^2. in the following manner. A solution of 20 grams of $AgNO_3$ in 100 ml. of H_2O is impregnated in the plaque and after drying, the plaque is immersed in a mixture of formalin and KOH, specific gravity of 1.30, in the proportion of 1 to 1 (by volume) for reduction, and thereafter washed with water and dried.

In order to improve the performance and prolong the service life of the thus treated plate as a gas ionizing electrode, the plate is immersed in a dispersion of polytetrafluoroethylene and after drying, subjected to a heat treatment at 350° C. in a nitrogen atmosphere. Oxygen generating electrode: A thin nickel plate of 30 x 40 x 0.05 mm. in size is used. Separator: A 40 x 40 mm. non-woven fabric of nylon (0.2 mm. in thickness) is used. Two oxygen ionizing electrodes (8), one oxygen generating electrode (7) and two separators (9), obtained in the manner described above, are assembled as shown in Figure 4.1f to compose the electrochemical element. This electrochemical element is disposed in the 20 ah cell. In this case, the quantity of an electrolyte is adjusted such that the lower ends of the separators are dipped in the electrolyte.

The cell is charged fully and after removing gases from the cell, air is introduced into the cell to form an air atmosphere therein. Use of oxygen is more advantageous over air for improving the gas absorbing ability but since it is conceivable that the interior of the cell is reduced to a pressure of 760 mm. Hg and further an actual product will encounter a problem of gas leakage, the air atmosphere was used in this example. After the cell is closed tightly with the air atmosphere, the oxygen ionizing electrode of the electrochemical element is connected with the cadmium negative electrode and the oxygen generating electrode with the nickel oxide positive electrode, and the cell is left to stand still. In the meantime, oxygen in the air forms hydroxyl ion at the oxygen ionizing electrode and the hydroxyl ion reacts with the negative electrode, $O_2 + 2H_2O + 2Cd \longrightarrow 2Cd(OH)_2$.

In this case, the inner cell pressure is reduced by about 15 mm. Hg which is the partial pressure of oxygen in air. The fact that Cd is converted into $Cd(OH)_2$ means discharge of the negative electrode. Therefore, when the cell is charged, oxygen gas is generated first from the positive electrode and the electrochemical element functions properly. Thus, the cell can be sealed. In order to observe the function of the cell, the inner cell pressure was measured by a pressure gauge provided on the cell, along with the values of i_2 (oxygen generating current) and i_3 (oxygen ionizing current), with the results shown in Figures 4.1b and 4.1c. Figure 4.1b shows the function of the cell when charged with a current of 4A (5 hour rate). As seen, the inner cell pressure does not change, and i_2 and i_3 remain 0 for 5 hours until the end of charging. This indicates that the charging current is entirely used for the charging of the active materials. As the cell becomes overcharged upon passage of 5 hours, the inner cell pressure begins to rise and at the same time the value of i_2 increases. The value of i_3 increases successively. The value of i_2 is slightly smaller than the value of i_3, because the difference i_1 therebetween is used for the charging of an uncharged portion or for the compensation of self-discharge current.

When the value of i_3 becomes equal to the value of i (4A), the inner cell pressure reaches an equilibrium state and in the meantime a state of $i = i_3 = i_2 = $ (4A) is produced. This state continues thereafter. This indicates that reactions represented by the formula $O_2 + 2H_2O + 4e^- \rightleftharpoons 4OH^-$ are taking place completely cyclically at the electrochemical element. Although in this example, use was made of an electrochemical element adapted for 5 hour rate charging, a shorter hour rate charging is also possible and the size can be made smaller by the use of an electrochemical element of higher performance. The electrochemical element behaves in substantially the same manner in the constant-voltage charging.

The discharge characteristic of the charged cell is shown in Figure 4.1c. As will be seen from the chart, the inner cell pressure, on commencement of discharge, quickly drops to -150 mm. Hg, that is, the state of air wherein the oxygen partial pressure is 0, as a result of a reaction represented by the formula

$$2Cd + (1/2)O_2 + H_2O \longrightarrow 2Cd(OH)_2$$

and the values of i_3 and i_2 also become 0. Namely, the cell shows a sufficiently satisfactory discharge characteristic. The fact that the values of i_3 and i_2 become 0 indicates that the cell is free of self-discharge. The example illustrated above is the most standard one. A deterioration of the oxygen ionizing electrode of the electrochemical element is mainly caused by the fact that the water repellency of that surface of the electrode which is in contact with the gases is lost and the passage (diffusion) of the gas through the electrode is blocked by the electrolyte attached to the surface.

For preventing such deterioration of the oxygen ionizing electrode and thereby prolonging the life of the same, it is preferable to provide a water-repellent layer (10) on that surface of the electrode which is brought into contact with gases as shown in Figure 4.1d. This may be attained by either attaching to the surface of the porous oxygen ionizing electrode a water-repellent synthetic resin powder, particularly a powder of fluorine-contained resin, e.g., tetrafluoroethylene-hexafluoropropylene copolymer, or attaching to the surface of the porous oxygen ionizing electrode with pressure a gas-pervious non-woven fabric made of a fiber of water-repellent synthetic resin, particularly fluorine-contained resin, for example, polytetrafluoroethylene or Polyflon Paper (Daikin).

By providing such a water-repellent layer on the surface of the oxygen ionizing electrode, the surface is prevented from being wetted and remains active, whereby a long service life of the electrode is assured. For the actual product of the closed type cell according to this process, it is one of the essential requirements that the electrochemical element only can be exchanged in a simple manner so as to use the cell over an extended period or upon occurrence of unexpected trouble. Such requirement may be met by connecting the electrochemical element integrally to a conventional vent plug as shown in Figure 4.1f. Referring to Figure 4.1f, a vent plug (11) to be fitted in a filling hole (12) formed in the lid (2) comprises a cylindrical plug body (16) having an inner flange (13), an outer flange (14) and an externally threaded portion (15) for engagement with an internal thread of the filling hole; a cylindrical inner cap (19) closed at the top end and resting on the inner flange (13), the cap having vent holes (17) which are closed with an annular rubber valve member (18); and an outer cap (21) having a pair of vent holes (20).

A lead (22) of the oxygen generating electrode (7) and a lead (23) of the oxygen ionizing electrodes (8) of the electrochemical element (6) pass through the inner flange (13) and the inner cap (19) and are led to the outside through the vent holes (20) respectively. Reference numeral (24) designates an adhesive consisting of an epoxy resin, which is used to seal the portions where leads (22) and (23) penetrate through the inner flange (13) and also to fixedly secure the inner cap (19) and the outer cap (21) with each other, and reference numeral (25) designates a packing. By connecting the electrochemical element (6) to the vent plug (11) integrally in the manner described, mounting or demounting or exchange of the electrochemical element (6) can be effected conveniently.

When the electrochemical element is to be connected to the vent plug (11) as described above, the size of the element undergoes a limitation. Therefore, it is preferable to shape the electrochemical element in a cylindrical configuration as shown in Figure 4.1e so as to increase the reaction area. With the vent plug (11) shown in Figure 4.1f, the gases interior of the cell, when the inner cell pressure has risen abnormally, move out of the inner cap (19) through the vent holes (17) upon forcibly expanding the rubber valve (18) and then are exhausted to the outside through the vent holes (20).

The sealed type cell is entirely free from the degradation of large current discharge characteristics which has been encountered previously as a result of closing the conventional cells, and the charging-discharging life is either equal to or even longer than that of the open type cells. In addition, this sealed type cell can be charged with a large current in a short period of time because the oxygen absorbing velocity is several tens times as high as that obtainable with a conventional negative electrode, e.g., cadmium electrode, owing to the oxygen ionizing electrode; further the cell is highly reliable in function since the electrochemical element closed in the cell functions solely with the oxygen interior of the cell.

Nickel Oxyhydroxide Electrode

T. Tsuchida, Y. Hayase, and Y. Fujisawa; U.S. Patent 3,674,561; July 4, 1972; assigned to Fuji Denki Kagaku, Japan describe electrodes for rechargeable alkaline cells. The electrode comprises a substratum, metallic nickel dendrite electrolytically cohered on the substratum, and nickel oxy-hydroxide exposed on the nickel dendrite. The process comprises the steps of preparing an electro-bath containing NO_3 ion in the range of about 0.16% to about 3.52%, SO_4 ion in the range of about 1.4% to about 18.1%, and Cl ion in the range of about 2.2% to about 11.5%, keeping the electro-bath weakly acidic, keeping the temperature of the electro-bath over 40° C., electrolytically cohering nickel dendrite layers to a substratum, and immediately electrolytically oxidizing the layers in an alkaline aqueous solution. A typical practicable example for producing a positive electrode of rechargeable alkaline cells according to the process is described as follows. Dendrite layers are electrolytically produced upon a substratum made of annealed nickel by the steps of dipping the substratum in dilute acid, corroding it so that the clear metal surface of nickel may be exposed, quickly washing it by passing through pure water, and inserting it into the plating solution, while passing current through the solution. The plating solution is a composiition consisting of 75% of $NiSO_4 \cdot 6H_2O$, 17.5% of $NiCl_2 \cdot 6H_2O$, and 7.5% of $Ni(NO_3)_2 \cdot 6H_2O$, the dissolving rate of the total salts by weight being 33%.

The temperature of the plating solution is kept to 90° C. and the pH value is regulated to 4. A positive electrode for rechargeable alkaline cells can be made by the steps of electro-depositing dendrite continuously upon a substratum for 10 minutes at the plating current density of about 20 a./dm.^{2} with respect to the apparent surface area of the substratum, drawing out the substratum immediately while continuously flowing current, washing down quickly any plating solution which clings to the layers in pure water, dipping the substratum in an aqueous solution of 37% of KOH, electrolytically oxidizing it for about 1 hour at the charging current density of about 15 ma./cm.^{2}, and cutting it off to a proper size.

OTHER CADMIUM CELLS

Porous Electrode Preparation

A process described by N. Marincic; U.S. Patent 3,607,406; September 21, 1971; assigned to P.R. Mallory & Co. Inc. involves forming a porous metal structure of controlled porosity, by starting with physical mixtures or fluid solutions containing a predetermined percentage of reactant materials to be active in the process, and then treating those reactant materials with other reactant materials to cause a reaction of metal substitution, whereby a porous structure will be formed of a desired metal of one of the reactants, with the undesirable materials left as soluble secondary byproducts which can be leached out of the porous structure, and then performing that operation of leaching or washing out the undesirable products in order to leave the desired porous structure as a finished unit.

Generally, the technique of forming the porous structure is based on the displacement reaction in which a less noble metal, that is, less electropositive, is used to reduce a salt of a more noble metal, that is, one more electropositive. The more noble metal is precipitated from its salt to form a desired porous deposit either on a solid substrate of the same metal, or on a solid substrate of a different metal. Foreign inert bodies can be incorporated into the initial mass to control the resulting structure either for porosity, or for mechanical strength, or for any other reason. The foreign bodies can be used during the preparation of the porous structure, and removed afterwards, for similar reasons, or as desired, thus permitting extreme flexibility procedurally and in end result.

Example 1: An expanded zinc plate was flooded with 0.2 mol $Pb(NO_3)_2$ and left alone for 1 hour. A porous lead structure resulted from the reaction.

Example 2: 5 grams of aluminum powder were placed between two layers of filter paper, forming a 2-millimeter thick layer of loose powder. The powder was flooded with 1 mol $CdCl_2$, and left for the reaction to complete, and then leached with distilled water. A porous cadmium structure resulted.

Example 3: A mixture was prepared, comprising 15 grams of zinc powder, 40 grams of $CdCl_2$ and 10 grams of sugar. Pellets were pressed applying 15,000 pounds per square inch. After full reaction, the structure was leached with distilled water. A porous cadmium electrode resulted.

Example 4: Three layers of expanded zinc plate were put together, degreased with acetone, and flooded with a 2 mol $CdCl_2$ solution. Elastic cadmium felt resulted, applicable as a metallic filter or as an electrode.

Example 5: A single layer of expanded zinc plate was placed on top of a sheet of solid silver foil. A solution of 2 mol $CdCl_2$ was poured over, while the expanded zinc was pressed gently with a wooden roller against the silver foil. A porous cadmium deposit on the silver substrate resulted, replicating exactly the pattern and the texture of the expanded zinc used. A microscopic investigation disclosed three types of porosity.

An ultimate structure within predesigned dimensional limits may thus be readily and simply achieved by arranging the displacement reaction in situ on an initial structure of substantially the same predesign dimensions. In Figure 4.2a, the operation of the treatment considered in example (5) is illustrated. A layer of expanded zinc or zinc mesh (10) is laid on a silver substrate (15), consisting of solid silver foil, and then the solution of 2 mol $CdCl_2$, as mentioned, poured over the structure, while the expanded zinc is gently pressed against the silver foil by a suitable tool, such as a wooden roller.

Microscopic investigation of the final product showed three types of porosity, as shown schematically in Figure 4.2b, after the soluble zinc chloride as a displacement reaction product was leached out. The cadmium electrode structure showed a microporosity (20) close to the surface, with decreasing microporosity (25) toward the axes of the original zinc mesh rods or filaments (10), and with a final network of canals (30) axially, but not necessarily continuous, along the axial regions of the original zinc rods (10).

FIGURE 4.2:   PREPARATION OF POROUS ELECTRODES

(BEFORE CADMIUM CHLORIDE POURED OVER)

a.

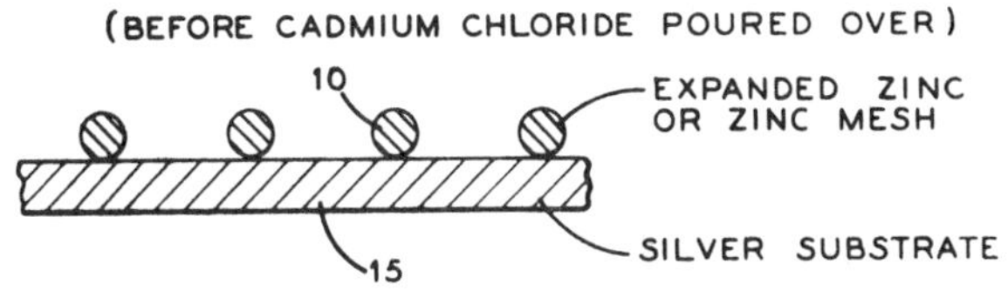

Zinc Overlaid on Silver Substrate

(continued)

FIGURE 4.2:   (continued)

b.

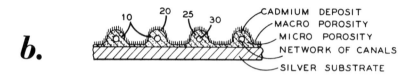

Porous Structure Formed After Treatment

c.

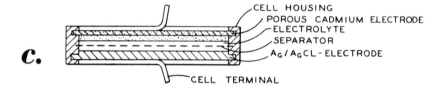

Sectional View of Silver–Cadmium Cell

Source: N. Marincic; U.S. Patent 3,607,406; September 21, 1971

The above cadmium electrode from Figure 4.2b was incorporated into an electro-chemical secondary cell, comprising this electrode, a Ag/AgCl electrode and 2 mol $CdCl_2$ plus 1 mol $NaClO_4$ as an electrolyte. The cell is conventionally rep-resented as follows: $Cd(Ag)/CdCl_2/AgCl/Ag$. Figure 4.2c represents the cell sche-matically. The cell was charged for one minute with 200 ma./cm.^{2} of the geometric area and discharged with 20 ma./cm.^{2} with 100% discharge efficiency.

Electrodes with Elongated Electrolyte Recess

M. Eisenberg; U.S. Patent 3,527,612; September 8, 1970; assigned to Electro-chimica Corporation describes an electrochemical cell including electrodes having a relatively longer dimension in the direction of current flow through the cell, i.e., thick electrodes, which are provided with elongate recesses to carry electrolyte to the interior portions of such thick electrodes to improve current distribution and hence the electrochemical efficiency of the cell. In conventional electrochemical cells, for example silver–zinc, silver–cadmium, or mercury–cadmium batteries, the cross sectional area available for current flow between the cathode and the anode electrodes is ordinarily relatively large compared to the electrode thickness, i.e., the dimension of the electrode taken in the direction of current flow through the cell.

Typically, electrodes having thicknesses of 0.030 inch to 0.060 inch are employed in cells where the cross sectional area is of the order of 1 to 10 or more square inches. Even in small button-type cells, electrode thicknesses, generally no greater than about 0.150 inch, may be used compared to an area for current flow of the order of 1/2 to 1 square inch. In certain applications, electrochemical cells capable of main-taining a predetermined amount of energy output are required; and such requirement necessitates an appreciable amount of active electrode material. If the cross sectional area available to accommodate such cell is limited, then the electrode may require to be of the above noted thick construction. An example of such a cell is illustrated in Figures 4.3a and 4.3b, Figure 4.3a being a lateral cross section to reveal certain details of the process. Such a cell (12) includes cathode enclosure (14) which houses active cathode material (16) and may, for example, consist of a relatively porous mass of a metal oxide, such as mercuric oxide, having a relatively long dimension (18) in the direction (20) of current flow through the cell.

Such an electrode, due to its relatively long dimension in the direction of current flow, is referred to as a thick electrode and includes an interior portion (22) remotely located from the other electrode or anode (24) which, for purposes of illustration, may comprise a porous mass of cadmium metal, also having a relatively long dimension (26) in the direction generally of the current flow line (20). As in the case of the cathode, anode (24) is housed in a suitable anode enclosure (28) adapted for secure engagement within one end of cathode enclosure (14), the two enclosures being insulated from each other by means of annular insulating gasket (30) which may be fabricated of rubber, neoprene, or other suitable material.

Confronting faces or front surfaces (32), (34) of the cathode and anode, respectively, have interposed between them a suitable porous separator (36) which also carries an amount of appropriate electrolyte, such as a solution of KOH in the alkaline mercuric oxide–cadmium cell being described here; and such electrolyte is present throughout most of the cell by virtue of the porosity of electrodes (16) and (24). During the discharge period of cell operation, the active materials proximate electrode faces (32), (34) are consumed; and thereafter current has to penetrate the porous structure of the electrodes deeper and deeper to maintain the electrochemical reactions supplying the power. At the same time, there is a buildup of reaction products which may tend to gradually block the passage of current.

With time, such buildup of reaction products usually results in an increase of internal cell impedance and increased electrode polarization due to the need for ions to diffuse further and further into the porous mass to support the electrochemical reaction. In conventional cells having electrodes of a particularly thick construction, such reaction should proceed toward and occur within the interior if such reaction is to be sustained; and this is relatively difficult if not impossible for the reasons just mentioned. An important feature of this process is that relatively large amounts of electrolyte are carried into the interior portion of a thick electrode sufficient to support a substantially complete electrochemical reaction with the electrode material.

In Figure 4.3a, this feature is achieved by means of cylindrical cavity (38) formed within cathode (16), such cavity having diameter (40) and length dimension (42), which is at least twice the diameter of the cavity and preferably greater than half the thickness dimension (18). Such cylindrical cavity is filled with and carries electrolyte to provide for improved current distribution and penetration within the interior portion (22) of the electrode. In an alternative variation not shown, the cylindrical cavity may be extended through the entire electrode; should extend to the interior portion of the electrode. The total mass of electrode material is selected so that after formation of cavity (38) the amount remaining is sufficient to support a substantially complete electrochemical reaction in accordance with the energy requirement required of the cell through its discharge cycle.

FIGURE 4.3:   ELECTRODE DESIGN

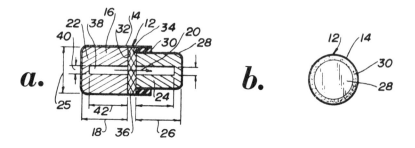

a.

Sectional View of Cell

b.

End View of Cell

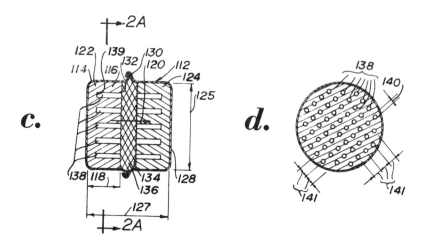

c.

d.

Sectional View of Button–Type Cell

Sectional View Along Line (2—2)
of Figure 4.3c

Source: M. Eisenberg; U.S. Patent 3,527,612; September 8, 1970

The construction described above may also be applied to cells of a somewhat different construction than the elongate cylindrical cell (12) shown in Figure 4.3a. More specifically, the process may also be applied to cells known as button-type cells exemplified in Figures 4.3c and 4.3d. A typical button-type cell derives its name from its generally flat cylindrical configuration. Such a cell (112) is shown having diameter (125) and housing therein electrodes (116), (124). Diameter (125) is substantially larger than the overall length (127), and hence the flat appearance; even though depth dimension (118) of cathode (116) in the direction generally of current flow through the cell indicated by arrow (120) is relatively long. Porous cathode electrode (116) is embedded in a suitable enclosure or housing (114), and anode (124) is encased in housing (128). Interposed between cathode front surface (132) and anode front surface (134) is micro-porous separator (136) wetted with a suitable electrolyte.

Cathode housing (114) is formed with a flared circumferential lip about which an overlapping portion of anode can (128) is secured as by crimping with insulating gasket (130) disposed therebetween to prevent electrical contact between the anode and cathode housings. The other details of the example shown in Figures 4.3c and 4.3d, for example the composition of the active electrode materials and the electrolyte utilized in the cell, are similar to those described in respect to Figures 4.3a and 4.3b above.

In the example of Figures 4.3c and 4.3d, the desirable results are achieved by means of a plurality of holes or cylindrical cavities (138) in electrode (116), each of such cavities forming an opening at surface (132) confronting anode surface (134) and having a terminus (139) proximate interior portion (122) of the thick cathode electrode (116). The array of cavities (138) may be seen in the sectional view of Figure 4.3d where, for sake of clarity, only several of such cylindrical cavities are identified by reference numerals.

As noted in respect to the description of Figures 4.3a and 4.3b, the relationship between the depth and the diameter of the cavities is maintained so that the latter penetrate to the region of the electrode proximate the interior portion thereby to carry the required amount of electrolyte into contact with electrode material of sufficient mass to support the desired degree of electrochemical activity for the particular operation that the cell is designed.

In Figures 4.3c and 4.3d, the longitudinal axis of the cylindrical cavities are maintained substantially parallel to the line of direction of current flow (120) through the cell and are spaced apart from each other at uniform intervals not less than about 1/2 nor greater than about 5 times the diameter of one of said cylindrical cavities.

LITHIUM AND ORGANIC ELECTROLYTE CELLS

ELECTRODES

Sulfur

A process described by D.A. Nole and V. Moss; U.S. Patent 3,532,543; Oct. 6, 1970; assigned to Aerojet-General Corporation relates to a battery utilizing a non-aqueous electrolyte and electrodes of lithium and sulfur.

Various cathode configurations have been investigated and it has been found generally desirable to employ the sulfur in combination with a conductive carbon to improve the conductivity of the cathode. While powdered sulfur may be pressed with particulate carbon and a suitable binder into a cathode configuration, this practice has not been usually followed as the resulting structure is not porous enough to allow good electrolyte penetration.

The preferred cathode configuration uses a porous carbon cloth structure upon which the active cathode material which may be either sulfur or a mixture of sulfur and a cathode composition such as cupric chloride have been applied. The active cathode ingredients are preferably applied to the carbon cloth support in mixture with conductive carbon and a suitable binder such as polyvinylchloride resin.

In one manner of application of the active ingredients to the carbon cloth support, a coating dispersion is prepared by placing 10 g. of a 70/30 mixture of powdered sulfur and powdered carbon black (previously ground and blended in a mortar) into 100 ml. of a 70/30 methylene chloride-carbon disulfide carrier containing 1.5% of polyvinylchloride resin. This mixture is then sprayed or dropped onto the carbon cloth support, and the solvent permitted to evaporate, leaving an electrode which is ready for use in the cell of the process. The sulfur cathode and the lithium anode are at least partially immersed and usually wholly immersed in the electrolyte solution of the cell.

Various techniques are available for the formation of the lithium anode. A particular simple structure involves the pressing of thin lithium sheets (circa 0.02" thickness) on either copper or stainless steel support screens. The cell is best constructed with an excess of lithium.

It is usually preferred to provide porous separators between the anode and cathode of the electrolytic cell and to this end a number of inert porous separator materials have been investigated with respect to their ability to pass current and to withstand the effects of the nonaqueous electrolytes. It will be understood that the separator is not an absolute necessity to the operation of the cell but a desirable element. Of the various porous separators tested, a polyethylene filter paper having about 60% porous area performed the best.

The battery of the process has a theoretical energy density of 1,130 watt-hr./lb. This energy density is approximately 30% greater than the best nonaqueous organic electrolyte system employing a lithium anode and copper fluoride CuF_2. During discharge of the cell of the process lithium metal of the anode is converted to lithium ion and in the process one electron is given up to the external circuit.

Simultaneously, the sulfur of the cathode obtains two electrons from the external circuit to form the sulfide ion. Upon charging the lithium ion receives one electron from the charged circuit and plates out as lithium metal on the anode while the sulfide ion gives up two electrons and is reconverted to sulfur at the cathode. The lithium-sulfur battery is characterized by a high energy density with a theoretical output of 1.3 to 2.0 times that of lithium batteries previously suggested. It is rechargeable, but however, may be utilized as a D-type cell.

In one configuration of the D-cell the cathode, anode and separator are cut into elongated strips, for example, strips of 1 1/2" x 10". The strips are laid with the anode on the bottom followed by the separator, then the cathode, and another separator. The strips are rolled around a carbon rod and inserted into a copper case of the D-cell type.

Desirably, internal cathode contact is improved by forming a slot in the carbon rod into which the cathode is inserted and bonded with a conductive carbon, sulfur, a volatile solvent such as methylene chloride and a polyvinylchloride binder. After an electrolyte is added to the D-cell, the cell is sealed with either epoxy resin or silicon rubber. In the preferred form a metal screw is located in the top of the carbon rod and upon sealing of the battery, the carbon rod per se is enclosed within the cell, leaving only the metal screw exposed for electrical contact. This type of cell may be loaded to about a 10 ampere-hour capacity.

In a second D-cell configuration, the lithium anode strip is inserted against the inside circumference of the copper case, a separator placed against the anode and a paste of cathode material cast into the remaining portion of the case. The paste comprises powdered conductive carbon, powdered sulfur, a volatile solvent and binder.

The volatile solvent is permitted to evaporate leaving a porous cathode structure. Prior to the drying of the paste through evaporation of the solvent, a carbon rod is inserted to provide electrical contact. After the cathode paste has dried, an electrolyte is added and the cell sealed in the fashion described previously. Again, it is desirable to completely seal the carbon rod within the battery case, leaving only a metal screw or other conductive surface exposed for electrical contact. The open circuit voltage of the cell is 2.8 volts. The initial internal resistance is substantially the same as that experienced in the first D-cell configuration described previously, but after a short period on closed circuit the cell resistance began increasing at a considerably faster rate. It is possible to load the cell of the second configuration as much as 20 ampere-hours capacity.

It has been found that the efficiency of the electric cell may be considerably improved by the use of a soluble sulfide material in the electrolyte, the sulfide material serving to boost the cell capacity and to extend the duration of discharge at the higher voltage. This improved electrolyte which may be described as a "reacted" electrolyte may be prepared by adding carbon disulfide to the electrolyte solvent and in a preferred embodiment is prepared by incorporating both finely divided lithium powder and carbon disulfide in the electrolyte which is then permitted to react at room temperature under constant stirring for a prolonged period, for example, for a period of a week, to prepare the reacted electrolyte.

In one manner of preparing the reacted electrolyte, finely divided lithium powder in the amount of 12 g./l. and carbon disulfide in the amount of 50 ml./l. is incorporated in a 50/50 mixture of dimethoxyethane and ethylene carbonate, along with one mol of lithium perchlorate. The mixture is permitted to stand at room temperature for a week with constant stirring.

Upon completion of the reaction, the mixture is filtered and the resulting reacted solution is then available for use in the cell of the process as the electrolyte. The use of the reacted electrolyte has been found to boost the discharge efficiency of a particular cell from 24 to over 50%. It is also noted that when one particular cell containing the reacted electrolyte is permitted to set over a weekend after being depleted to 41.2% of its capacity, its open circuit voltage returns to the normal 2.55 volts and an additional 0.17 ampere-hour discharge is obtained, resulting in a total discharge efficiency of 73.4%.

The cell may be discharged (and recharged) to essentially 100% of theoretical capacity; however, because of internal resistance, the last 20 to 25% of capacity is normally not considered profitable. Attempts have been made to identify precisely the active products in the reacted electrolyte, but as yet this has not been conclusively established although it is evident that the materials are soluble sulfide. It is believed that the compositions are likely polysulfides containing both lithium and sulfur. Carbon disulfide may also be present. It is believed that the reacted electrolyte minimizes or substantially prevents the formation of polysulfides on the cathode during charge and discharge.

Cupric Sulfide

G.M. Gerbier and V.L. Dechenaux; U.S. Patent 3,655,446; April 11, 1972; assigned to Societe Des Accumulateurs Fixes et de Traction (Societe Anonyme), France describe a method for preparing a cupric sulfide electrode in which substantially stoichiometric amounts of copper and sulfur are used and which eliminates limitations as to size of batches capable of being processed at a time. The process comprises, a sulfuration stage more especially including compressing the heated resultant products of a presulfuration stage in a mold and heating the product at a temperature within the 120° to 160°C. range, and a final processing stage or step at a temperature within the 120° to 160°C. range outside of the mold, where previous to the sulfuration stage or step a presulfuration stage or step is carried out, consisting in heating at a temperature within the 120° to 160°C. range a first mixture formed by the whole stoichiometric amount of required copper and about 25 to 35% of the stoichiometric amount of required sulfur, then in adding to the thus processed mixture of the presulfuration stage the complement of the required stoichiometric amount of sulfur or a slightly larger amount to provide a second mixture, carrying out the sulfuration stage of the second mixture and subsequently after removal from the mold effecting a further heating of the electrode resulting from the sulfuration stage.

Advantageously, after the presulfuration stage a material capable of increasing the electric conductivity of the desired electrode is added to the mixture of the presulfuration stage at the same time as the complement of sulfur in making the second mixture for the following reason. During the discharge of a cell provided with a cupric sulfide cathode, the electrochemical reduction of cupric sulfide is effected in two distinct steps: first the cupric sulfide is reduced to cuprous sulfide according to the reaction:

$$2 \ CuS \ + \ 2e^- \ \longrightarrow \ Cu_2S \ + \ S^=$$

then the cuprous sulfide is reduced to copper according to the reaction:

$$Cu_2S \ + \ 2e^- \ \longrightarrow \ 2 \ Cu \ + \ S^=$$

Both cupric sulfide and copper are electrically conductive but cuprous sulfide is not. Due to this fact, at the end of the first reduction step, the electrode mixture is not sufficiently conductive. This would cause an increase of the internal resistance of a cell provided with such an electrode, this resistance increasing, e.g., from 1 to 3 ohms. This is why it is advantageous to provide in the desired electrode a material capable of increasing its electrical conductivity.

This material may be constituted by acetylene black the content of which may be about 2% by weight of the total amounts of sulfur and copper. This 2% content of acetylene black is sufficient to decrease by half the aforesaid variation of internal resistance; a larger amount does not bring any substantial improvement at the usual discharge rates.

Nickel flakes also may be added together with the acetylene black, the nickel flake content being about 2 to 5% by weight of the total stoichiometric amounts of sulfur and copper. Cells built with resulting electrodes display improved performance at high rates; thus the energy density of a cell the cathodes of which comprises acetylene black 2% and nickel flakes 2.5% reaches 205 wh/kg. whereas the energy density of a cell of the same type without these improvements is 185 wh/kg. for a 40 hour discharge.

Also, nickel powder can be used instead of acetylene black. However, for obtaining a similar result, the amount of nickel powder added in making the second mixture must be about 10% by weight of the total stoichiometric amount of sulfur and copper, so that the weight of the resultant electrode is increased and this may be a drawback in some cases.

In order to make the electrode according to the process it is advantageous to use a copper powder made of dendritic particles the size of which are less than 50 microns and the apparent density of which is within the 1 to 2 range. Such particle shape promotes the sulfuration and the sintering of sulfide particles during heating, thus ensuring the mechanical strength of the desired electrode. A powder with scale shaped particles may also be used, the grain size being between 1 and 40 microns and the apparent density being about 1. The following examples illustrate the process.

Example 1: A first batch of copper powder of the type described and powdered sulfur in which a stoichiometric full amount of copper and from 25 to 35% by weight of the required stoichiometric amount of sulfur are mixed together to form a first mixture. The copper powder is preferably in the form of dendritic particles whose size is less than 50 microns with an apparent density within the 1 to 2 range. This particle shape promotes sulfuration and sintering of sulfide particles during heating, thus ensuring the mechanical strength of the electrode to be formed. In the alternative, the copper powder may be in the form of scale shaped particles whose grain size ranges from between 1 to 40 microns and with an apparent density of about 1.

This first mixture of the full stoichiometric amount of such copper powder together with 25 to 35% by weight of the required stoichiometric amount of sulfur is subjected to a presulfuration stage or step by heating the mixture to a temperature of from 120° to 160°C. for 2 to 5 hours.

The product of this presulfuration stage is then mixed with a complemental amount of the total stoichiometric amount of sulfur increased up to 5% by weight of the total required stoichiometric amount of sulfur and this second mixture is deposited in a mold and compressed therein under a pressure ranging from 0.25 to 2 torr/cm.^{2} and preferably at a pressure of 0.5 torr/cm.^{2}.

The compressed second mixture in the mold is then heated in a sulfuration step to a temperature ranging from 120° to 160°C. for about 15 hours. Thereafter, the product of this second sulfuration step in the form of a molded electrode is removed from the mold and cooled. This electrode after cooling is heated again outside of the mold for about 10 to 15 hours at a temperature ranging again from 120° to 160°C. During such final heating, the excess sulfur eliminates itself. Cooling prior to the final heating may be omitted if desired. The resulting finished electrode is cupric sulfide.

There is no limitation in this procedure as to the size of the batch treated as all possibilities of ignition of the sulfur likely to occur if the original copper-sulfur mixture were to contain the full stoichiometric amounts both of copper and sulfur in a mass exceeding a weight of about 20 g., are prevented by the process.

Example 2: The same presulfuration stage of the same first mixture as in Example 1 is carried out for the same length of time at the same temperature. The product of this presulfuration stage is then mixed with additional sulfur in the same amounts as in Example 1 to provide the second mixture to which is added a conductivity increasing material such as acetylene black in the amount of about 2% by weight of the total stoichiometric amounts of sulfur and copper, or else a mixture of this latter amount of acetylene black with nickel flakes the latter in the amount of from 2 to 5% by weight of the stoichiometric amounts of sulfur and copper, or else, nickel powder instead of acetylene black may be added to the second mixture in the amount of 10% by weight of the total stoichiometric amount of copper and sulfur.

The second mixture containing at least one of the conductively increasing materials is then placed in a mold and compressed therein under a pressure ranging from 0.25

to 2 torr/cm.^{2}. This compressed mixture in the mold is then heated in a sulfuration step to a temperature ranging from 120° to 160°C. for about 15 hours. Thereafter, the product of this sulfuration step in the form of a molded electrode is removed from the mold and cooled, if desired.

The electrode after removal from the mold is again heated outside the mold for 10 to 15 hours at a temperature again from 120° to 160°C. During such final heating excess sulfur is eliminated. The resulting finished electrode as in Example 1 is cupric sulfide whose conductivity is increased by the added conductivity increasing material. Again, there is no limitation as to size of batch treated for the same reasons as in Example 1.

Example 3: An electrochemical cell is constructed with a unit comprising at least one cupric sulfide electrode of Examples 1 or 2 together with a lithium sheet electrode positioned at at least one face of such cupric sulfide electrode with an insulating separator interposed between the respective electrodes, the separator being, for example polypropylene felt.

This electrode-separator unit is positioned in a casing and a nonaqueous electrolyte comprising, for example, tetrahydrofuran organic solvent and lithium perchlorate solute is added to the casing. The electrochemical cell thus constructed is a lithium-cupric sulfide electrochemical cell with nonaqueous electrolyte. Other electrolytes also may be used. The lithium electrodes likewise may be replaced by other suitable electrodes.

Cuprous Chloride

In a process described by <u>G.M. Gerbier and V.L. Dechenaux; U.S. Patent 3,661,648; May 9, 1972; assigned to Societe des Accumulateurs Fixes et de Traction (Societe Anonyme), France</u> electrodes for electrochemical cells are prepared from porous copper-containing carrier bodies by immersion at ambient temperature of such bodies in cupric chloride dissolved in an organic nonaqueous solvent such as methanol and subsequently washed in an organic compound capable of dissolving cupric chloride, e.g., methanol to remove excess cupric chloride and then dried in vacuo.

The porous carrier bodies may be prepared by sintering arborescent copper powder selected from the group consisting of acicular form or dendritic form having an apparent density of between 1 and 2 first molded to desired shape then heated to 700°C. in a reducing atmosphere for one-half hour. A conductive copper or copper plated support of metal gauge or perforated metal sheet may be incorporated in the powder during molding.

In the alternative the copper carrier body can be molded from such copper powders agglomerated with an organic binder selected from the group consisting of polystyrene, polyethylene and polytetrafluoroethylene and this molded body immersed in cupric chloride dissolved in organic solvent such as methanol, with subsequent washing in organic compound such as methanol to remove excess cupric chloride and ultimate drying in vacuo. The following examples illustrate the process.

Example 1: A methanolic solution of cupric chloride consisting of a quantity of cupric chloride comprised between 500 and 600 g. is dissolved in 1 liter of methanol. A porous sintered copper carrier is dipped or immersed in this solution for about one-half an hour at ambient temperature, it being possible, however, to extend this immersion time to about three-quarters of an hour. The reaction which occurs during the immersion time is as follows: part of the copper forming the original porous metal carrier is corroded, by the nonaqueous cupric chloride solution to form the active material in the form of cuprous chloride in the carrier, thus providing the desired electrode.

The so-treated electrode is then removed from the nonaqueous solution and washed by an organic compound capable of dissolving the cupric chloride, for example, methanol, to remove the excess cupric chloride and then the washed electrode is dried in vacuo.

In the resulting electrode, the copper content of the active material in the form of cuprous chloride corresponds practically to about twice the corroded copper of the carrier, since this cuprous chloride comprises as much copper derived from the metallic copper as that supplied by the nonaqueous cupric chloride corroding solution.

By this method, in a single operation, about a third of the original copper of the original carrier becomes corroded to cuprous chloride. If this process is repeated by again immersing the electrode which has already been treated once in the nonaqueous methanolic solution, the further amount of corroded copper resulting is very slight. Thus, a repetition of several operational immersion cycles, it is found produces no more than 5% of extra cupric chloride.

Example 2: The porous copper carrier to be treated in the manner of Example 1 can be made as follows. Copper powder in arborescent form selected from the group consisting of dendritic or acicular particles is preferably used, because such a powder has very great entangling qualities, while having a structure such that it can easily be corroded by cupric chloride. Relatively light copper powders of this type having an apparent density practically between 1 and 2, can be used.

Firstly, such powder is molded into the required shape of the electrode, then the corresponding molded powder cakes are elevated to a temperature of about 700°C. for about half an hour in a reducing atmosphere. A suitable conductive support, in the form of metal gauze or of perforated metal sheets, for example, can be incorporated during molding into these electrode shapes, being located preferably in the median plane. These conductive supports may advantageously be made of copper, or also can be made of copper-plated steel.

Example 3: In an alternative procedure, the copper carrier for treatment in the manner of Example 1 can be made as follows. Copper powder of the same kind as in Example 2 is agglomerated with an organic binder. This organic binder may be selected from the group consisting of polystyrene, polyethylene, or polytetrafluoroethylene. In effecting this carrier product, the copper powder and organic binder are agglomerated and molded into desired carrier shape. This shaped carrier is then subjected to the immersion procedure in nonaqueous cupric chloride methanol solution according to the procedure of Example 1.

In the electrode, it has been found that the corroded cuprous chloride fraction of the metal copper in the molded carrier is no longer about one-third but rather about one-fourth or one-fifth of the metal copper, thus enabling the production of electrodes having a metallic carrier of improved conductance.

The electrodes produced by the procedures of any of the examples are of utility particularly in electrochemical generators of primary type with a lithium anode and a nonaqueous electrolyte. They are useful, too, in other known types of electrochemical generators with different anodes and electrolytes.

Fluorinated Carbon

N. Watanabe and M. Fukuda; U.S. Patent 3,536,532; October 27, 1970; assigned to Matsushita Electric Industrial Co., Ltd., Japan describe an electric current producing primary cell of high energy density which is composed of a negative electrode having a light metal as active material, a nonaqueous electrolyte and a positive electrode having a solid fluorinated carbon as active material, the solid fluorinated carbon being represented by the formula $(CF_x)_n$ where x is not smaller than 0.5 but not larger than 1 and obtained by the fluorination of a crystalline carbon, such as graphite. The utility of the positive electrode active material is high and nearly 100%, the flat characteristic of discharge voltage is excellent and the shelf life is long owing to the chemical stability in the electrolyte of the fluorinated carbon used as active material.

The process for preparing the fluorinated carbon is as follows. A 200 mesh graphite powder was charged in a reactor made of nickel and the reactor was heated externally in an electric furnace while bleeding air, until it reaches a predetermined temperature. When the temperature had reached about 450°C., fluorine was slowly introduced into the reactor and the reaction between the fluorine and graphite was continued for about 2 hours while maintaining the fluorine pressure at 0.8 atmosphere.

The particle size, the reaction temperature and the fluorine pressure can suitably be selected from the standpoint of economy. As an example, the relationship between the amount of fluorine absorbed by the graphite and the temperature is shown in Figure 5.1a. As seen, the reaction starts at a temperature higher than 300°C. and a temperature of 450°C. is optimum for the reaction, but when the temperature exceeds 450°C. the fluorinated graphite formed is further combined with fluorine and flown away in the form of gaseous CF_4, etc.

Industrially and from the standpoint of production safety, the reaction for producing the fluorinated carbon is preferably carried out at a temperature of 350° to 450°C. for 2 to 5 hours in a fluorine gas atmosphere at a pressure not greater than the atmospheric pressure, particularly preferably at a pressure of 0.5 to 0.8 atmosphere, although the reaction temperature is variable depending upon the reaction time.

The powder of fluorinated graphite obtained in the manner described above was mixed with an electrically conductive agent consisting of acetylene black and a binder consisting of polyethylene tetrafluoride powder at the weight ratio of 1:0.2:0.2 to prepare a positive electrode active material. Since the mixture is highly moldable, a positive electrode can be produced simply by molding the mixture with a nickel screen.

The size of the molded electrode was 40 x 40 x 1 mm. and the theoretical capacity was about 2 ah. The negative electrode used in combination with the positive electrode was 40 x 40 x 0.4 mm. in size and had nickel leads. The electrolyte used was a solution of 1 M lithium perchlorate $LiClO_4$ in 1 liter of propylene carbonate. As separator, a sheet of polypropylene nonwoven fabric having a thickness of 0.2 mm.

was used. The elements described previously were disposed in a polyethylene case and sealed to obtain a battery. The assembly of the battery was effected in a dry argon atmosphere.

The discharge characteristic of the battery when discharged at 100 ma. is represented by the curve (1) in Figure 5.1b. The curves (2) and (3) represent the discharge characteristics of batteries in which use is made of AgCl and CuF_2 as active material for the positive electrode respectively.

FIGURE 5.1:   PRIMARY CELL EMPLOYING FLUORINATED CARBON ELECTRODE

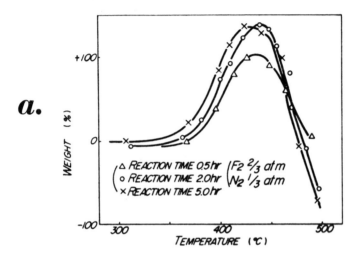

Percent Fluorine Absorbed by Graphite as a Function of Temperature

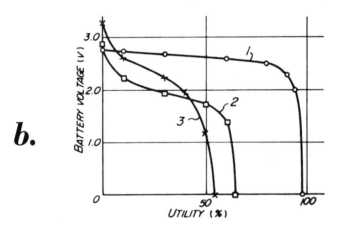

Discharge Characteristics of Batteries

Source: N. Watanabe and M. Fukuda; U.S. Patent 3,536,532; October 27, 1970

The open circuit voltage used was 3.3 to 3.6 volts for the battery, 2.85 volts for the battery of the characteristic curve (2), and 3.53 volts for the battery of the characteristic curve (3). From this chart, it will be seen that the discharge characteristic of this battery is superior to those of the conventional ones.

Namely, the battery is characterized by the fact that the utility of the active material is nearly 100% and that the flat characteristic is excellent. The discharge voltage of the battery is somewhat lower than that of the battery comprising CuF_2, in the initial stage of discharge operation, but this battery is superior to the latter in respect of utility of the active material and flat characteristic. With reference to self-discharge, the active material of this process showed substantially no deterioration even after storage of the battery for about 6 months.

In the example described above the production process was described as a general practice but, when the solid fluorinated graphite is used in a battery designed for low-rate discharge, blending of a metal powder or carbon powder, which is normally used as an electrically conductive ingredient, is not particularly needed, since solid fluorinated graphite possesses the property of carbon, i.e., electrical conductivity, in a considerable degree.

This is advantageous in increasing the theoretical quantity of electrolyte charged in the battery. Such property of solid fluorinated graphite is exactly contrary to the original presumption that the property of fluorine-contained resin, e.g., ethylene tetrafluoride, would generally appear as the value of x in the formula $(CF_x)_n$ becomes larger, and is one of the features of the solid fluorinated carbons produced by the process.

Metal Chromates

A process described by A.N. Dey; U.S. Patent 3,658,592; April 25, 1972; assigned to P.R. Mallory & Co., Inc. provides an electric cell comprising a light metal anode, a metal chromate cathode, and an organic electrolyte. Light metal anodes suitable for the process may be composed of lithium, sodium, potassium, beryllium, calcium, magnesium, aluminum, and the like as defined in U.S. Patent 3,413,154. The preferable anode material is lithium.

The depolarizers used are the metal salts of the oxyacids of chromium. The chromates and dichromates of copper, silver, iron, cobalt, nickel, mercury, thallium, lead and bismuth and mixtures are particularly suitable as depolarizers.

Suitable electrolyte may be made by dissolving organic or inorganic salts of light metals in the organic solvents. For example, 1 to 2 molar solutions of lithium perchlorate or lithium aluminum chloride in tetrahydrofuran solvent constitute a suitable organic electrolyte. Other light metal salts such as perchlorate, tetrachloroaluminate, tetrafluoborate, chloride, hexafluophosphate, hexafluoarsenate, etc. dissolved in organic solvents like propylene carbonate, dimethyl sulfite, dimethyl sulfoxide, N-nitrosodimethylamine, gamma-butyrolactone, dimethyl carbonate, methyl formate, butyl formate, acetonitrile and N,N-dimethylformamide, can be used as electrolytes.

The metal chromate cathodes are fabricated by mixing a mixture of powdered metal chromate and graphite as a conductive diluent in 7:3 weight ratio with 3% by weight of an aqueous dispersion of polytetrafluoroethylene sometime referred to as colloidal

Teflon R which acts as a binder. Sufficient amount of an organic solvent such as isopropanol is then added to the mix to form a paste. The paste is then thoroughly mixed to form an easily pliable dough (2). Metal chromate cathodes are molded on an expanded nickel current collector (4), by placing the above dough on the current collector which in turn is placed in a rectangular die, and then pressing the dough at pressures of 70 to 80,000 psi.

The process is schematically shown in Figure 5.2a. The excess isopropanol is squeezed out of the dough and it results in a compact rectangular cathode (8) with adequate mechanical integrity. The cathode is then dried in air and cured at a temperature of 300°C. for 2 hours. The curing process enhances the mechanical integrity of the cathode even further. The electrical conductivity of the cathode is found to be quite adequate. In the above mentioned method of electrode fabrication, the weight ratios of metal chromate and graphite can be varied from 24:1 to 1:1. The preferred ratio is 7:3.

FIGURE 5.2: LITHIUM-METAL CHROMATE CELL

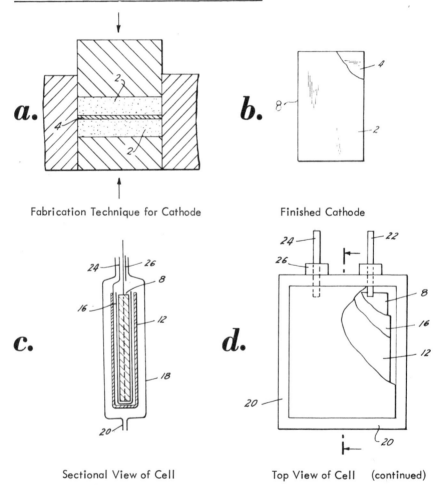

a. Fabrication Technique for Cathode

b. Finished Cathode

c. Sectional View of Cell

d. Top View of Cell   (continued)

FIGURE 5.2: (continued)

e.

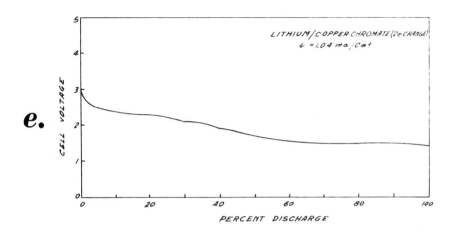

Discharge Curve for Lithium–Copper Chromate Cell

f.

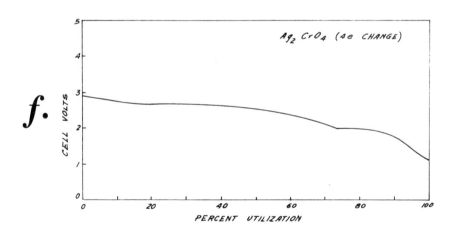

Discharge Curve for Lithium–Silver Chromate Cell

Source: A.N. Dey; U.S. Patent 3,658,592; April 25, 1972

A corner of the metal chromate cathode, made according to the above procedure, is scraped to bare the current collector to which the tab is spot welded for electrical connection. The finished cathode (8) is shown in Figure 5.2b. The cathode is used in conjunction with two lithium anodes (12) prepared by pressing two rectangular pieces of lithium metal, 0.02 inch thick, on expanded stainless steel, to construct a lithium-metal chromate cell, in parallel plate configuration as shown in Figure 5.2c. A layer of filter paper (16) is interposed between the lithium anode and the metal chromate cathode, to provide a separator and electrolyte absorber. The sectional view of the cell is shown in Figure 5.2c. The cell is enclosed in a foil-laminate enclosure (18) by seals (20).

3 cc of 1 molar solution of $LiClO_4$ in tetrahydrofuran is then added to the cell and the cell is heat sealed permanently. The top view of the cell with partial cut-outs is shown in Figure 5.2d. The cells are cathode limited, i.e., the cathode has a lower capacity than the anode.

The lithium-metal chromate cells are discharged using a constant current of 4.5 ma. corresponding to a current density of 1 ma./cm.^{2} based on the geometric area of both sides of the cathode. This would correspond to a 20 to 30 hour rate. The operating characteristics of the several lithium-metal chromate cells, under the above type of load, is shown in the following table.

Cell	Open Circuit Voltage	Initial Operating Voltage	Average Operating Voltage
Li/CuCrO$_4$	3.6	2.3	2.0
Li/Ag$_2$CrO$_4$	3.5	2.8	2.4
Li/HgCrO$_4$	3.5	2.5	2.3
Li/PbCrO$_4$	3.1	2.3	1.6
Li/CoCrO$_4$	3.0	2.3	1.4

All the cells showed exceedingly high open circuit and operating voltages compared to conventional alkaline cells. The cell voltage versus fraction of the cathode capacity discharged plot of the Li/CuCrO4 cell is shown in Figure 5.2e. The cathodic reaction assumed for the calculation of the cathode capacity is:

$$CuCrO_4 \ + \ 2Li^+ \ + \ 2e \ \longrightarrow \ Cu \ + \ Li_2CrO_4$$

It is evident from Figure 3.2e that the cathode utilization efficiency is close to 100%. The theoretical energy density of the above cell, based on the open circuit voltage and the above cell reaction is 400 wh/lb. The similar discharge curve for the Li/Ag2CrO4 cell is shown in Figure 5.2f. The cathodic reaction is assumed to be:

$$Ag_2CrO_4 \ + \ 4Li^+ \ + \ 4e \ \longrightarrow \ 2Ag \ + \ CrO_2 \ + \ 2Li_2O$$

The cathode utilization efficiency is again found to be 100%. The system is also remarkable for its exceptionally high operating voltage.

Sulfur Dioxide as Cathode Depolarizer

D.L. Maricle and J.P. Mohns; U.S. Patent 3,567,515; March 2, 1971; assigned to American Cyanamid Company describe a primary and secondary electrochemical cell effective without electrode separator which comprises in combination an anode of a metal capable of reducing sulfur dioxide, a cathode particularly of a high surface area material, an electrolyte salt solution containing sulfur dioxide as the major cathode depolarizer, with the use of a cosolvent if desired.

Example: Suitable cells are constructed in conventional fashion using a cylindrical glass vessel closed with a rubber stopper through which are sealed connectors for the anode and cathode and means for admitting sulfur dioxide and inert gas. The clean, dry cell is placed in an argon-filled dry box where the electrodes are prepared and the dried electrolyte added to the cell.

The cosolvent, if employed is then introduced into the cell. The cell is then removed from the dry box and sulfur dioxide gas is passed through the solution until the desired amount has collected in the cell. The performance of the cell is tested by connecting a voltmeter to the output terminals and an ammeter and variable load resistor in series with the cell.

A cell is prepared employing 0.5 molar tetrabutylammonium perchlorate in propylene carbonate, the solution being saturated with sulfur dioxide together with a 2.15 cm.^{2} lithium ribbon anode and a 1 cm.^{2} platinized platinum cathode. This cell exhibits an open circuit voltage (called OCV) of 2.9 volts and delivers 10 ma./cm.^{2} (cathode current density, called CCD) at 1.6 volts.

Cold Extrusion of Lithium Anode

A. Blondel and J.-F. Jammet; U.S. Patent 3,663,721; May 16, 1972; assigned to Societe des Accumulateurs Fixes et de Traction (Societe Anonyme), France describe electrochemical cells utilizing lithium anodes which are prepared by cold extrusion into sheet form. The extruded sheets are severed into prescribed band lengths and the bands are pleated in zig-zag form to provide multiple folds or pleats.

Thin cathode plates are positioned between pleats, being insulated from the pleated lithium anode by suitable separating means. The dimension of the cathode plates parallel to the width of the bands is such as to leave a zone along at least one edge of the pleated anode band that has no opposite cathode plate counterpart.

Electrical connections are secured to the pleated band in the zone and are in the form of tabs, for example, of stainless steel, spot welded to an expanded metal strip which latter in turn is fastened to the lithium anode band in the zone as by ultrasonic welding. The separating means may be in the form of an enclosing sheath or envelope of insulating material into which the lithium band is inserted prior to pleating or the cathode plates may be individually sheathed in separator material.

Referring to Figure 5.3, the anode which is to be constituted by a zig-zag folded or pleated sheet of lithium is advantageously made by cold extrusion of a lithium ingot to form a lithium band by means of a suitable extruding apparatus (E) as diagrammatically shown in Figure 5.3a.

This apparatus (E) comprises a die holder or tube (1), a carrier (2) for the die holder (1) and the die-plates (3) as shown in Figure 5.3a. A lithium ingot (4) is placed in the die holder (1) and is pressed towards the die-plates (3) by means of a suitably operated extruding piston (5) guided in its movement by a guide member (6).

FIGURE 5.3:   LITHIUM ANODE CELLS

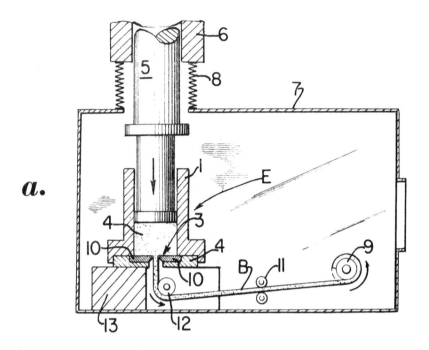

Anode Extrusion Apparatus

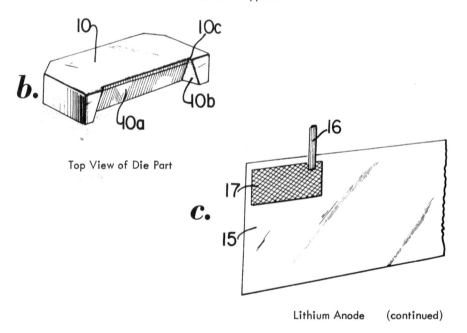

Top View of Die Part

Lithium Anode   (continued)

FIGURE 5.3:   (continued)

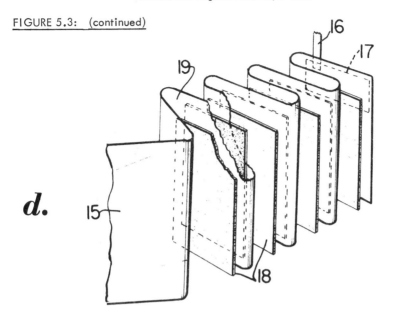

Perspective View of Zig-Zag Construction

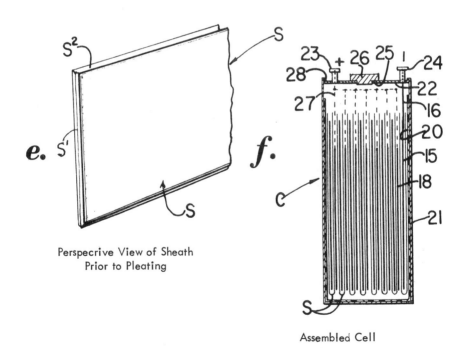

Perspecrive View of Sheath
Prior to Pleating

Assembled Cell

Source: A. Blondel and J.-F. Jammet; U.S. Patent 3,663,721; May 16, 1972

The extrusion operation on the lithium ingot (4) is carried out at ambient temperature and wholly under dry argon atmosphere in a hermetic container (7) so that the band (B) when issuing from the passage in the die plate (3) remains clean on its surface. Bellows (8) connect the container (7) to the guide member (6) used for guiding the piston (5) which is operating for extrusion of band (B).

The lithium band (B) issuing from the passage of the die plate (3) must be drawn in order to be kept straight upon its issuance. This problem is solved by a winch (9) whose windup is effected by suitable means (not shown).

Before windup the lithium band (B) passes through smoothing rollers (11), after changing its direction around a guiding roller (12). The extrusion apparatus rests on support (13). With an extrusion pressure of about 0.4 ton/cm.^{2} it is possible to obtain a lithium band 61 mm. wide and about 1.2 mm. thick. Die-plate (3) is obtained by assembling two half-dies parts (10) in the carrier (2), each half-die part (10) being similar to that shown in Figure 5.3b.

It should be noted that the outward and downward slant of the bevelled faces (10a) and (10b) of these half-die parts (10) is pronounced in order to prevent lithium from adhering to them. This kind of die part is provided with a slot (10c) and in assembly the opposite faces of these slots are parallel with trued surfaces, defining the orifice of die (3).

The extruded lithium band (B) subsequent to windup is unrolled from winch (9) and is then cut to the desired plate length for making an anode plate (15) (Figure 5.3c). Electrical connection means (16) and (17) are then fastened to edge zones of the anode plate (15), Figure 5.3c. In this example, these means comprise a stainless steel tab (16) having the shape of a grid for lightness sake, which is spot-welded to an expanded metal strip (17) made, e.g., of stainless steel.

The fastening of strip (17) to the edge zone of the anode plate (15) is effected by ultrasonic welding of the strip (17) to one edge zone of the lithium plate (15). For this purpose, the edge of the plate (15) is placed on the anvil of an ultrasonic welding machine. Ultrahigh frequency sounds are directed to the strip (17) and anode plate (15) for a time which depends on the strip surface area (about 6 seconds for 8 cm.^{2}) at a power of about 1000 watts.

This ultrasonic welding is also carried out in a dry argon atmosphere. For a correct application of the above described process it is necessary that:

> the lithium surface of plate (15) be very clean (therefore the welding apparatus is maintained in a dry argon atmosphere);
> the anvil be made of trued tempered steel so that lithium is not welded to the anvil;
> the titanium sonotrode (not shown) must be covered, e.g., by a polyethylene terephthalate sheet to prevent lithium from being welded to titanium of the sonotrode;
> the thickness of the lithium plate should be at least equal to 0.6 mm.

An anode (15) according to the process is shown in Figure 5.3c, being provided with its electrical connection means (16) and (17). The lithium anode (15) bears the stainless steel tab (16) spot welded to strip (17) made of an expanded metal such as

stainless steel and the strip (17) itself welded by ultrasonic welding to anode (15).
Upon completion of attachment of the electrical connecting means (16), (17) to
anode (15), a separator is then placed around the anode. According to the process,
this separator (S) (Figure 5.3e) constitutes a sheath obtained by lengthwise folding
of a sheet, for example, made of polypropylene felt or the like, and by heat welding the opposite end edges of the sheet as at (S^1).

Advantageously this sheet is teaseled on its external faces which are not to be in
contact with the anode. The lithium anode (15) provided with its electrical connection means is slipped into the separator sheath through its open mouth (S^2). Then
the assembled anode and sheath are zig-zagwise folded or pleated on a master gauge
and cathode plates are placed between the adjacent pleats resulting from the zig-
zag folding as illustrated in Figure 5.3d.

Figure 5.3d diagrammatically shows a perspective view with broken-away portions
of part of the unit of anode and cathode plates; however, the sheath shaped separator
(S) enclosing the anode (15) and the electrical connection means associated with the
cathodes have not been shown in this figure for clarity's sake.

The anode (15) provided with its electrical connection means (16), (17) is made by
folding or pleating the plate (15) of Figure 5.3c enveloped in sheath (S) upon itself.
Between each pair of pleats is placed a cathode plate (18) having the shape of a
thin substantially rectangular plate. A cell may thus be constituted by folding the
anode eight times in order to constitute seven folds and placing seven cathode plates
in the pleats between these folds.

As seen in Figure 5.3d, the height of the cathode plates (18) is smaller than the
width of anode (15). Thus, a longitudinal zone (19) is created along the upper edge
of anode (15), such zone being without opposite cathode surface and therefore not
being electrochemically operative. The electrode connection means (16), (17) are
fastened on the anode in the zone (19). Thus, current is correctly collected until
the end of the discharge.

As an illustration, the width of band (15) being 61 mm., the width of zone (19) may
be from 5 to 10 mm. This assembly is introduced into a pocket (20), Figure 5.3f,
made of insulating material such as polyethylene, which is then placed in a metal
casing (21), e.g., made of tin plated iron sheet. The pocket (20) is mainly used for
insulating the electrodes (15) and (18) from the metal casing (21). The cover (22)
of the casing is provided with positive and negative terminal posts (23) and (24) and
with a perforation (25) used for pouring the electrolyte in the casing, which can be
hermetically closed as by a plug (26).

The connections (27) of the cathode plates and the connecting tab (16) of the anode
(15) are respectively welded to the positive and negative terminal posts (24) and (23)
of the cover (22). Such assembly operations are carried out in a closed container
under a dry air flow.

The cells (C) thus constituted prior to joining of the cover (22) to casing (21) are
placed in vacuo and then under an argon atmosphere prior to soldering at (28) of
the cover (22) to the casing (21), e.g., by tin soldering. Due to the fact that the
cell (C) comprises only one anode (15) made of a folded length of lithium band instead of several distinct anode plates, time is gained in the manufacture of the cell

since only one electrical connection is to be welded for collecting the current of the lithium electrode (15). The electrolyte for cell (C) may be constituted, for example, by a lithium perchlorate solution in a mixture of tetrahydrofuran and 1,2-dimethoxyethane. It must necessarily be free from any trace of water. As an illustration this electrolyte may have the following composition by volume.

	Percent
Tetrahydrofuran	62
1,2-dimethoxyethane	27
Lithium perchlorate	11

ORGANIC ELECTROLYTES

Tetrabutylammonium Chloride in Propylene Carbonate

A.J.-C. Caiola, H.R. Guy and J.-C. Sohm; U.S. Patent 3,658,593; April 25, 1972; assigned to Societe des Accumulateurs Fixes et de Fraction (Societe Anonyme), France describe electrochemical cells utilizing lithium negative electrodes capable of discharge at high rate in which the lithium electrode operates in substantially insoluble manner utilizing a nonaqueous electrolyte in which lithium and its oxidation products formed during discharge are insoluble, in which the lithium is not corroded and which has acceptable conductivity and in which excessive electrode polarization and passivation is prevented.

The composition of the nonaqueous electrolyte comprises at least a solution of a tetrabutylammonium salt in propylene carbonate and advantageously a solution of tetrabutylammonium chloride $(C4H9)4NCl$ in propylene carbonate mixed with tetrabutylammonium perchlorate. The concentration of tetrabutylammonium salts in propylene carbonate is preferably in the range of 0.5 to 1 mol/liter and advantageously the ratio of tetrabutylammonium perchlorate concentration to tetrabutylammonium chloride concentration expressed in mols/liter does not exceed 10.

Organic Aprotic Solvent and Lewis Acid

In a process described by M. Eisenberg; U.S. Patent 3,660,162; May 2, 1972; assigned to Electrochemica Corporation a relatively lightweight galvanic cell is constructed by using aluminum, magnesium or beryllium, either in their relatively pure metallic form, in combinations with each other, or as alloys, as electrode supports, and an electrolyte comprising organic aprotic solvents. Such cells have been found to perform well where such support materials support lithium and calcium anodes as well as cathodes comprising metal chlorides, sulfides, fluorides and bromides.

Examples of such anodes are those in which the active material comprises calcium, strontium and barium and the alkali metals of group Ia of the periodic system of elements. Examples of such cathodes are those in which the active material comprises copper chloride, copper bromide, copper fluoride, copper sulfide, silver fluoride, silver difluoride, silver chloride, silver bromide and the halide compounds of nickel, cobalt and manganese.

The organic aprotic electrolytes comprise a solute which may be a Lewis acid, preferably with a coordinating salt, or any other kind of a salt sufficiently soluble in the organic aprotic solvent to yield an electrolyte with a specific conductivity of at least 5×10^{-4} mhos/cm. and yet not participate appreciably in the electrode reactions of the cell. Examples of Lewis acids are aluminum fluoride, aluminum chloride, aluminum bromide, boron halides such as boron chloride, boron fluoride and boron bromide, antimony fluorides and antimony chlorides. Examples of such other kinds of salts are lithium perchlorate, sodium perchlorate and magnesium perchlorate.

The Lewis acids, or salts are dissolved in aprotic solvents. The Lewis acids may be dissolved either with or without alkali metal halide or earth alkali coordinating compounds such as lithium fluoride, lithium chloride, lithium bromide, sodium chloride or sodium fluoride. Examples of the aprotic solvents successfully used include pentacyclic esters, aliphatic ethers, cyclic ethers, nitroparaffins, cyclic ketones, aliphatic nitriles; chlorinated esters, cyclic and aliphatic amines and amides. Where desired, compatible combinations of the just listed aprotic solvents may be used such as those described in U.S. Patent 3,468,716. Thus, the particular solvent and solute is largely a matter of choice so long as their combination produces an electrolyte that is aprotic in nature and has sufficient specific conductivity.

The electrode support members are not, of course, completely free from corrosive attack by some organic electrolytes. Those combinations of support materials and electrolytes in which corrosion occurs more rapidly have nevertheless been found quite suitable for use in reserve type batteries, that is primary batteries in which the electrolyte is introduced just prior to battery operation which operative period does not ordinarily exceed 1 week.

A 20 x 20 mesh (0.010-inch thick wire) aluminum screen type 1100 in an aluminum chloride solution in nitromethane-propylene carbonate solvent mixtures will corrode at a rate rendering its use limited to such a reserve type battery where there is an insufficient amount of coordinating salts, such as lithium chloride, in the electrolyte.

In general, the closer the proportionality between the Lewis acid and coordinating salt to a ratio of 1:1 by weight, the lesser the rate of support corrosion. Thus, one part by weight of aluminum chloride, a Lewis acid, to one part by weight of lithium chloride, a coordinating salt, provides for minimal corrosion. For more acidic electrolytes, that is those containing a relatively low proportion of coordinating salt, the life of the support can be substantially extended, where desired, by electroplating it with nickel. The proportion of Lewis acid to coordinating salt by weight should be between 1:1.5 and 4:1.

Example 1: A 20 x 20 mesh aluminum screen alloy, No. 1100, which is commercially pure aluminum with a minimum content of 99% aluminum, to which an aluminum contact tab has been welded on, is used to make a lithium anode by pressing on in an inert atmosphere of argon, a piece of lithium metal. A similar piece of aluminum screen is used as a support grid for the cathode in which a mixture of copper chloride with small additions of graphite as a conducting additive and a suitable binder is pressed to yield a cathode.

Using six such cathodes and seven anodes interleaved with a nonwoven polypropylene separator a cell is assembled with an electrolyte comprising by volume 80% nitromethane, 20% propylene carbonate containing 3 mols/liter of aluminum chloride and 0.3 mols/liter of lithium chloride. The open circuit voltage of this cell, the construction of which is described more fully in U.S. Patent 3,468,716, was found

to be 3.10 volts at an ambient temperature of 20°C. Two cells thus constructed
have been discharged at the same time with two control cells in which the grids were
screens of the same mesh, i.e., 20 x 20 mesh, and wire thickness, but made of nickel.
As can be seen from the following table, in which the comparison is given, the
aluminum equipped cell yields about the same amount of power, but because of a
lighter weight results in a substantial increase of energy density as expressed in
watt-hours per pound.

Example 2:   Two cells are built using aluminum type 1100 grids of the same mesh as
in the preceding example, but electroplated with nickel. These cells are also discharged at 12 amp. and in the same electrolyte as Example 1, yielding comparable
cell voltages and capacities as the other cells. However, due to the lower weight
they were capable of significantly larger energy densities as illustrated in the fourth
column of the following table.

Example 3:   Cells are constructed using perforated sheet of magnesium alloy type
AZ31B (96% magnesium, 3% aluminum, 1% zinc), a commercial alloy with a specific gravity of 1.77. Cells are constructed using these perforated magnesium
sheets as supports and filled with the same electrolyte as given in Example 1. The
energy density yield obtained is 15 to 18% greater than those of the aluminum cells
of the following table.

Energy Density Yields of Organic Electrolyte Cells with Nickel and with Aluminum Support Grids*

	Nickel	Aluminum Type 1100 (Example 1)	Aluminum Type 1100 Plated with Nickel (Example 2)
Cell weight, No. lb.	0.320-0.322	0.240-0.245	0.245-0.250
Average cell voltage (under load)	2.17-2.20	2.14-2.18	2.12-2.17
Percent faradaic efficiency (% yield of theoretical capacity)	66-67	65-67	68-72
Ah delivered to 1.0 v. cutoff	5.6-5.7	5.5-5.7	5.9-6.0
Energy density yield, wh/No. lb.	37.8-39.2	49-54	50-53.2

*5 ah nominal capacity cells discharged at room temperature at 12 amp.

SOLID STATE CELLS

SOLID ELECTROLYTES

Tetramethylammonium Iodide–Silver Iodide Reaction Product

D.M. Smyth, C.H. Tompkins, Jr. and S.D. Ross; U.S. Patent 3,567,518; Mar. 2, 1971; assigned to Sprague Electric Company describe an electrical component comprising a pair of electrodes separated by a solid electrolyte comprising the quaternary ammonium iodide–silver iodide reaction product of a Z–quaternary ammonium iodide and silver iodide, where the molar ratio of silver ion to the quaternary ammonium ion is greater than 2:1. The Z–quaternary ammonium ion is a member of the group consisting of R_4N^+, N—R substituted pyridinium, N—R substituted quinolinium and N—R substituted isoquinolinium. The R group is an alkyl, aryl, alkaryl or aralkyl group. The R groups of R_4N^+ are the same or any combination of the groups.

When the electrical component is a battery, the negative electrode comprises silver and the positive electrode comprises a source of iodine vapor. A particularly convenient source of iodine vapor is the triiodide salt of a Z–quaternary ammonium ion. The Z–quaternary ammonium ion of the triiodide has the same definition as the Z–quaternary ammonium ion of the Z–quaternary ammonium iodide. In a preferred battery cell the negative electrode consists essentially of a mixture of silver powder and the quaternary ammonium iodide–silver iodide reaction product in the weight ratio of from 6:1 to 1:1.5. The positive electrode consists essentially of a mixture of (1) conductive carbon, (2) a source of iodine vapor and (3) the quaternary ammonium iodide–silver iodide reaction product, in the ratio of from 1:4:0 to 1:22:80 respectively.

The use of the quaternary ammonium iodide–silver iodide reaction product in the negative and positive electrodes serves to reduce the contact resistance between these electrodes and the electrolyte. A particularly preferred battery cell employs as the interelectrode the solid electrolyte reaction product of tetramethylammonium iodide and silver iodide. The negative electrode consists essentially of 1 to 2 parts by weight of the solid electrolyte and 1 to 2 parts by weight of silver. The positive electrode consists essentially of from 1 to 3 parts by weight each of (1) conductive carbon, (2) tetramethylammonium triiodide and (3) the solid electrolyte.

Example 1:   A battery cell was prepared as follows: 2.078 grams of tetramethylammonium iodide, $(CH_3)_4NI$, and 9.705 grams of silver iodide, AgI, were mixed and
ground together and then sealed in an evacuated quartz tube. The mixture was
heated in a furnace to 300°C. and held there for 1/2 hour. The reactants became
molten and complete mixing was assured by agitating the tube. The reaction product
was quenched by removing the tube immediately to room temperature. The reaction
product was removed from the tube and reduced to a fine powder. Employing
50,000 lbs./in.^{2}, a disc of the reaction product 1/2 inch in diameter and 25 mils
thick was formed. The compressed disc had a resistivity of 65 ohm-cm.

As the anode, 2 parts by weight of silver powder to 1 part by weight of the powdered
reaction product was formed into a disc having the same dimensions as the solid
electrolyte disc. As the cathode, 1:20:35 parts by weight of carbon black, tetramethylammonium triiodide and the solid electrolyte reaction product respectively,
were mixed and compacted into a disc having the same dimensions as the anode and
solid electrolyte discs. Platinum foil contact sheets, 10 mils thick, were spring
pressure contacted to the anode and cathode. This cell had an open circuit voltage
of 0.62 volt and an internal resistance of 50 ohm.

In view of the fact that a comparable cell employing $RbAg_4I_5$ and RbI_3 would cost
more than ten times that of the cell of the foregoing example, the results are outstanding. This, coupled with the fact that the electrolyte of the subject battery is
far more stable at temperatures below room temperature than that of the rubidium
system, strongly illustrates the advantages of the process.

To illustrate the greater stability of the tetramethylammonium iodide-silver iodide
reaction product over $RbAg_4I_5$, samples of each electrolyte were pressed between
electrodes consisting of a mixture of powdered silver and the reaction product. The
samples were stored at 25°, 20° and 0°C. under both ambient and dry conditions.
Resistance readings at 1,000 Hz. were taken for each sample at frequent, but
irregular intervals.

The initial resistance values for the reaction product were approximately ten times
higher than those of the $RbAg_4I_5$ and these values remained effectively constant
throughout the 120 day evaluation period. All the $RbAg_4I_5$ samples stored in ambient
moisture increased in resistance from three to four orders of magnitude in the same
time period, thereby greatly surpassing the resistance of the reaction product samples.
Under dry storage a similar increase for the $RbAg_4I_5$ was noted at 0°C., but not
at the higher temperature. Other samples of the reaction product were stored at
-78°C. and these also have shown no increase in resistance over a 120 day storage
period.

The following tabulation shows resistivity versus specific 7-quaternary ammonium
iodide-silver iodide reaction products. The samples were prepared by heating the
indicated proportions at 200° to 350°C. in an evacuated quartz tube until the
reaction was complete. The reaction product was removed from the tube and reduced
to powder form. Resistivity was determined by pressing the powder into pellets and
measuring resistance as a function of the solid electrolyte thickness. The electrodes
employed were pressed mixtures of the solid electrolyte reaction product and powdered silver. In the table, resistivity is measured in ohm-cm.

Composition	Reaction product		
	(1)	(2)	(3)
Moles AgI			
Moles Z-quaternary ammonium : iodide			
1:1	2.0×10^5		
2:1	360	3.7×10^6	3.7×10^4
3:1	120	1,600	470
4:1	65	230	180
6:1	55	140	190
9:1	70	280	270

[1] Tetramethylammonium iodide-silver iodide reaction product.
[2] p-Methoxytrimethylbenzylinium iodide-silver iodide reaction product.
[3] 4-vinyl, N-methylpyridinium iodide-silver iodide reaction product.

It is believed that good advantage over pure silver iodide can be obtained with a molar ratio of silver ion to quaternary ammonium ion of as high as 20:1. The process conditions necessary to form the solid electrolyte include heating the Z-quaternary ammonium iodide with silver iodide at a temperature between 250° to 350°C. in a closed system having a comparatively inert atmosphere, (e.g., helium or a partial vacuum) for a period between 5 to 60 minutes.

In the foregoing example, in place of tetramethylammonium iodide, the following iodides may be reacted with silver iodide to form a solid electrolyte reaction product within the scope of the process: N-methylpyridinium iodide, 4-vinyl-N-methyl-pyridinium iodide, poly-4-vinyl-N-methylpyridinium iodide, N-methylquinolinium iodide, trimethylanilinium iodide and p-methoxybenzyltrimethylammonium iodide.

Example 2:   Another battery cell was prepared as in Example 1 except that the electrolyte was prepared by reacting 1.060 grams of poly-4-vinyl-N-methylpyridi-nium iodide and 4.638 grams of silver iodide in an evacuated quartz tube at 300°C. for 1/2 hour. The powdered reaction product had a resistivity of 250 ohm-cm.

Potassium Silver Iodide

W.R. Hruden; U.S. Patent 3,598,654; August 10, 1971 describes compounds which are capable of ionic conduction (ionophores) having the form M_2AgI_3 where M is a univalent electropositive element or radical used in solid state electrochemical devices such as batteries.

Although any member of the family of compounds having the form M_2AgI_3 may be used, for the purposes of this discussion, K_2AgI_3 (dipotassium silver triiodide) will be employed as a representative member. All of the members of the family and specifically Rb, Cs and NH_4 should be directly substitutable for K in the rest of the disclosure with appropriate change in the names the compounds formed and with comparable results. As with its related compounds of the family, dipotassium argentous triiodide is orthorhombic and may conveniently be prepared from aqueous solution in accordance with the reaction:

$$2KI + AgI \longrightarrow K_2AgI_3$$

This compound has been found to have a specific ionic resistivity of ca. 37 ohm-cm. In the preparation of this compound two mols of KI are dissolved in the minimum amount of water to give a saturated solution. The solution is then warmed to approximately 80°C. and one mol of AgI is dissolved in the KI solution with stirring. The K_2AgI_3 is then obtained by crystallization. Several solid state electrochemical energy-producing cells have been prepared by employing an anode of powdered silver pressed into a pellet, an electrolyte of dipotassium argentous triiodide, similarly in the form of a pellet and pelletized cathode having the following composition:

	Percent
RbI_3	50
Acetylene black carbon	25
K_2AgI_3	25

These cells were capable of delivering 1 ma./cm.^{2} with an electromotive force of 0.65 volt. The solid state cells were produced by placing the anode and cathode pellets on opposite sides of a pellet of K_2AgI_3 electrolyte. The sandwich assembly thus formed was held together under sufficient pressure to assure adequate electrical contact between the pellets.

Lithium-Lithium Iodide and Silver-Silver Iodide

A process described by M.L.B. Rao; U.S. Patent 3,455,742; July 15, 1969; assigned to P.R. Mallory & Company Incorporated provides a solid electrolyte cell comprising two metal/metal halide voltaic half-cells in combination. High energy density for the system is realized by employing a light weight, high voltage anode, an alkali metal and its halide in combinations, such as the lithium/lithium iodide half cell.

The cathodes selected for the cells are metal/metal halide voltaic half cells, particularly silver/silver halide, such as silver/silver iodide. Other metal/metal halide cathodes suitable for the process are silver/silver mercuric halide, such as silver/silver mercuric iodide (Ag/Ag_2HgI_4), lead/lead halide and copper/copper halide. The system employs a combination of two different salts as electrolyte, such as lithium iodide in combination with silver halide, silver mercuric halide, lead halide or copper halide.

The major problem with solid electrolyte cells is the selection of suitable electrolytes. The ion or ions of the electrolytes should involve the ionic transport of the major part of the current. These and other requirements are satisfied by the following system, constituting the preferred form of the process: Li/LiI/AgI/Ag.

It will be noted that the cell is a high energy density solid electrolyte system comprising two metal/metal halide voltaic half-cells in combination. Practical cells of this kind have been assembled by employing pressed pellets (1 cm.^{2} x 1 mm.) of each of the electrolyte salts, appropriately sandwiched with lithium and silver metal. Current potential measurements of the cell have been carried out after providing suitable electrical connections and wrapping the cells in Teflon tape. A cell of the described character is illustrated in Figure 6.1.

Referring to Figure 6.1a, reference numeral (10) denotes the anode, which may be lithium metal. Anode (10) is in contact with one face of a solid electrolyte layer (12), such as one of lithium iodide, the other face of which is in contact with cathode (14). The cathode may be composed of silver iodide. A contact layer (16) of high electrical conductivity, such as one of silver foil, is applied to cathode (14). Lead wires (18) and (20) are connected to anode (10) and to cathode contact (16), respectively, and constitute the electrical terminals of the cell. For protection,

FIGURE 6.1:   HIGH ENERGY DENSITY SOLID ELECTROLYTE CELLS

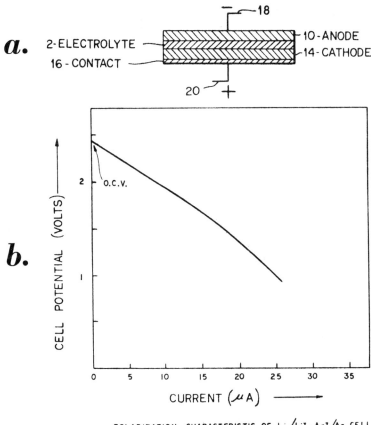

POLARIZATION CHARACTERISTIC OF Li/LiI-AgI/Ag CELL

(a) Sectional View of Solid Electrolyte Cell
(b) Operating Characteristics of Solid Electrolyte Cell

Source: M.L.B. Rao; U.S. Patent 3,455,742; July 15, 1969

the cell may be wrapped in Teflon tape, leaving the lead wires outside of the tape wrap. Figure 6.1b gives the open circuit and the operating characteristics of typical cells assembled as described in the foregoing. The open circuit voltage of the cell is 2.2 to 2.45 volts. The cell potential varies linearly with the current drawn, indicating that internal resistance dominates the cell performance. Cells of similar performance characteristics may be made by replacing the silver iodide cathode with other silver/silver halide cathodes. Furthermore, cathodes of silver/silver mercuric halides, lead/lead halides and copper/copper halide may be combined with lithium/lithium iodide anodes to provide satisfactory high energy density cells. The performance of all of the above mentioned cells may be improved by reducing the thickness of the electrolyte layer, by adding impurity salts to enhance ionic migration of charge carriers, by operating the cells at elevated temperatures or by any combination of these expedients. A plurality of the cells may be connected in series or parallel, or both, to obtain batteries of higher voltage or capacity. Also, the cells of the process may be used as secondary or rechargeable cells.

Of the various alkali metal halides, lithium iodide is the one which possesses a specific resistance of the order of 5×10^6 ohm-cm. at room temperature and is thus eminently suitable for cell applications. In the preferred form of the process, a lithium/lithium iodide half cell is combined with a silver/silver iodide half cell. However, the process also contemplates a cell in which the lithium/lithium iodide anode is combined with an iodine cathode (Li/LiI/I). A suitable iodine cathode may be prepared by absorbing iodine on charcoal (0.75 gram of iodine per gram of charcoal). While in combination with a lithium/lithium iodide anode both a silver/silver iodide cathode as well as the iodine cathode can provide a practical and operative solid electrolyte cell, there are certain essential differences between the characteristics and the mode of operation of the two types of cells:

(a) When using a silver/silver iodide cathode between current collectors, there is a combination of two different salts LiI/AgI as compared to single electrolyte of a cell employing iodine cathode.

(b) The cell employing a silver/silver iodide cathode operates with high energy density whereas the cell employing an iodine cathode operates with low energy density.

(c) The cell employing a silver/silver iodide cathode operates on a voltaic cell principle of dissolution and deposition of metals at the metal/metal halide interface. In contrast to this, the mechanism of the cell employing an iodine cathode involves redox potentials.

Generally speaking, the cells employing a silver/silver iodide cathode are superior to those having an iodine cathode. The former dispense with the use of iodine which requires a binder and a conductor for the half cell operation. This is unnecessary when the cathode is silver/silver halide, for example, silver/silver iodide, and where the depositing metal itself, e.g., Ag from AgI, serves as conductor, the cathodes requiring no additional conductor or binder. In contrast to iodine, the metal/metal halide cathodes are noncorrosive and hence qualify the cells for applications wherein the presence of corrosive materials, such as iodine, cannot be tolerated. For the above mentioned reasons the use of silver/silver halide cathodes greatly simplifies the manufacture and assembly of both single cells, as well as of stacked cells.

Quaternary Ammonium Silver Iodide Salts

A process described by B.B. Owens; U.S. Patent 3,476,606; November 4, 1969; assigned to North American Rockwell Corporation relates to solid state electric cells in which the conductivity-imparting component of the solid electrolyte is an organic ammonium silver iodide salt.

Solid ionic conductors are known and are of particular utility as the electrolyte in a solid state electric cell. The silver halides have been found useful as such solid electrolytes. One device employing silver iodide as a solid electrolyte is described in U.S. Patent 2,689,876. These solid state cells are advantageous compared with conventional cells and batteries with respect to shelf life stability, leak-free properties, freedom from pressure buildup during the electrochemical reaction, and flexibility with respect to construction design and miniaturization. However, the usefulness of such devices, particularly at room temperature, is limited principally by the low ionic conductivity of the solid electrolyte. For example, the ionic conductivity at room temperature of the silver halides is 10^{-6} (ohm-cm.)$^{-1}$, resulting in solid state cells having too high an internal resistance for many applications. Pressed silver iodide pellets have been reported as having an ionic conductivity at room temperature as high as 2.7×10^{-4} (ohm-cm.)$^{-1}$.

Preparation of the quaternary ammonium silver iodide compound $N(CH_3)_4Ag_2I_3$ was reported by Kuhn and Schretzmann in, Angew. Chemie, 67, 785 (1955). The preparation of this compound was also reported by Bradley and Greene in Trans. Faraday Soc., 63 (2), 424 (1967), who found no evidence of any substance with a high conductivity. Other work shows alkali metal silver iodide ionic conductors having a room temperature conductivity of 0.2 (ohm-cm.)$^{-1}$. However, for certain applications the need exists for other solid ionic conductors having a conductivity at least greater than that of the silver halides yet possessing other advantageous properties such as enhanced low temperature conductivity and lower cost of production.

In this process, there are provided ionically conductive solid compositions of matter and solid state electrochemical devices utilizing these compositions as the solid electrolyte element where the electrolyte compositions have an ionic conductivity greater than that of silver iodide and contain at least 75 cationic mol percent, suitably between 75 and 97.5 cationic mol percent silver cations, preferably between 80 and 90 mol percent.

The conductivity-imparting components of these compositions are organic ammonium silver iodide salts which may be expressed by the empirical formula $QI \cdot nAgI$, n having any value between 3 and 39 inclusive and Q being an organic ammonium cation having an ionic volume between 30 and 85 cubic angstroms. Where the substituents on the nitrogen atom of Q are aliphatic groups, e.g., methyl, ethyl; or aralkyl groups, e.g., benzyl, then Q must be a quaternary ammonium ion, i.e., four carbon atoms are attached to the nitrogen atom. A preferred composition range is from QAg_4I_5 to QAg_9I_{10} where Q preferably is a quaternary ammonium cation. The nitrogen of the organic ammonium cation complex may be attached to separate organic groups or may form part of a cyclic structure.

Specifically preferred conductive compositions of matter are tetramethylammonium octasilver nonaiodide $N(CH_3)_4Ag_8I_9$, tetraethylammonium octasilver nonaiodide $N(C_2H_5)_4Ag_8I_9$ and pyridinium octasilver nonaiodide $HNC_5H_5Ag_8I_9$. Solid state electrochemical devices utilizing the ionic conductive compositions as solid electrolytes preferably utilize organic ammonium polyiodide salts for the associated electrode acting as electron acceptor and utilize a silver-containing composition for the electrode acting as electron donor. A preferred solid state cell includes a silver-containing anode, a solid electrolyte element comprising $N(CH_3)_4Ag_8I_9$, and a polyiodide-containing cathode comprising $N(CH_2H_5)_4I_3$.

Example 1: The preparation of conductive compositions with varying silver content is as follows. Tetramethylammonium silver iodide was prepared with varying silver content using a melt-anneal technique according to the following equation:

$$N(CH_3)_4I \ + \ nAgI \ \longrightarrow \ N(CH_3)_4Ag_nI_{n+1}$$

the value of n being varied from 2 to 20 (from 67 to 95 mol percent silver cation). The tetramethylammonium iodide and varying quantities of silver iodide were intimately mixed together as fine powders to give total weights of 10 to 20 grams for each sample. The mixtures were melted at a temperature between 200° and 300°C., quenched at room temperature, pelletized and then annealed, at 165°C. The conductance of selected 2 gram samples in the form of pellets to which silver electrodes were attached was measured at room temperature. For $N(CH_3)_4Ag_2I_3$ (i.e., n = 2) the conductivity value was 10^{-7} (ohm-cm.)$^{-1}$, essentially nonconductive. The conductivity values obtained for the samples were essentially as shown in curve (1) of Figure 6.2a, a maximum 0.03 (ohm-cm.)$^{-1}$ being obtained for a silver content of 86 cationic mol percent (n = 6).

FIGURE 6.2:   SOLID STATE ELECTROCHEMICAL DEVICE

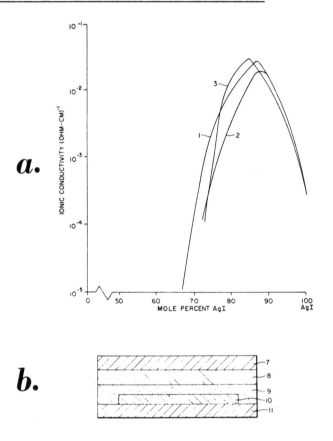

(a) Variation of Ionic Conductivity as a Function of Silver Content
(b) Cross-Sectional View of Cell

Source: B.B. Owens; U.S. Patent 3,476,606; November 4, 1969

Similar results were obtained when the starting material used was tetraethylammonium iodide, as well as pyridinium iodide, corresponding essentially to the results shown in Figure 6.2a for curves (2) and (3), respectively. The maximum conductivity value for tetraethylammonium silver iodide was shown for a silver content of 88 cationic mol percent (n = 8). For pyridinium silver iodide the maximum conductivity value was observed for n = 6.

A paste method of synthesis was also used for the preparation of the tetramethyl and tetraethyl conductive compounds. Desired amounts of the quaternary ammonium iodide and silver iodide were intimately mixed and ground together, sufficient water being added to form a thick paste. The mixtures were then heated to near dryness in an oven at 80°C. and then further dried in vacuum. The resulting product was ground and pelletized followed by annealing in an argon atmosphere at a temperature of 125°C. for 8 hours to several days. It was found that samples prepared in this manner showed a maximum in the conductivity curve at a slightly higher silver content corresponding to 88 cationic mol percent (n = 8).

Tetramethylammonium silver iodide was also prepared by mixing 2 grams $N(CH_3)_4I$ and 14 grams AgI (1:6 molar ratio) in 60 grams water. The reactants were boiled together for 1 hour, a change in the color of the solid phase from yellow to white indicating that the silver iodide was being consumed. The solution was then cooled, the supernatant liquid was decanted, and the precipitate was washed with acetone and filtered over vacuum. X-ray analysis of the product showed that two conductive phases corresponding to n = 8 and n = 4 were present.

Example 2:   The preparation of solid state electric cells with organic ammonium silver iodide electrolyte is as follows. Test cells were prepared having a configuration essentially similar to that shown in Figure 6.2b. For the composite anode a copper wafer made contact with 1 gram of an anode mix consisting of silver powder containing dispersed carbon and conductive electrolyte material $RbAg_4I_5$.

The composite cathode consisted of a titanium wafer in contact with 1 gram of a cathode mix containing RbI_3 as electron acceptor, together with carbon and $RbAg_4I_5$. The organic ammonium silver iodide electrolyte composition was 3 grams in weight. All of the conductive compositions listed in the table functioned as suitable electrolytes in the solid state cells. For the cell containing pyridinium heptasilver octaiodide electrolyte, the cathode mix contained $(C_2H_5)_4NI_3$ as electron acceptor in place of RbI_3.

Electrochemical Cells (Ag/Electrolyte/RbI3)

Electrolyte Empirical Formula	Open Circuit EMF (volts)	Flash Current (milliamperes)
$(C_2H_5)_3NCH_3Ag_4I_5$	0.65	19
$(C_2H_5)_3NC_3H_7Ag_4I_5$	0.64	2.5
$(CH_3)_3NC_6H_5Ag_4I_5$	0.66	20
$C_2H_5N(CH_3)_3Ag_4I_5$	0.65	50
$(CH_3)_4NAg_4I_5$	0.66	120
$(CH_3)_4NAg_6I_7$	0.66	100
$C_5H_5NHAg_7I_8$	0.66	650

Example 3:   Solid State Electrochemical Timer — A solid state coulometer for use as a timing device was assembled using as electrolyte element a conductive composition having the empirical formula $N(C_2H_5)_4Ag_4I_5$ equivalent to 80 cationic mol percent silver ion. The device was built in tabular form with the other timing electrode consisting of a titanium tube 0.610 in. long, 0.210 in. diameter and having a wall thickness of 0.025 in. The inner wall of the titanium tube was lined with 0.3 gram of carbon-polycarbonate mixture. The inner counter-electrode, concentric with the outer electrode and used as a reservoir of silver, was a 0.3 gram blend of silver, carbon, and $RbAg_4I_5$. The solid electrolyte 0.6 gram, was disposed between the two electrodes and in intimate contact with each.

In operating the device, a fixed amount of silver was first transported from the counter-electrode to the timing electrode by passage of a constant preselected current through the device for a preselected time. During timing operation, a constant current, flowing in a reverse direction to the setting current, is used to strip the silver from the timing electrode, resulting in a marked increase in voltage across the device. This voltage increase is used as a signal-actuating mechanism. Thus, the foregoing device was set at 500 microamp., for 3 seconds with the positive power lead connected to the counter-electrode. A voltage drop of 115 millivolts was recorded across the device during setting.

Stripping was accomplished with the positive power lead connected to the timing electrode and using a 10 microamp current. An initial voltage drop of 8 millivolts was recorded across the device. The voltage drop started rapidly at 140 seconds, reaching cutoff voltage of 630 millivolts at 149 seconds. The accuracy of the device is seen from a comparison of the initial input of 1500 microamp.-sec. (500 microamp. for 3 sec.) with a timing output of 1490 microamp.-sec. (10 microamp. for 149 sec.).

Alkali Metal Silver Iodides

G.R. Argue and B.B. Owens; U.S. Patent 3,443,997; May 13, 1969; assigned to North American Rockwell Corporation describe an electrochemical device which includes a solid electrolyte element comprising an ionically conductive composition of matter where the conductivity-imparting component has the formula MAg_4I_5 where M is a univalent ion selected from the class consisting of K, Rb, NH_4, Cs, and related combinations, Cs being present only as minor constituent of M, i.e., less than 50 ion percent of M, together with means for providing a flow of ions through the electrolyte element.

Illustrative of such electrical devices are solid state electrical timers, coulometers, and adaptive computer components, as well as solid state electrical cells and batteries. In each of these devices there is a flow of electric current by a movement of ions through the solid electrolyte element, an associated electrode acting as an electron acceptor and another associated electrode acting as an electron donor.

As an example of a solid state timer, a silver film anode and an inert cathode, e.g., platinum, are provided, with a solid electrolyte element disposed between. An applied voltage strips off a thin film of silver from the anode and exposes an underlying inert metal, suitably platinum. Upon depletion of the active metal film of the anode, the cell becomes polarized, and the voltage across the timer changes markedly. This change in voltage can serve to actuate a signal device such as a relay, light, or alarm. The time interval for the timer may be readily predetermined by the current passing through the cell and the amount of active metal on the film anode.

Example 1: The electrical cell utilizing conductive composition having empirical formula KAg3I4 is as follows. A cell was constructed consisting of a 0.125 mm. thick silver foil anode, a 6 mm. thick pellet of the conductive composition having the empirical formula KAg3I4 as electrolyte, and a 20% iodine–80% carbon pellet as the cathode. A current density of 20.5 ma./cm.^{2} was obtained across a 50 ohm load at 0.40 volt when the cell was maintained at a temperature above 35°C. This current is approximately 40 times that reported for Ag3Si cells and 20,000 times better than that reported for any other solid state battery. An open circuit voltage of 0.68 volt was obtained in agreement with a theoretical value of 0.687 volt. The cell was operated for 7 hours at an average current density of 1.2 ma./cm.^{2}. A 6 cell battery was assembled having an open circuit voltage of 4.2 volts, and was used successfully to operate a commercial transistor radio.

Example 2: The preparation of electric cell having composite electrodes is as follows. Silver metal for the mixed anode layer is prepared either by reduction of silver nitrate by copper or by reduction of silver oxide by carbon. The silver is then intimately mixed with equal amounts of carbon and of electrolyte material. The mixture is heated to at least the melting point of the electrolyte, cooled and then ground to a fine powder. The electrolyte is sieved through a 250 mesh U.S. Standard sieve before use in the cell. The cathode is prepared by mixing carbon and the electrolyte in equal proportions, heating to the melting point of the electrolyte, quenching and then grinding the material together while adding iodine.

For a typical 100 milliampere-hour cell, 0.5 gram I_2 + C + electrolyte material is placed in a one inch stainless steel die and pressed at 18,000 lbs. The resultant pressed disk is then placed in a second insulated die, and an appropriate amount of electrolyte material is added to the die. By using a slightly oversized die and pressing onto the cathode disk at 18,000 lbs., a cup of electrolyte is formed around the cathode. Then an appropriate amount of anode material (0.5 gram Ag + C + electrolyte is equivalent to a 100 milliampere-hour cell) is placed into the die on top of the previously pressed electrolyte and pressed at 18,000 lbs. The cell is formed. Tantalum foils are placed over the anode and cathode for convenience in effecting electrical contact. The complete assembly is then preferably encapsulated in an epoxy-type resin so as to give both a cell of rugged construction and one that is protected from atmospheric corrosion.

Example 3: The improvement by use of electric cell with composite electrodes is as follows. The following electric cells were prepared essentially as shown for Example 2, using composite cathodes and with both individual and composite anodes, and were found to have the following characteristics:

Cell	Composite Anode	Electrolyte	Composite Cathode	Internal Resistance, ohms	Silver Utilization, %
1	1 g. Ag	2 g. (RbAg3I4)	0.5 g. I2, 1 g. C, 0.5 g. (RbAg3I4)	1.1	12
2	1 g. (RbAg3I4), 0.5 g. Ag, 1 g. C	3 g. (RbAg3I4)	0.5 g. I2, 1 g. C, 3.0 g. (RbAg3I4)	0.2	70

As may be noted from a comparison of the two cells using the conductive composition having the empirical formula RbAg3I4 as electrolyte, upon incorporation of elec- trolyte also in the silver anode, the internal resistance of the cell is reduced mark- edly, in part by decreasing the anode electrolyte contact resistance. However, more importantly, the utilization of silver is almost sixfold greater in the composite anode cell compared to the other type of cell.

Example 4:   The preparation of $Cu/(RbAg3I4)/I2$ cell is as follows. A cell was also constructed in which the silver anode was replaced by a copper anode, following essentially the procedure shown in Example 1. The cell has an open circuit voltage of 0.68 volt and a flash current of 15 ma./cm.^{2}. A continuous current of 0.1 milliampere per square centimeter was drawn for several hours.

Rubidium Silver Iodide

T. Takahashi and O. Yamamoto; U.S. Patent 3,558,357; January 26, 1971 describe a laminated assembly of a thin solid electrolyte of $RbAg4I5$, or $Ag3SI$ sandwiched between a thin anode of silver or a material containing silver and a thin cathode of tellurium, selenium, or a material containing tellurium or selenium. The electrolyte and cathode can be easily formed by vacuum evaporation deposition. A layer-built solid-electrolyte cell can be readily fabricated by successive repetitions of the vacuum evaporation procedure to build up a plurality of unit cells in stacked forma- tion.

Example 1:   9.3 grams of silver iodide and 2.1 grams of rubidium iodide were mixed and ground. The resulting mixture was sealed under a vacuum within a glass tube, caused to react for 2 hours at a temperature of 500°C., then cooled abruptly, and again caused to react for 15 hours at 160°C. thereby to synthesize $RbAg4I5$.

A mixture of 0.5 gram of this $RbAg4I5$ and 0.3 gram of silver powder of 200 mesh particle size to form an anode, 0.5 gram of the $RbAg4I5$ to form an electrolyte, and a mixture of 0.5 gram of the $RbAg4I5$, 0.2 gram of tellurium, and 0.1 gram of silver telluride to form a cathode were stacked in laminar arrangement and pressed into a tablet of a diameter of 1.2 cm. A silver wire and a gold wire were connected as lead wires to the anode and cathode side, respectively, of the tablet, and the entire tablet was bonded with an epoxy resin thereby to form a cell.

At room temperature, this cell exhibited an open circuit voltage of 0.21 volt and a voltage of 0.14 volt when discharging with an internal resistance of 1 ohm and 40 ma. When the cell was discharged with a current of 10 ma., the capacity up to the instant at which the cell voltage became 0.1 volt was 40 ma.-hour. At 100°C., the cell exhibited an open circuit voltage of 0.23 volt and a voltage of 0.17 volt when discharging with an internal resistance of 1 ohm and 40 ma. This cell was rechargeable. When it was charged with a current of 10 ma., and a charge voltage of 0.33 volt for 4 hours and then discharged, it exhibited the same discharge curve as that of the first discharge.

Example 2:   A cell was fabricated in accordance with procedure set forth in Ex- ample 1 except for the use of selenium instead of tellurium and silver selenide in- stead of silver telluride. At a temperature of 25°C., this cell exhibited an open circuit voltage of 0.25 volt and an internal resistance of 3 ohms. When discharging with a current of 10 ma., the cell exhibited a voltage of 0.14 volt. The capacity of the cell when discharged with 5 ma. up to the instant at which the cell voltage became 0.1 volt was 30 ma.-hour.

Example 3:   A mixture of 0.7 gram of silver iodide and 1.0 gram of 200 mesh silver powder was molded into a tablet of a diameter of 1.2 cm. to be used as an anode. 1.5 grams of silver iodide was molded into a 1.2 cm. diameter tablet to be used as an electrolyte. A mixture of 0.7 gram of silver iodide and 0.5 gram of tellurium was molded into a tablet of the same diameter to be used as a cathode. These three tablets were then stacked laminarly in the order of anode-electrolyte-cathode and held in pressed state in a stainless steel holder.

At a temperature below 147°C., the modification temperature of silver iodide, this cell exhibited a high internal resistance and could not discharge with a high current density. At a higher temperature, however, the internal resistance became less than 1 ohm, and discharge with a high current was possible. At 250°C., the open circuit voltage was 0.26 volt. When, with a current of 50 ma. and an initial voltage of 0.15 volt, the cell was discharged continuously for 3 hours, the cell voltage became 0.12 volt. Recharging with a current of 50 ma. was also possible, during which the charge voltage became 0.33 volt. When, after recharging thus for 3 hours, the cell was discharged, the same discharge curve as that of the initial discharge was obtained.

While the solid electrolyte cell according to the process exhibits a lower open circuit voltage than that of a cell in which a halogen is used as the cathode material, the cell of the process is capable of discharging with a current density of several ma./cm.^{2} evan at room temperature. Moreover, tellurium and selenium do not melt at temperatures up to 350° and 217°C., respectively, and because of the low vapor pressures, fully gastight cells can be readily fabricated.

Another feature of the process is that, when tellurium is used, thin films can be formed in a simple manner by vacuum evaporation, whereby the entire cell can be made extremely thin. Since the electrical conductivity of selenium is substantially lower than that of tellurium, the discharging characteristic of the cell when one or more selenium thin films are used is poorer than that when tellurium thin films are used.

Zinc Salt Complex

F.E. Swindells; U.S. Patent 3,669,743; June 13, 1972; assigned to Melpar, Inc. describes an electrochemical cell which has a zinc electrode and a silver electrode between which is positioned a solid nonhygroscopic electrolyte in the form of a complex compound of a zinc salt with ammine, hydrazine, hydroxylamine, or aniline; such as zinc chloride ammine. A further layer of silver chloride may initially be disposed between the solid complex and the silver electrode, or may subsequently be formed in that location upon charging of the cell. The preferred salts to be converted to the complex are zinc chloride, zinc bromide, and zinc sulfate.

In a typical procedure, conversion to the diammine is accomplished by exposing anhydrous zinc chloride to ammonia gas. The resulting zinc chloride ammine, $ZnCl_2 \cdot 2NH_3$ is then dried, pressed into pellets, and the pellets conditioned in a high relative humidity atmosphere. Finally, the conditioned pellets are compressed between electrode strips of silver and zinc to form the solid state cell. Although it is in a discharged state where the immediately preceding procedure is followed, the cell may be conveniently and rapidly charged by application of a low level DC voltage, which converts a portion of the zinc salt to the metal and forms an equivalent amount of silver chloride (AgCl). A substantial amount of the energy thus stored is recoverable upon discharge of the cell, and the cell may be recycled through several successive charging and discharging periods without apparent deterioration.

In one example, the solid electrolyte composed of a zinc salt complex is provided in the form of a flexible sheet structure in which a fibrous or porous membrane is impregnated with the solid electrolyte. In the following example an electrochemical cell having the structure $Ag|AgCl|ZnCl_2 \cdot X|Zn$ was formed, where X is the compound forming the complex with the zinc salt.

Example 1:   In the preparation of the $Ag|AgCl$ half cell, a sheet of 1.5 ml. Ag was cut in the shape of a circular electrode (1) (Figure 6.3a) with an extension (2) suitable for making conductive connection to the cell. The electrode was then cleaned by immersion in benzene in an activated ultrasonic cleaning unit, after which it was subjected to successive rinses of acetone, distilled water, and grain alcohol.

To provide the necessary rigidity for further treatment, a layer of dielectric tape (e.g., Scotch brand, No. 470) was applied to the back of the electrode, and the front side was masked with the same tape [in the region designated by reference numeral (3)] to expose a circular area of 6.45 cm.^{2} (approximately 1 1/8 in. in diameter) on which to provide a layer (4) of AgCl. To that end, the masked electrode was placed in a dilute (approximately 0.2 N) HCl bath for conversion of 15μ of the silver to 37μ of AgCl by electrolysis at a rate of 3.88 ma./cm.^{2}, thereby producing a 25 ma.-hour cell capacity. The resulting $Ag|AgCl$ half cell was then immersed in distilled water for several hours, removed, and dried.

The Zn electrode (5) (Figure 6.3b) was prepared by cutting and cleaning a sheet of 2.0 mil Zn in the same manner as that described above for the Ag electrode. Thereafter, the Zn electrode was amalgamated by rubbing with a 10% solution of $HgCl_2$. Regidification was again accomplished with the aid of a layer of the dielectric tape.

The electrolyte was prepared by dipping 100μ thick filter paper (e.g., No. 50, made by W. & R. Balston, Ltd.) into a solution of $ZnCl_2$ and H_2O (100 grams $ZnCl_2$, and 50 ml. H_2O), after which is was exposed to ammonia gas for a period of 72 hours while being dried in a desiccator containing $CaSO_4$ (e.g., Drririte brand desiccant). The complex compound thus obtained was found to possess high electrical AC resistance, which dropped to 10 ohms or less following exposure to an atmosphere of high humidity (93% relative humidity in this example) for a period from 1 1/2 to 2 hours. After the latter step, the electrolyte was found to have experienced a weight gain of 10.6%, corresponding to the addition of one molecule of water. The complex compound formed by this process has $ZnCl_2 \cdot Zn(OH)_2$ as a major component.

In assembling the cell, one or two layers of electrolyte prepared as set forth above was placed between the $Ag|AgCl$ half cell and the Zn (Hg) electrode. In a typical example, an electrolyte (7) having a disk shape as shown in Figure 6.3c, was

FIGURE 6.3:   RECHARGEABLE CELL WITH SOLID ZINC SALT COMPLEX
ELECTROLYTE

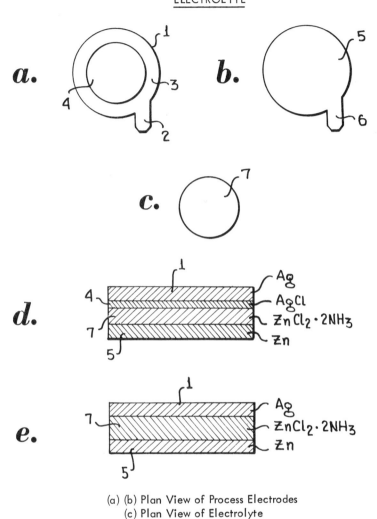

(a) (b) Plan View of Process Electrodes
(c) Plan View of Electrolyte
(d) (e) Sectional View of Assembled Cells

Source: F.E. Swindells; U.S. Patent 3,669,743; June 13, 1972

composed of a sufficient number of layers (again, usually one or two) to provide it
with a composite thickness of 0.008 in., while the positive electrode (i.e., the
Ag|AgCl half cell) was 0.002 in. thick and the negative (Zn) electrode was also
0.002 in. thick. Electrolyte (7) was placed in direct contact and overlying relationship with AgCl region (4) (Figure 6.3c), and the Zn electrode then placed
against the electrolyte, the three sharing a common axis and the two electrode

extensions (2) and (6) lying in spaced-apart relationship at a common end of the cell. Finally, a moisture barrier was applied, in the form of another layer of dielectric tape to each electrode and the entire cell dipped in a krylon solution for coating thereby. Immediately after fabrications, the cell was tested to determine its electrical characteristics. The open circuit voltage was approximately 1.0 volt, with the Zn|ZnCl$_2$·X half cell contributing a potential of 0.768 volt and the Ag|AgCl half cell a potential of 0.22 volt. The AC resistance was under 10 ohms, ranging from 4 to 10 ohms in most of the units fabricated. The theoretical capacity of a cell fabricated in the manner set forth above (25 ma.-hour), is limited by the thickness of the AgCl.

Such an electrochemical cell is rechargeable, and can be subjected to several cycles of discharging and charging without substantial degradation. The number of cycles is increased by overcharging the cell, provided the charging potential has a magnitude below that which would cause short circuiting of the cell. The latter value was found to range from 1.4 to 1.6 volts for many of the cells fabricated in the manner above.

Example 2:   A cell essentially similar to that of Example 1 was fabricated using a different porous membrane in place of the filter paper. In particular, Metricel, type SL having a pore diameter of 0.8μ was used instead of filter paper, and dipped in the zinc chloride solution. The remaining steps were the same. The resulting cell had a DC resistance of 20 ohms, and was quite similar to that fabricated by Example 1 in all other electrical properties.

Example 3:   Using the same procedure and materials as in Example 1, a cell was fabricated with an active area of 5 cm. by 10 cm. Its flexible character and operability in a deformed (from a flat or planar configuration) condition were demonstrated by wrapping the cell around a 1 1/8 in. diameter tube of Lucite and taping it in place with pressure sensitive tape. The cell had an open circuit voltage of 0.92 volt in this condition.

Vapor Deposition Technique for Silver Halides

A process described by G.M. Goldberg; U.S. Patent 3,554,795; January 12, 1971; assigned to Technical Operations, Incorporated provides ultra-thin film solid electrolyte devices for use as batteries, sensors, and the like, comprising electrodes and solid electrolytes all formed by vacuum deposition. Special fabrication methods are shown for depositing a silver halide electrolyte in a very thin layer without the occurrence of voids in the electrolyte by first depositing a layer of silver which is converted to silver halide in halogen gas atmosphere and then completing the electrolyte by silver halide vapor deposition techniques.

The preferred procedure for fabricating a void-free cell (for example, a Pt/AgH/Ag cell) is generally as follows (Figure 6.4a). First evaporate by conventional techniques a layer of silver, representing the silver electrode, upon a suitable substrate (quartz, for example). The silver layer may vary widely in thickness, however, it has been found that a thickness in the order of 1μ is satisfactory. The silver layer is then placed in an iodine chamber for approximately 24 hours to ensure the conversion of most of the silver to silver halide. A second layer of silver-iodide is then deposited by evaporation over the original ultra-thin silver halide film which acts as a wetting layer offering nucleii for the formation of uniform and

FIGURE 6.4:   VAPOR DEPOSITION TECHNIQUE FOR ULTRA-THIN SOLID
ELECTROLYTE

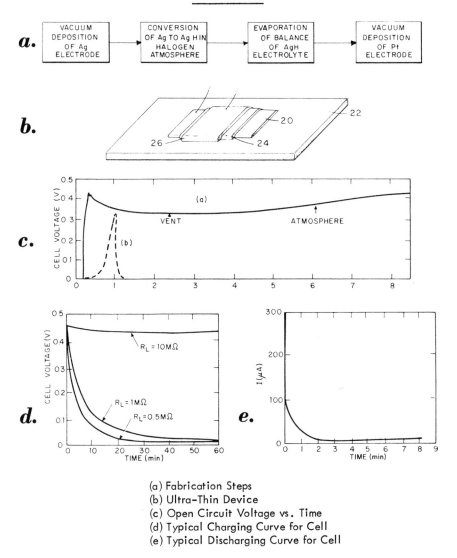

(a) Fabrication Steps
(b) Ultra-Thin Device
(c) Open Circuit Voltage vs. Time
(d) Typical Charging Curve for Cell
(e) Typical Discharging Curve for Cell

Source: G.M. Goldberg; U.S. Patent 3,554,795; January 12, 1971

void-free deposits of silver halide. The film of silver iodide may be deposited from a tungsten boat at a pressure of 3×10^{-5} to 5×10^{-5} mm. Hg. The boat temperature is kept constant during the evaporation inthe range of 550° to 680°C., preferably at 560°C. for AgI. The total thickness of the electrolyte is preferably in the range of 4 to 12μ. Finally, the platinum electrode (20) is deposited on the electrolyte (22) by sputtering, care being taken to mask off the silver electrode to prevent short circuiting of the cell. A CVC AST-100 low energy sputtering unit may be employed with deposition pressure preferably at 2×10^{-3} mm. Hg. The target voltage is preferably 600 volts with the receiving surface placed approximately 6 inches from the target.

Alternatively, the cell may be formed by first sputtering a platinum electrode on a substrate to a thickness of 500 to 1000 A. An ultra-thin film of silver, e.g., about 1000 A. thick is evaporated upon the platinum electrode. Assuming, that a silver iodide electrolyte is desired, the laminate is placed in an iodine atmosphere until the silver is substantially completely converted to silver iodide. As in the first described method, silver iodide is then evaporated upon the iodided silver film, followed by deposition of the second electrode, in this case the silver electrode.

Although each of the above methods yields operable cells, the first described method is perhaps preferred, for the reasons: (1) one less step is required in the fabrication process, and (2) there exists the possibility that when the ultra-thin silver film is deposited upon the platinum electrode, an alloy may be formed which would have a saturating effect on the platinum electrode and may cause a lack of uniformity in electrical characteristics from cell-to-cell.

Electrical contact may be established with the thin-film electrodes by cementing fine electrical wires to the electrodes with conductive Eccobond Solder No. 56-C. As suggested, the above techniques are adaptable for the formation of electrolytes of silver iodide, silver bromide and other materials; however, in spite of the greater susceptibility of silver iodide to the formation of voids, this material is preferred over other potential electrolytes, having relatively low resistance to the conduction of ions, and exhibiting relatively great reluctance to release its halogen component as a result of photolytic and/or electrolytic action.

It has been found that recording of the open circuit voltage (OCV) of a cell during its formation in a vacuum makes it possible to study the electrical properties of thin-film cells as a function of time during deposition. Figure 6.4b shows typical open circuit voltage (OCV) vs. time curves during deposition of the silver electrode for both sound and shorted cells. Figure 6.4c shows a typical transient charging curve for a cell formed as just described. Figure 6.4d shows a series of discharge curves for the same cell for loads of 0.5 MΩ, 1 MΩ, and 10 MΩ.

In related work, J.I. Masters, P. Vouros and J.P. Clune; U.S. Patent 3,575,715; April 20, 1971; assigned to Technical Operations, Incorporated describe ultra-thin film solid electrolyte devices for use as batteries and sensors comprising electrodes and solid electrolytes all formed by vacuum deposition. Evaporated electrolyte layers of silver bromide, silver iodide, and potassium silver iodide (KAg_4I_5) are described. Special fabrication methods are shown, especially for minimizing ionic and interface resistance in such devices.

Pressed Pellet Containing Charge Transfer Complex and Water

I.D. Blackburne, G.C. Morris, L.E. Lyons, and R.G. Hoare; U.S. Patent 3,582,404; June 1, 1971 describe a solid state electrical cell of the type having a compressed pellet containing a charge transfer complex between two electrodes, the pellet incorporating a polar liquid, preferably water.

The process is based on the fact that in this type of cell the addition of a polar liquid, such as water, to the solid electrolyte pellet profoundly affects the properties of the cell, in particular by raising the short-circuit current available by a factor of the order of a thousand.

Some experimental results are given showing the effects of addition of water to various cells. In each case the electrolyte pellets were made by dry-mixing the constituents and pressing them at 4,000 to 8,000 psi pressure. Where there was more than one electrolyte component, the components were mixed in equal quantities of weight. Electrodes in each case, unless otherwise stated, were magnesium.

Electrolyte	Voltage (volts)	Short-circuit current	Comments
(a) Iodine	0.8	10 μA.9	After disiccation.
	1.14	31 μA.	After 2 minutes exposure to atmosphere.
	1.35	60–80 ma.	1 drop water added.
Active electrode Al	0.83	60 μa.	Dry.
Active electrode Sn	0.11	3.3 μa.	Dry.
(b) Iodine	1.48	32 μa.	Dry.
	1.60	150 ma.	1 drop water added.
	1.55	50 ma.	After 5 miutes.
Pyranthrone	1.48	5.4 ma.	15 minutes vacuum desiccation.
	1.50	200 ma.	Second drop of water added.
	1.50	94 ma.	After 4 minutes.
(Active electrode Al)	0.86	1.49 ma.	Dry.
(c) Iodine	1.45	60 μa.	
	1.50	55 μa.	30 minutes ageing.
		100 ma.	1 drop water added.
Di-bromo anthracene		45 ma.	After 45 seconds.
(d.b.a.)	1.42	2.5 ma.	30 minutes vacuum desiccation.
		190 ma.	Second drop of water added.
		100 ma.	After 120 seconds.
(d)	1.32	1.2 ma.	Dry.
		380 ma.	Several drops of water added.
KI-I$_2$-d.b.a.	1.50	10 ma.	2 hours ageing.
	1.15	250 ma.	Further water added.
		80 ma.	3½ minutes ageing.
(e) I$_2$-paraffin	0	0	Addition of water had no effect.
(f)	1.35	0.4 μa.	After desiccation.
	0.94	18 μa.	12 minutes ageing.
I$_2$-naphthalene	1.72	12 μa.	In moist atmosphere.
	1.5	160 ma.	1 drop water added.
		45 ma.	9 minutes ageing.
(g)	1.10	10.6 μa.	After desiccation.
	1.25	12.5 μa.9	Moist atmosphere.
	1.30	13.5 μa.	2 minutes ageing.
I$_2$-methylene blue	1.50	150 ma.	1 drop water added.
	1.66	150 ma.	7 minutes ageing.
	1.58	150 ma.	20 minutes ageing.
(h)	1.25	20.5 μa.	After desccication.
	1.35	45 μa.	In moist atmosphere.
I$_2$-polyvinyl pyridine	1.60	90 ma.	1 drop water added.
	1.00	38 ma.	15 minutes ageing.

NOTE.—μa. = microamps ; ma. = milliamps.

From the above examples it will be seen that addition of 1 drop (about 0.15 cc) of water raises the short-circuit current spectacularly except in case (e) where the electrolyte remained inactive, probably because the paraffin prevented any penetration. Cases (f) and (g) indicate that while exposure to a moist atmosphere has some effect, this is small compared to that caused by addition of liquid water.

The general pattern shown by cases (b), (c), (d), (f), (g) and (h) indicates that ageing decreases the first affect of water addition, but the short-circuit current does not drop to anywhere near the dry current. Later further addition of water raises the current again and ageing returns it to a new value higher than the first aged value. Case (g) is particularly favorable. The iodine-methylene blue electrolyte gives a high short-circuit current and a nearly constant voltage during ageing. While water has been used as the additive in all the above examples, other polar liquids such as dimethylformamide or tetrahydrofuran may be used instead, and give similar results.

Rare Earth Fluorides

A.C. Lilly, Jr. and C.O. Tiller; U.S. Patent 3,657,016; April 18, 1972; assigned to Phillip Morris Incorporated describe a solid state battery comprised of an oxygen-impermeable casing containing an electrochemical cell comprised of a cathode, an anode spacedly disposed with respect to the cathode and comprised of a metal forming stable fluorides, and a solid rare earth fluoride electrolyte in contacting relation with the anode and the cathode. Batteries are preferably constructed by thin film deposition and exhibit an operating characteristic where the output voltage existing between the anode and cathode temporarily decreases to zero where external load resistance decreases below a predetermined level.

By way of introduction to the process, a battery comprising a casing of silicon dioxide containing an electrochemical cell having a bismuth anode, a lanthanum fluoride electrolyte and a gold cathode will be considered in connection with the electrochemical reactions presumed to occur in the cell. A theoretical open circuit voltage of approximately 0.57 volt exists between the bismuth anode and the gold cathode and upon loading, the following reactions are presumed to occur:

$$\text{In the electrolyte:} \quad LaF_3 \rightleftharpoons LaF_2^+ + F^-$$
$$\text{At the anode:} \quad Bi + 3F^- \rightleftharpoons BiF_3 + 3e^-$$

This lanthanum difluoride cation is substantially immobile, and the gold cathode merely collects electrons. The battery electrode requirements are evidently that the anode form stable fluorides and that the cathode be an inert conductor. Thus, examples of various metals which are usable as battery electrodes include, for the cathode, gold, platinum, rhodium and palladium, and for the anode, silver, zinc, bismuth, beryllium, cadmium, rubidium, lanthanum, iron and lead. Since the battery is responsive to oxygen, as will be discussed, it is required that oxygen be evacuated from the anode, cathode and electrolyte upon assembly, or that such assembly be performed in an oxygen-free enviroment, and that the assembly then be encased in an oxygen-impermeable casing. Usable casing materials include silicon dioxide, tantalum oxide, aluminum oxide and silicon oxide.

The rare earth fluorides, i.e., the fluorides of scandium, yttrium, lanthanum and of the metals of the lanthanide series (atomic numbers 58 through 71), e.g., cerium, praseodymium, neodymium and erbium, are employed as the battery electrolyte.

Underlying the need for the oxygen-impermeable battery casing are the following considerations. The rare earth fluorides exhibit a large density of Schottky defects, i.e., crystalline lattice vacancies created by the removal of an ion from its normal site and placing same on or near the crystal surface. At room temperature, in excess of 10^{19} fluorine vacancies per cm.^{3} are provided, and the vacancies are of sufficient size to provide mobility for the relatively large oxygen anions, if present. Furthermore, the dielectric constants of the rare earth fluorides have been found to accommodate electron capture by oxygen, where an interelectrode potential difference of 0.2 volt and above exists between the anode and cathode contacting the electrolyte. Under such circumstances, the following further reactions are presumed to occur in such unencased oxygenated battery:

$$\text{At the cathode:} \quad O_2 + e^- \longrightarrow O_2^- \quad \text{(electron capture)}$$

At the anode: $4Bi + 3O_2^- \longrightarrow 2Bi_2O_3 + 3e^-$

Battery output current, desirably dependent exclusively on the electrodes, electrolyte and load resistance, will be seen to be further dependent on ambient oxygen, which is presumed to effectively decrease the internal resistance of the battery. In environments whose oxygen concentration is variable or uncontrolled, spurious change in battery performance characteristics will undesirably result. Whereas oxygen is thus deleterious to consistent battery performance, the presence of water vapor does not affect same to any substantial effect.

Referring to Figures 6.5a, 6.5b and 6.5c one structural arrangement of the battery comprises an oxygen-impermeable insulative support member (10), a thin film of anodic metal (12) overlying the support member, a thin film of solid electrolyte (14) overlying the anodic film and the support member, a thin film of cathodic metal (16) overlying the electrolyte and the support member and a thin oxygen-impermeable insulative film (18) overlying the cathodic film and the support member.

In this arrangment cell elements (12), (14) and (16) are substantially circular films concentrically disposed relative to the center of the battery. Film (14) is in contiguous electrical contact with films (12) and (16). Thus all conductivity between anodic film (12) and cathodic film (16) occurs through electrolyte film (14). While these films are all in further contacting relation to support member (10), since this member is insulative no interfilm conductivity is provided thereby. The same is true for impermeable film (18) which is electrically insulative.

Anodic film (12) includes a strip portion (12a) extending through films (14) and (16) (Figure 6.5c) to the exterior of the battery, the strip terminating in anode pad (12b) by which electrical connection may be made to the anodic film. To preserve battery geometry as respects anode-cathode conductivity through electrolyte film (14), and to prevent direct short-circuiting of films (12) and (16), an insulative layer (20) of oxygen-impermeable material is arranged in overlying relation to anodic film strip portion (12a) and support member (10) to enclose strip portion (12a) and electrically insulate same from films (14) and (16).

Electrical connection is made to cathodic film (16) through strip portion (16a) and pad (16b). As strip portion (16a) extends exteriorly of the sensor, same contacts only insulative impermeable film (18) and insulative support member (10). Since no contact is made with underlying enclosed films (12) and (14), no insulative member need be employed in conjunction with strip portion (16a).

In manufacture of the battery, the various films are deposited in vacuum through appropriate masks onto support member (10). Typically a glazed alumina sheet, Alsimag 614, 1/32 inch in thickness with a 743 glaze of borosilicate, is loaded into a vacuum deposition chamber together with evaporation boats each containing material for one of the battery films. The support member is selectively positioned in target positions above the respective material-containing boats by a turntable or carousel.

Evaporation masks are supported between the boats and targets on a second carousel which can be raised to bring a mask into contact with the support member. The chamber is evacuated by means of an ion absorption pump to 2.5×10^{-8} torr and pressure is maintained throughout the deposition process at 2 to 4 times 10^{-7} torr.

The support member is heated to in excess of 125°C. to remove water and other absorbed contaminants. Thereupon the anodic film evaporation mask is raised against the support member and the anodic film material body is moved into registration with the substrate and mask. The body is then heated with current flow predetermined to give sufficiently rapid deposition, the rate of deposition of each material being selected to insure a smooth and even layer.

Upon completion of deposition of the anodic film, the evaporation boat containing the material constituting insulative film (20) is moved into target position and the evaporated mask corresponding to desired insulative film geometry is raised against the anodic-filmed support member. Then film (20) is deposited over the limited anodic film area illustrated in Figure 6.5a.

Upon completion of deposition of insulative film (20) upon the support member and anodic film (12), the evaporation boat containing electrolyte film material is moved into the target position and the electrolyte film evaporation mask is raised against the support member in place of the anodic mask and deposition of this film ensues. As particularly illustrated in Figure 6.5c, the deposited electrolyte film is spaced at a limited portion of the circular periphery from anodic film (12) by insulative layer (20). Substrate temperature is maintained at approximately 125°C. and evaporation is maintained at a relatively slow rate. Such method parameters provide for a reduced number of grain boundaries, low internal electrolyte resistance and avoidance of amorphous electrolyte structure.

Deposition of cathodic film (16) is next performed by movement of the cathodic film material evaporation body and evaporation mask into the target position. As illustrated in Figure 6.5a, the cathodic film mask defines strip portion (16a) and pad (16b) in addition to the interior circular cathodic film proper.

FIGURE 6.5:   SOLID STATE BATTERY

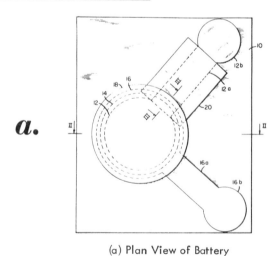

a.

(a) Plan View of Battery

(continued)

FIGURE 6.5: (continued)

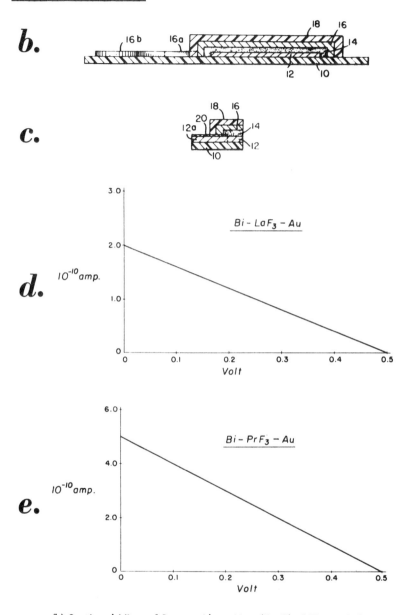

(b) Sectional View of Battery Along Line (II—II) of Figure 6.5a
(c) Partial Sectional View of Battery Along Line (III—III) of Figure 6.5a
(d) (e) Performance Characteristics of Batteries

Source: A.C. Lilly, Jr. and C.O. Tiller; U.S. Patent 3,657,016; Apr. 18, 1972

As illustrated in Figure 6.5b, the strip pad portions (16a) and (16b) are deposited directly upon support member (10) whereas all remaining portions of the cathodic film are deposited upon electrolyte film (14) and insulative layer (20) (Figure 6.5c). Finally the deposition of oxygen-impermeable film (18) is accomplished by movement of appropriate evaporation boat and mask into the target position. This film is deposited over the entire exposed surface of cathode film (16) except for pad (16b) and most of strip portion (16a). This film contacts insulative film (20) as is shown in Figure 6.5c.

Example 1:   A battery having the structure of Figures 6.5a through 6.5c is constructed by employing bismuth anodic film material, lanthanum fluoride electrolyte film material and gold cathodic film material. The respective film thicknesses are: 4320 A., 8635 A., and 2015 A. An open circuit voltage of one-half volt is measured between anodic film pad (12b) and cathodic film pad (16b). Upon connection of a load resistance of 1.0×10^9 ohms to the film pads, an output voltage of 0.15 volt is provided, indicating a load current of 1.5×10^{-10} amp. With constant loading for 12 hours, a decrease in output voltage of 0.018 volt occurs.

These and other performance characteristics of the batter are shown in Figure 6.5d, where the relationship between terminal voltage of the battery and output current is indicated as the load resistance is varied from 10^{10} to 0 ohms. From this plot, the effective internal resistance of the battery may be seen to be approximately 2.5×10^9 ohms, and it will be noted that the battery is self-limiting as respects maximum allowable output current. Thus, at an output current of 2.0×10^{-10} amp., output voltage falls to zero.

Example 2:   A battery having the structure of Figures 6.5a through 6.5c is constructed by employing bismuth anodic film material, praseodymium fluoride electrolyte film material and gold cathodic film material. The respective film thicknesses are: 3174 A., 2744 A., and 804 A. An open circuit voltage of one-half volt is measured between anodic film pad (12b) and cathodic film pad (16b). Upon connection of a load resistance of 1.0×10^9 ohms to the film pads, an output voltage of 0.25 bolt is provided, indicating a load current of 2.5×10^{-10} amp. With constant loading for 12 hours, a decrease in output voltage of 0.027 volt occurs.

These and other performance characteristics of the battery are shown in Figure 6.5e, where the relationship between terminal voltage of the battery and output current is indicated as the load resistance is varied from 10^{10} to 0 ohms. From this plot, the effective internal resistance of the battery may be seen to be approximately 1.1×10^9 ohms, and it will be noted that the battery is self-limiting as respects maximum allowable output current. Thus, at an output current of 5.0×10^{-10} amp., output voltage falls to zero.

Two Contiguous Thin Film Layers of Halides

F.E. Swindells and W.R. Lanier; U.S. Patent 3,547,700; December 15, 1970; assigned to Melpar, Incorporated describe a solid state battery which includes an electrolyte composed of two contiguous thin film layers of halides of the respective metals utilized as the electrodes of the battery. One of the electrolyte thin film layers is a halide salt of a divalent or higher valence metal doped with a halide salt of a monovalent metal. The monovalent metal has an ionic radius that most closely approximates, relative to the other monovalent metals, the ionic radius of the divalent or higher valence metal in the aforementioned one of the electrolyte layers. The electrodes are, like the electrolyte layers, of thin film construction. According to the process, a solid state thin film battery or cell of the type shown in Figure 6.6a is fabricated as follows. The substrate (10), which is preferably composed of borosilicate glass, quartz, ceramic, or plastic film such as polyester or polyimide, is initially outgassed at approximately 400°C. for from 20 to 30 minutes to render its surface more likely to promote adhesion of and direct continuous contact with a thin film electrode (12). Suitable electrode materials include silver, magnesium, lead, cadmium, calcium, barium, and beryllium.

In a first example, silver is used as the material for electrode (12) and is preferred for its electrical conductivity and its capability of strongly adhering to any of the aforementioned substrate materials. The silver layer (12) is deposited to a thickness of approximately 1000 A. from a tantalum boat in a vacuum environment of 10^{-5} torr, with the substrate temperature maintained at approximately 150°C.

FIGURE 6.6:   SOLID STATE BATTERY CELLS

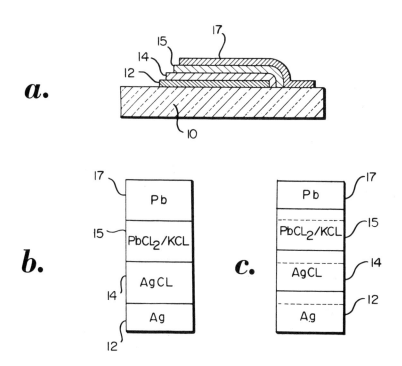

a.

b.   *c.*

(a) Cross-Sectional View of Thin Film Form
(b) Schematic Indicating Boundary Layer Positions — Charged Cell
(c) Schematic Indicating Boundary Layer Positions — Discharged Cell

(continued)

FIGURE 6.6:   (continued)

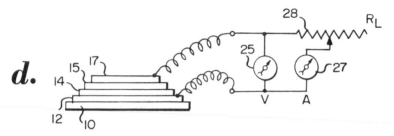

Measurement Circuit

Source: F.E. Swindells and W.R. Lanier; U.S. Patent 3,547,700; Dec. 15, 1970

An electrolyte layer (14) of silver chloride is then deposited to a thickness of approximately 2500 A. from a crucible composed of a material such as quartz, thoria or
zirconia to prevent decomposition of the silver chloride during the evaporation
process. It is desirable, for this reason, that the deposition of silver chloride be
carried out from a crucible composed of material of the type specified, rather than
from a tantalum or tungsten boat. The deposition of the AgCl layer (14) is also
performed under vacuum conditions at the above specified pressure, with the substrate temperature maintained at 125°C.

Layer (15) is an electrolyte layer which may vary in specific composition, but which
is, according to the process, composed of a halide of a divalent or higher valence
metal doped with from 0.01 to 50% of monovalent metal halide by weight of total
composition (in contrast with mol percent) of the electrolyte layer to be doped,
with the dopant distributed through the layer. Layer (15) is composed of lead
chloride doped with 1% by weight of potassium chloride. This layer is evaporated
on silver chloride film (14) to a thickness of from 5000 to 15,000 A. The lead
chloride-potassium chloride composition may be deposited from a crucible or from
a tantalum boat. In any event, electrolyte layer (15) is preferably substantially
thicker, e.g., up to ten times thicker, than electrolyte layer (14), because of
power considerations to be examined presently.

The cell is completed by depositing a layer (17) of electrode material, such as lead
in the case of lead chloride as the primary constituent of electrolyte layer (15),
evaporated from a tantalum boat under the specified vacuum pressure, to a thickness
of 1000 A. with the substrate temperature maintained at 75°C. It will be noted
that the substrate temperature is reduced during deposition of each successive layer,
and this will be generally true for any specific composition of cell layers. It is
essential, of course, that the two electrodes be separated throughout by the electrolyte and/or the substrate, as shown.

Upon completion of the above process, the thin film doped cell is in a charged state
and may be stored for subsequent use, or utilized immediately without additional
charging. Alternatively, the cell may be fabricated in an uncharged state by simply
omitting the AgCl layer, which will appear in any event during electrolyte reaction

accompanying charging of the cell. Referring to Figure 6.6b, the relative positions
of the boundaries of the several layers of a charged cell are shown for the exemplary
electrode and electrolyte materials described above.

The thickness of each individual layer is exaggerated in both Figures 6.6b and 6.6c,
for the sake of clarity. When the cell is connected in an electrical circuit such
that a path is completed from the cathode (Ag) (12) to the anode (Pb) (17) via an
appropriate load, current flows through the circuit and there is a shift of the layer
boundaries toward the anode end of the cell, as the cell discharges. This phenomenon is illustrated by the difference in location between the dashed and solid boundary
lines in Figure 6.6c. The dashed lines represent the original boundaries between
layers in a fully charged cell, and the solid lines represent the boundaries between
layers of the cell when discharged.

This boundary shift is explained by oxidation of the lead to lead chloride, with an
accompanying reduction of the silver chloride to silver. As is to be expected,
charging the cell reverses this shift of the boundaries back toward the silver (cathode)
electrode. In the charging process, the silver is oxidized to silver chloride and the
lead chloride is reduced to lead, the silver then becoming the anode and the lead
the cathode.

It should be clear that the capacity of the battery cell expressed in ampere-hours is
dependent upon the amount of material available for transport, according to Faraday's
law. In the example given above, the lead chloride layer and silver chloride layer
must be sufficiently thick to provide the desired capacity, and this will hold for
any materials of which the electrolyte layers may be composed. The required
thickness of the layers to provide a desired cell capacity may be determined in
accordance with the stoichiometric relation

$$1/2Pb \ + \ AgCl \ \longrightarrow \ 1/2PbCl_2 \ + \ Ag$$

where one gram equivalent of the reactants is equivalent to 26.8 ampere-hours
capacity. The same approach to determine layer thickness may be followed for
other cell materials. Other examples of suitable compositions of electrolyte layer
(15) are as follows. Magnesium chloride may be doped with from 0.01 to 50% of
lithium chloride, by total weight of this part of the electrolyte. This will require
the use of magnesium in place of silver for electrode layer (17).

In general, it may be stated that the metal in the compound of which electrolyte
layer (15) is composed is the metal to be used as the electrode (17) adjacent that
electrolyte layer. Cadmium chloride doped with up to 50% by weight of sodium
chloride, or calcium chloride doped with up to 50% by weight of sodium chloride,
or barium chloride doped with up to 50% by weight of potassium chloride, or
beryllium chloride doped with up to 50% by weight of lithium chloride, may alternatively be used for electrolyte layer (15). In general, the monovalent metal whose
salt is to be utilized as the dopant is selected on the basis of its having an ionic
radius most closely corresponding to the ionic radius of the divalent metal or higher
valence metal whose halide is the principal constituent of electrolyte layer (15).

Tests were performed on the cells using apparatus shown in Figure 6.6d. A voltmeter
(25) is connected between electrodes (12) and (17), and a microammeter (27) is
connected through a variable high resistance (28) across the voltmeter. Representative

electrical characteristics of a typical cell in which magnesium chloride doped with lithium chloride was used as the electrolyte layer (15), are as follows:

R_L (28)	V (across 25)	I (through 27)
A (open)................................	1.80	0
1.00 megohm.........................	1.75	1.5µa
0.50...................................	1.69	3.2
0.00...................................	1.50	15.0
0.15...................................	1.49	29.6
0.01...................................	1.13	112.0

Because of sensitivity of the cells to ambient moisture the cells may be wrapped in protective insulative film (e.g., polyester), or may be encapsulated immediately after fabrication.

ANODES

Dispersion of Iodine Reactive Material in Electrolyte or Anode

B.B. Owens and J.R. Humphrey; U.S. Patent 3,661,647; May 9, 1972; assigned to Gould Ionics, Incorporated have found that resistance build-up results in solid state electric cells from the diffusion of a mobile oxidant such as iodine diffusing through the solid state electrolyte to the anode where the oxidant reacts with a reductant present in the anode to form a resulting highly resistant layer on the anode surface. This process is directed to a resulting cell where the mobile oxidant such as the iodine is prevented from reaching the surface of the anode to react with the reductant such as Ag to form the highly resistive layer on the anode surface adjacent the electrolyte.

The process thus comprises disposing a material in the electrolyte which will tie up the mobile oxidant through a reaction with it or complexing it to prevent it from reaching the surface of the anode. For example, when silver is dispersed throughout the solid electroltye, silver iodide then results from the diffusion of the iodine through the electrolyte where it encounters the silver particles. Thus the iodine is prevented from reaching the anode surface to form a highly resistive AgI layer. There are additional various materials that can be added to the electrolyte to ac-complish the desired result of tying up the mobile oxidant before it reaches the anode. Through the use of a material added to the electrolyte to tie up the mobile oxidant, it has been found that the resulting solid state electric cell will experience little or no increase in resistance over long periods of storage.

The process is particularly directed to solid state electrochemical devices where the solid electrolyte element utilized has unusually high ionic conductivity. Examples of such electrolytes are described in U.S. Patent 3,443,997 where the electrolyte has a general formula MAg_4I_5 where M represents K, Rb, NH_4, or Cs, and in U.S. Patent 3,476,606 where it contains an organic ammonium ion. In the following examples the anode element is prepared by a method described in U.S. Patent 3,503,810 while the cathode is prepared according to U.S. Patent 3,476,605.

Example 1: A reference cell not incorporating the concept of this process was pre-pared having an anode comprised in weight percent, 53% silver, 42% $RbAg_4I_5$ and 5% carbon. The anode as well as the cell had a diameter of one-half inch and a thickness of 20 mil. The electrolyte layer was comprised in weight percent,

90% RbAg$_4$I$_5$ and 10% of a thermoplastic polycarbonate resin produced from the reaction of bisphenol A and phosgene. The electrolyte pellet was formed by mixing the resin with the RbAg$_4$I$_5$ and pressing the mixed powder at a load of 1,000 lbs. to achieve the pellet. The resin served as a filler which aids in the processing and to occupy some of the area of the cracks and crevices between the grain boundaries of the RbAg$_4$I$_5$.

The cathode is comprised, in weight percent, of 77% (CH$_3$)$_4$NI$_9$, 10% carbon and 13% RbAg$_4$I$_5$. The assembly was then pelletized at 10,000 lbs. and encapsulated in an epoxy type resin to give both a cell of rugged construction and one that is protected from atmospheric corrosion. When a cell of the above composition and construction was tested at 70°F., it was found to have an initial resistance of 5.4 ohms; after 1 year had elapsed at the foregoing temperature the resistance of the cell increased to 8.7 ohms. At 160°F. a cell of the aforegoing construction had an initial resistance of 4.4 ohms which increased to 10.9 ohms at the end of 1 year.

Example 2:   Cells were made identical to that described in Example 1 having the same dimensions utilized in the same anode and cathode compositions. The electrolyte layer however contained 5 weight percent of silver. The silver was introduced by blending the electrolyte powder of Example 1 with the appropriate amount of silver oxalate, and then heating the mixture to 165°C. for 15 minutes to decompose the silver oxalate. The final composition contained 5% silver, 9.5% resin, and 85.5% RbAg$_4$I$_5$.

Test cells were then built by the same techniques as used in Example 1. At 70°F. a cell made in accord with this example had an initial resistance of 39.6 ohms. After 1 year the resistance decreased to 38.5 ohms. More dramatic results achieved at an elevated temperature of 160°F. for the initial resistance of a cell having the composition of the example was 26 ohms. After 1 year the resistance had decreased to 9.7 ohms.

The initially high resistance of these cells was due to partial disproportionation of the electrolyte RbAg$_4$I$_5$ into AgI and Rb$_2$AgI$_3$. These resistive salts caused the observed high resistance, and although there is some reduction in cell voltage during discharge, it was found that when resistive materials are dispersed in the electrolyte they do not severely degrade cell performance. If the resistive material is formed only on the anode surface then severe polarization results during low temperature discharge. It has been found that the silver can be introduced into the electrolyte without obtaining this initial high resistance by using powdered silver, or annealing the resistive material or keeping the reactants cooler during the pyrolysis step. The resistance of the cells stored at 160°F. decreased to 9.7 ohms because the resistive salts were recombining to form RbAg$_4$I$_5$.

Example 3:   In this example a graded electrolyte was utilized comprised of two separate pellets, the first pellet comprising anode alone while the second pellet comprised silver dispersed in electrolyte adjacent to the anode. In the cells made in accordance with this example the anode and cathode were of the same composition and made in the same way as described in Example 1. The cell and components had the same dimensions also as that set forth in Example 1. The difference however in this cell resides in the structure and composition of the electrolyte between the cathode and normal anode. The first pellet was formed of the electrolyte composition which was the same as that set forth in Example 1. This pellet had a thickness of 12 mils. A second pellet of 11 mils thickness was formed and disposed between the electrolyte pellet and the anode. The second pellet contained 74 weight percent $RbAg_4I_5$ and 26 weight percent of silver and did not contain any resin. The initial resistance of the cell of this example at 70°F. was 6.6 ohms. After 1 year the resistance slightly increased to 7.6 ohms. The initial resistance of a cell made in accord with this example at 160°F. was 4.8 ohms. After 1 year the resistance had increased to 6.4 ohms.

In related work B.B. Owens and J.E. Oxley; U.S. Patent 3,663,299; May 16, 1972; assigned to Gould Ionics, Incorporated describe a solid state electric cell having a mobile oxidant such as iodine in the cathode, a solid state electrolyte and a suitable anode containing a reductant material such as silver which is capable of reducing the mobile oxidant, where at least the anode additionally contains a material which is capable of reacting with the primary reaction product of the mobile oxidant and reductant to form a resulting product in situ that has improved ionic conductivity as compared to the primary reaction product.

Silver, Carbon and Electrolyte Matrix

In a process described by I.J. Groce; U.S. Patent 3,503,810; March 31, 1970; assigned to North American Rockwell Corporation an anode composition for a solid electrolyte power cell is prepared by blending a compound of silver, preferably Ag_2O, with carbon and with a solid electrolyte material. The resulting blend is then heated to reduce the silver compound and give finely divided silver metal powder in a matrix of carbon and electrolyte.

Example 1: Chemically pure Ag_2O powder, 21.5 grams, was blended by means of a spatula with 22.6 grams of AgI powder, 7.4 grams of RbI crystals and 7.5 grams of carbon (acetylene black) in a glass beaker. The mixture was further mixed and pulverized by feeding it twice through a disc pulverizer, following which it was placed in a 3 inch diameter stainless steel retort. The retort was evacuated, backfilled with argon, and heated to 450°C. in an electrically heated furnace. Initiation of the reduction reaction was evidenced by a rapid increase in the pressure inside the retort after 10 minutes. The retort was then cooled by lowering it into a container of water. This completed the first half of the reduction.

The reaction product was removed from the retort and blended with a second charge of 32.3 grams of Ag_2O and again pulverized through the disc pulverizer. This mixture was next placed into the retort and heated to 450°C. after evacuating and filling the retort with argon. The second reduction was also evidenced by a rapid increase in pressure inside the retort but required 15 minutes to initiate. The retort was cooled in the same manner as above and the product removed. This completed the reduction, yielding 84 grams (98% recovery) of product with a nominal weight composition of $1.0 Ag + 0.6 RbAg_4I_5 + 0.15 C$.

The anode composition was then compacted at 5 tons per square inch and heated in an open beaker for 20 minutes at 400°C. When the compact was cooled it was broken and crushed in a mortar and pestle to give a random distribution of particle sizes up to 10 mesh. Although no sieve analysis was made, very little fine powder (-200 mesh) was observed. Test cells fabricated from this batch, having the representation $Ag/RbAg_4I_5/I_2+C$, averaged 109 ma.-hr. (20 ma. current drain) out of a theoretically available 125 ma.-hr. This comprised a utilization of 88% of the silver from an anode containing 59% Ag.

Example 2: The procedures utilized for blending and pulverizing were the same as in Example 1 except that the batch size was increased and the ratios changed as follows:

	Grams
Ag_2O	107.5
AgI	48.0
RbI	15.0
Carbon	20.0

After pulverizing these in the disc pulverizer, the material was added to the reduction retort. The second reduction was prepared by combining the lots from the first reduction with 107.5 grams of Ag_2O, and blending and pulverizing in the disc pulverizer. The final product was again compacted into pellets at 5 tons per square inch and heated for 20 minutes at 400°C. Product yield was over 90%. The cooled and crushed product was tested in 125 ma.-hr. cells, giving an average of about 116 ma.-hr. or 93% silver utilization in an anode blend of 1.0 Ag + 0.3 electrolyte + 0.08 C which contained 72.5% silver.

In related work with solid ionic conductors based on alkali metal silver iodides G.R. Argue and B.B. Owens; U.S. Patent 3,701,686; October 31, 1972; assigned to North American Rockwell Corporation describe a solid state electric cell having a solid electrolyte disposed between an anode and a cathode, and where solid electrolyte material is dispersed in at least the anode, and preferably in both the anode and the cathode. Preferably the dispersed electrolyte material is of substantially the same composition as the material of the solid state electrolyte.

Lithium Coated with Oxygen-Containing Lithium Compound

S.R. Ovshinsky; U.S. Patent 3,615,835; October 26, 1971; assigned to Energy Conversion Devices, Incorporated describes a room temperature DC voltage generating device comprising a solid body of lithium metal carrying on and in direct contact with at least a part of its surface a solid layer of a composition consisting essentially of at least one oxygen-containing lithium compound such as lithium nitrate or lithium sulfate, or lithium chloride or lithium bromide.

The solid layer is effective in preventing direct contact of a water-moisture-containing gaseous environment with the body of lithium metal in the area of the solid layer. The process provides an electrical circuit including electrical contacts respectively connecting the body of solid lithium metal, and the solid layer, and means providing a water-moisture-containing gaseous environment whereby to moisten the layer with generation of DC voltage occurring on contact of the moisture with the layer.

Referring to Figure 6.7a, one form of the voltage-generating device of this process is generally designated at (10). It includes a body of lithium metal in the form of a sheet or disc or layer (11) which has been coated with a solid layer or film (12) of any of the above noted compositions. This layer or film (12) of lithium compounds may also include iodine. A needle electrode (13) is inserted into the lithium metal

FIGURE 6.7:   SOLID STATE DC GENERATING DEVICE

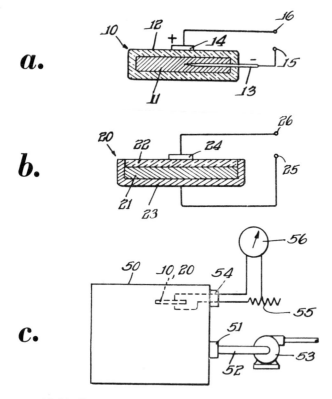

(a) (b) Illustrations of Two Forms of the Voltage Generating Device
(c) Vacuum Measuring Device

Source: S.R. Ovskinsky; U.S. Patent 3,615,835; October 26, 1971

body (11) so as to be in contact therewith and an electrode (14) is suitably secured to and contacts the outer surface of the lithium compound coating or layer or film (12). The electrodes (13) and (14) are connected respectively to terminals (15) and (16). The solid layer (12) of the composition prevents contact of a water-moisture-containing atmosphere with the body (11) of lithium metal.

When the device (10) of Figure 6.7a is subjected to a gaseous moisture-containing environment, which may be moist air or another water-moisture-carrying gas which may be a gas in which free nitrogen is essentially absent, for instance, argon, carbon dioxide, neon, or the like, the coating or film or layer (12) is moistened and it takes up moisture from and loses moisture to the environment in accordance with the moisture content of the environment. The lithium metal body (11) and the moistened coating or film or layer (12) generates a substantial DC voltage which is transmitted from the electrodes (13) and (14) to the terminals (15) and (16), the terminal (16) being positive with respect to the terminal (15). The value of the DC voltage appearing at the terminals (15) and (16) depends upon the moisture content of the coating or film or layer (12) and hence upon the moisture content of the gaseous environment.

When the device (10) is subjected to an environment, such as atmospheric air, having a high moisture content, say 60 to 80% RH, the device (10) operates to produce DC voltages across the terminals (15) and (16) and, depending also upon the partic- ular lithium compound or mixtures utilized, may be in the volt range and may even exceed 3 volts. As the moisture content of the gaseous environment decreases, the voltage generated by the device (10) correspondingly decreases and, therefore, the device (10), in addition to being a simple and powerful DC voltage generator, may also operate as a device for producing a DC voltage in accordance with the moisture content of the gaseous environment so as to provide a means for measuring the moisture content of the gaseous environment.

Another form of the voltage-generating device is generally designated at (20) in Figure 6.7b. It includes a solid body, in the form of a layer or sheet or disc (21), of lithium metal inserted in the bottom of a cup- or dish-shaped member (23) which is formed of electrically conducting material, such as metal or the like, the lithium metal (21) making electrical contact with the member (23). A solid layer or coating or film (22) of a composition containing at least one lithium compound, or mixtures, of the kinds described above, is supplied to the body (21) of lithium metal, it being applied in any suitable manner as indicated previously.

The compositions may also be tamped or pressed into the member (23) against the body (21) of lithium metal. An electrode (24) is suitably applied to the outer sur- face of the layer (22) of the composition, and the electrode (24) and the body (21) are connected by leads to terminals (26) and (25). The member (23) and the layer (22) of the composition prevent contact of the water-moisture-containing atmosphere with the body (21) of lithium steel. The outer surface of the composition layer is exposed to and contacted by the water-moisture-containing gaseous environment and the device (20) of Figure 6.7b operates to generate substantial DC voltage in the same manner as discussed above in connection with the device (10) of Figure 6.7a.

The voltage-generating devices of Figures 6.7a and 6.7b may be stored before use by encapsulating them in a water-moisture-proof container so that they will not be subject to water-moisture until ready for use. When ready for use, they are re- moved from such container and subjected to the gaseous environment to generate substantial DC voltages in accordance with the water-moisture content.

Figure 6.7c illustrates the manner of utilizing the voltage-generating devices (10) or (20) of Figures 6.7a and 6.7b for measuring and indicating vacuum conditions. Here, a container or tank (50) carries a moisture-containing gas and it is connected through a coupling (51) and a pipe (52) to a vacuum pump (53) which operates to produce a vacuum in the tank or container (50). As the pump (53) is operated, a moisture-containing gas is pumped from the tank or container (50) to produce a vaccum therein. In so doing, moisture is withdrawn from the tank along with the gas and the moisture content of the gas remaining in the tank is proportional to the vacuum drawn in the tank. A voltage-generating device (10) or (20) of the kinds illustrated in Figures 6.7a and 6.7b is placed in the tank so as to respond to the moisture content of the gas in the tank. The device (10) or (20) has leads extending through a suitable coupling (54) in the tank which are connected through an adjustable resistance (55) to a meter (56). Thus, the device (10) or (20) generates a voltage in accordance with the moisture content of the gas in the tank (50) and hence the vacuum condition and this voltage so produced is indicated by the meter (56). The meter (56) may be calibrated to indicate the degree of vacuum in the tank (50) and the calibration of the meter may be effected by the adjustable resistance (55).

The DC voltage generating devices are most desirably operated at room or ambient temperatures. They can, however, be operated at lower temperatures as well as at higher temperatures. In general, it is particularly preferred that they be operated at temperatures in the range of 20° to 60°C.

CATHODES

Quaternary Ammonium Polyiodides

B.B. Owens; U.S. Patent 3,476,605; November 4, 1969; assigned to North American Rockwell Corporation describes the use of an organic ammonium polyiodide cathode component for a solid state electric cell containing a silver anode and an ionically conductive silver-containing solid electrolyte. The polyiodide compositions utilizable as electron-acceptor cathode component are defined as QI_n where Q is an organic ammonium cation, preferably a quaternary ammonium cation, and n has a value ranging from 2 to 11, inclusive. Specifically preferred polyiodide compositions are tetramethylammonium heptaiodide $N(CH_3)_4I_7$ and the tetraethyl-ammonium triiodide $N(C_2H_5)_4I_3$ and heptaiodide $N(C_2H_5)_4I_7$.

The electron-acceptor cathode component may be utilized in any solid state electric cell having a conductive-anode electron donor, preferably silver, and a solid electrolyte where the current preferably is transported by silver cations. The cathode compositions QI_n provide a lower iodine activity compared with that of pure iodine and generally also the inorganic polyiodides, resulting in greater cell stability, longer shelf-life, and less corrosion.

Example 1: The preparation of quaternary ammonium polyiodides is as follows. A molar portion of tetramethylammonium iodide was reacted with varying molar portions of iodine in a closed vessel at a temperature of 65°C., the molar amounts of iodine used varying from 1 to 3 mols I_2 per mol of tetramethylammonium iodide. Polyiodide compositions were obtained in all instances, having an empirical formula ranging from $N(CH_3)_4I_3$ to $N(CH_3)_4I_7$, the equilibrium vapor pressure of the polyiodide products increasing with increasing iodine content but being substantially below that of pure iodine. Similar polyiodide products were prepared by performing the reaction at 120°C. followed by quenching of the reaction product. These techniques of preparation were also used to prepare other polyiodide compounds QI_n. Several typical preparations are shown as follows.

To 12.16 grams of $(C_2H_5)_3NCH_3I$ was added 12.7 grams of I_2. As soon as the I_2 was added, the contents changed to a black liquid. After 1 1/2 hours no smell of I_2 was detectable and the contents started to harden.

To 13.56 grams of $(C_2H_5)_3NC_3H_7I$ was added 12.7 grams of I_2, the contents becoming liquid upon I_2 addition. After 1 1/2 hours the contents were hard and black in color. No odor of I_2 was detectable.

To 13.16 grams of $(CH_3)_3NC_6H_5I$ was added 12.7 grams of I_2. The compounds were blended together, forming a silver blue-green product. After 1 1/2 hours, a slight odor of I_2 was noted.

To 14.96 grams of $C_2H_5N(C_3H_7)_3I$ was added 12.7 grams of I_2, the contents becoming liquid. After 1 1/2 hours the contents were almost hard, but still wet. No odor of I_2 was noted.

To 15.66 grams of $N(C_3H_7)_4I$ was added 12.7 grams of I_2. The contents were blended together, no liquid being formed. After 1 1/2 hours the mix turned green in color and had a strong pungent odor.

To 10.05 grams of $N(CH_3)_4I$ was added 12.7 grams of I_2. The contents were blended together, no liquid being formed. After 1 1/2 hours the contents turned green in color. No odor of I_2 was detectable.

Example 2: The preparation of cathode blend is as follows. Cathode blends of 1 gram electrolyte, 1 gram carbon, and 3 grams polyiodide [varying from $N(C_2H_5)_4I_3$ to $N(C_2H_5)_4I_9$] were prepared by melting $RbAg_4I_5$ electrolyte together with the carbon, blending in tetraethylammonium iodide, and heating to 120°C. with iodine. While a suitable blend was obtained, the utilization of I_2 was only 75%. Utilization approaching 100% was obtained when carbon was first blended with an aqueous solution of tetraethylammonium iodide, and the blend was then dried and powdered and added to an acetone solution of $RbAg_4I_5$. The acetone was vaporized off from the mixture, the recovered product was dried and powdered, and the desired amount of iodine was then added. The mixture was heat cured for 18 hours at 115°C. followed by 48 hours at 65°C.

Example 3: The preparation of an electric cell using organic ammonium polyiodide cathode composition is as follows. Electric cells were prepared having a structure corresponding to that shown in Figure 6.8.

FIGURE 6.8:   CROSS-SECTIONAL VIEW OF SOLID STATE CELL UTILIZING AN ORGANIC AMMONIUM POLYIODIDE CATHODE

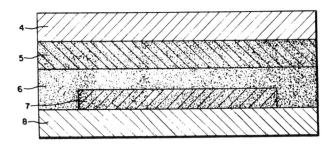

Source: B.B. Owens; U.S. Patent 3,476,605; November 4, 1969

The anode composition consisted of a blend of silver, carbon and $RbAg_4I_5$. The electrolyte element was $RbAg_4I_5$. The cathode composition was prepared essentially as described for Example 2 and consisted of carbon, $RbAg_4I_5$ and the organic ammonium polyiodide. The measured cell voltages are shown in the following table.

EMF of Electric Cells ($Ag/RbAg_4I_5/QI_n$)

QI_n in Cathode	Available I_2, wt. %	Open Circuit EMF (volts)
$(CH_3)_4NI_3$	56	0.655
$(CH_3)_4NI_4$	64	0.643
$(CH_3)_4NI_5$	73	0.652
$(CH_3)_4NI_6$	76	0.667
$(CH_3)_4NI_7$	80	0.667
$(CH_3)_3NC_2H_5I_3$	54	0.652
$CH_3N(C_2H_5)_3I_3$	51	0.64
$(C_2H_5)_4NI_3$	50	0.640
$(C_2H_5)_4NI_5$	67	0.643
$(C_2H_5)_3NC_3H_7I_3$	48	0.633
$C_2H_5N(C_3H_7)_3I_3$	46	0.614
$(C_3H_7)_4NI_3$	45	0.604
$(C_3H_7)_4NI_5$	63	0.625
$(C_3H_7)_4NI_7$	72	0.657
$(CH_3)_3NC_6H_5I_3$ (trimethylphenyl-ammonium triiodide)	49	0.639
$C_9H_7NHI_3$ (quinolinium triiodide)	50	0.655
$C_9H_7NHI_5$ (quinolinium pentaiodide)	66	0.657
$C_9H_7NCH_3I_3$ (1-methylquinolinium triiodide)	48	0.648
$C_9H_7NCH_3I_5$ (1-methylquinolinium pentaiodide)	65	0.650
$C_5H_5NCH_3I_3$ (1-methylpyridinium triiodide)	53	0.658
$C_5H_5NCH_3I_5$ (1-methylpyridinium pentaiodide)	70	0.661

For a representative cell, using $N(C_2H_5)_4I_7$ as electron-acceptor cathode component, the total cathode weight (grams) of the constituents used for the cathode blend is:

$$0.17 \; N(C_2H_5)_4I \; + \; 0.5 \; I_2 \; + \; 0.1 \; C \; + \; 0.27 \; RbAg_4I_5 \; = \; 1.04 \; g./100 \; ma.-hr.$$

For a 200 ma.-hr. cell, there is utilized 2.08 grams cathode, 1.40 grams anode, and 1.0 gram electrolyte, or 4.48 g./200 ma.-hr., calculating to a power density of 13.0 watt-hr./lb.

Example 4:　The preparation of an electric cell using organic ammonium compositions for electrolyte and cathode is as follows. A test cell of structure similar to that of Figure 6.8 was prepared having an anode composition (1 gram) consisting of Ag, C, and $RbAg_4I_5$; pyridinium heptasilver octaiodide, $C_5H_5NHAg_7I_8$ (3 grams), as electrolyte; and a cathode composition (1 gram) consisting of carbon, $RbAg_4I_5$ and $N(C_2H_5)_4I_3$. The open circuit voltage obtained was 0.66 volt with a flash current of 650 ma. After seven months storage, the open circuit voltage was 0.64 volt and the flash current was 150 ma. Similar cells were also prepared using organic ammonium silver iodide electrolytes, but using RbI3 as the electron-acceptor in a composite cathode. After seven months storage, flash currents for the best of these cells had decreased from an initial value of 120 to 0.2 ma., and from an initial value of 100 to 1 ma., thereby demonstrating the enhanced stability obtained by use of an organic ammonium polyiodide composition as electron-acceptor compared with RbI3.

The quaternary ammonium polyiodide cathode materials are advantageous compared with pure iodine in providing a lower iodine activity resulting in reduced cell corrosion and longer shelf-life. They are further advantageous over other inorganic polyiodides such as RbI3 or CsI3 in providing a lower iodine activity, a higher weight percent of available iodine, lower cost, nonreactivity with the electrolytes RbAg4I5 or KAg4I5, longer cell shelf-life, and greater compatibility with organic ammonium silver iodide electrolytes.

Organic Iodine Charge Complexes

J.R. Moser; U.S. Patent 3,660,163; May 2, 1972; assigned to Catalyst Research Corporation describes a solid state primary cell having a lithium anode, an iodine cathode and a solid lithium halide electrolyte. The lithium-iodine cells of this process have a high output voltage, typically an open circuit voltage of 2.7 to 3.0 volts depending primarily on cell design and the cathode material. The system has a theoretical energy density of 213 watt-hours per pound and energy densities as high as 136 watt-hours per pound have been obtained during discharge of encapsulated cells at room temperature.

Referring to Figure 6.9a, the cell comprises a metal anode contact member or current collector (2), a sheet lithium anode (4), a cathode (6), and a cathode current collector consisting of a metal screen (8) that is preferably spot-welded to a metal sheet (10). The cell components are stacked and the stack is compressed suitably at pressure of 1,000 psig, to form a complete cell.

The anode contact member has holes (12) punched by a nail or other conical punch, giving projecting sharp edges that dig into and secure the contact member to the lithium anode. A metal screen or gauze may also be used to advantage as an anode contact member. The cathode material, suitably finely divided, is compacted when forming the cell into a solid disc having the cathode contact screen (8) embedded. If desired, the cathode material may be compacted into a pellet of the desired size prior to assembly into the cell.

As shown in Figure 6.9b, the resulting cell, which has a plurality of laminated layers in intimate contact, is then encapsulated in an inert potting compound (14), suitably a polyester, that serves to protect the cell from degradation or exposure to the atmosphere.

A lithium iodide electrolyte is formed in situ by reaction of the iodine in the cathode with the lithium anode. It is equally satisfactory, and in some instances preferable, to form a film of lithium salt electrolyte on the anode surface abutting the cathode prior to cell assembly, most conveniently by exposing the anode surface to dry air or argon atmosphere containing halogen gas or vapor. It will be recognized that

FIGURE 6.9:   LITHIUM-IODINE CELL

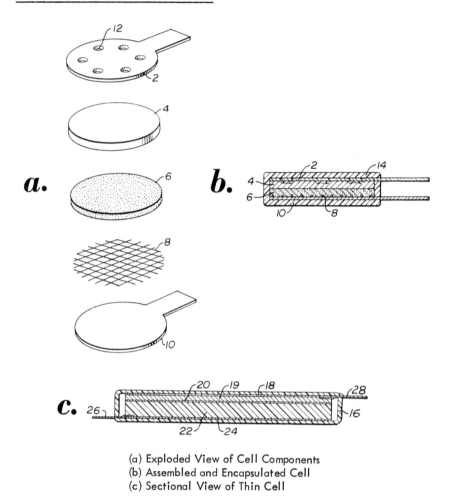

(a) Exploded View of Cell Components
(b) Assembled and Encapsulated Cell
(c) Sectional View of Thin Cell

Source: J.R. Moser; U.S. Patent 3,660,163; May 2, 1972

additional lithium iodide electrolyte is formed by the electrochemical reaction of
the cell. The cell of Figure 6.9c may be made in the form of a thin sheet, as little
as 15 to 20 mils thick, of any size. The cell is enclosed in plastic envelope (16),
suitably polyvinyl chloride, such as Process 40, Teflon or other plastic film impervious to iodine and ordinary atmospheres, that is oxygen, nitrogen and water
vapor. A thin metallic anode current collector (18), suitably a nickel plate deposited on the plastic envelope by vacuum deposition or electroless plating, abuts
a thin lithium electrode (19), suitably 1 to 10 mils thick. The lithium is most conveniently in the form of a foil, but it may also be deposited on the current collector
by vacuum deposition, electroplating or other conventional methods. The initial

thin film of lithium iodide electrolyte (20), may, as previously described, be performed on the lithium surface before assembling the cell or formed spontaneously when the clean lithium anode surface is brought into contact with the cathode. The cathode (22) may be a compacted powder or is more conveniently applied to the anode as a paste of cathode material or cathode material and binder, for example, vinyl dispersions such as O Kun's Liquid Vinyl. The cathode current collector (24) may be a thin metal foil or, as in the case of the anode current collector, a metal coating deposited on the plastic envelope. Metal leads (26) and (28), for external circuit connections, are connected to the anode and cathode current collectors respectively and tightly sealed through openings in the plastic envelope.

The cells are adversely affected by atmospheric moisture, so cells are assembled and encapsulated in a dry atmosphere, suitably in dry rooms or enclosures having a relative humidity less than 2%, using substantially anhydrous and/or dried components. All of the cell assemblies and tests of nonencapsulated cells described herein were performed in such a dry room or in dry boxes having even dried air or argon atmospheres.

Example 1:   Primary cells were made using various electronically conductive cathodes containing iodine available for electrochemical reaction. Cathodes were made by mixing iodine with the other powdered cathode components, graphite and/or an organic material, and the mixture was compacted at 5,250 psig into a pellet 1.25 cm. in diameter and about 1 mm. thick. The organic materials used were those that react spontaneously with iodine to form a charge transfer complex.

In some cases, as indicated, an organic I_2 charge transfer complex was separately prepared by conventional methods and then compacted into cathode pellets, either alone or admixed with graphite. The cathode pellet was placed between a 1.23 cm.^{2} lithium anode and a nickel foil collector, wrapped with Teflon tape and compressed under slight pressure with a clamp. The lithium anode disc was cut from lithium ribbon that had been cleaned with petroleum ether and scraped. The cell was tested for open circuit voltage (OCV), current output at various voltages ($\mu A/V$), and short circuit current (SCC). The cells were then coated with lacquer and retested in 24 hours. Typical results are shown in Table 1.

TABLE 1

Battery number	Cathode composition	Initial test				24 hour test			
		O.C.V.	S.C.C. (μA)	μA/V	μA/V	O.C.V.	S.C.C. (μA)	μQ/V	μA/V
1	99% iodine, 1% carbon	1.61	24	0.6/1.54	24/0.04	0.49	21	0.19/0.51	20/0.03
2	95% iodine, 5% carbon	2.85	4,600	1.05/2.85	1,430/2.35	2.80	680	.93/2.40	470/0.75
3	90% iodine, 10% carbon	2.90	6,300	1.10/2.90	1,530/2.50	2.85	1,100	1.05/2.85	760/1.25
4	50% iodine, 50% polyethylene	2.90	122	1.05/2.75	110/0.20	2.80	1,450	1.10/2.80	720/1.15
5	50% iodine, 50% polypropylene	2.90	145	1.00/2.80	130/0.20	2.85	710	1.10/2.85	500/0.80
6	47½% iodine, 47½% polypropylene, 5% carbon	2.85	470	1.00/2.85	380/0.65	2.75	1,650	1.07/2.75	790/1.30
7	Pyrene 2 I_2 [1]	2.95	630	1.10/2.95	450/0.80	2.95	150	1.06/2.90	135/0.30
8	50% pyrene 2 I_2 [1] 50% iodine	3.00	480	1.10/3.00	370/0.70	2.90	305	1.06/2.85	255/0.45
9	Pyrene 2 I_2 [2]	3.00	1,600	1.10/3.00	770/1.35	2.95	3,100	1.10/2.95	1050/1.85
10	50% pyrene 2 I_2 [2] 50% iodine	3.00	2,900	1.10/3.00	860/1.45	2.85	1,100	1.07/2.85	600/1.00
11	50% iodine, 50% phenothiazine	2.95	1,200	1.07/2.95	680/1.15	2.90	1,450	1.09/2.90	740/1.25
12	47½% iodine 47½% phenothiazine 5% carbon	2.90	2,700	1.02/2.90	1,050/1.85	2.85	1,750	1.08/2.85	820/1.35
13	2 phenothiazine 3 I_2 [3]	2.85	118	1.07/2.85	104/0.20	2.85	102	1.09/2.75	93/0.15
14	Perylene I_2	2.01	2,700			2.80	3,700		

[1] Prepared by melting together pyrene and iodine.
[2] Prepared by precipitation from CCl^4 solution.
[3] Prepared by precipitation from benzene solution.

Other organic materials used with iodine in cells with similar performance, either with or without 5 to 10% added graphite, include nylon, Lucite, Lucite paste in dichloroethylene, pyrrole, polypyrrole, naphthalene, dimethyl glyoxime, phenol-phthalein, phthalimide, erythrosine, methylene blue, urea, brominated pyrene, Teflon and o-tolidene.

Example 2: When a powdered electronic conductor, such as carbon or powdered metal is incorporated in the cathode to improve conductivity, it is preferred to form a film of electrolyte on the lithium anode before assembly of the cell to diminish internal short circuiting. The electrolyte film is formed by exposing the lithium surface to a dry air or argon atmosphere containing a vapor reactive with lithium to form a conductive salt of lithium, preferably the halogens, I_2, Cl_2, or Br_2, although other reactive vapors may be used, such as methanol or ethyl ether. To illustrate, cells were made in accordance with Example 1 having a cathode of 47.5% $C_{16}H_{10} \cdot 2I_2$, 47.5% iodine and 5% powdered carbon; one cell had a clean lithium anode and a second cell had a lithium anode coated with an LiI film. The results from performance tests as in Example 1 were as follows:

TABLE 2

	Initial Test		24 Hour Test	
Anode	OCV (volts)	SCC (ma.)	OCV (volts)	SCC (ma.)
Li	2.20	3,200	2.70	1,000
Li coated with LiI	3.10	3,500	3.00	3,100

Example 3: Six cells one-half inch in diameter and 0.215 inch thick were made in accordance with Figure 6.9a in which the anode contact was nickel foil, the anode was lithium, the cathode contact was 60 mesh nickel gauze and the cathode was a pellet of 2 phenothiazine $\cdot 3I_2$ charge transfer complex prepared by mixing together and heating phenothiazine and iodine in the indicated proportions. The lithium anode surface abutting the cathode was coated with a film of LiI formed by exposing the surface to iodine vapor in an argon atmosphere. The cell components were stacked and compressed at 1,100 psig to form the completed cell assembly. The cells were continuously discharged at a current of $25\mu A$ at 33°C. Table 3 shows the voltage at $25\mu A$ current drains and the short circuit current at various time intervals.

TABLE 3

Time (hr.)	Volts at $25\mu A$	SCC (μA)
0	2.6-2.7	1,100-1,600
25	2.3-2.4	250-400
49	1.3-1.5	50-80

In related work, A.A. Schneider and J.R. Moser; U.S. Patent 3,674,562; July 4, 1972; assigned to Catalyst Research Corporation has found that a mixture of iodine with a poly-2-vinylpyridine $\cdot I_2$ or a poly-2-vinylquinoline $\cdot I_2$ charge transfer complex is an improved cathode material of a plastic state and in conjunction with a metal anode, for example lithium, provides primary cells with improved capacity and performance characteristics.

A.M. Hermann, F. Gutmann, and A. Rembaum; U.S. Patent 3,660,164; May 2, 1972; assigned to California Institute Research Foundation describe primary cells capable of operating essentially in the solid state and utilizing the reaction of a halogen with a metallic anode. The cathode is a charge transfer complex in which the acceptor component is the halogen and the donor component is an organic compound, typically aromatic or heterocyclic. Preferred anode materials include magnesium, calcium and barium.

The maximum current produced by such cells can be increased by surrounding them with an atmosphere containing more than the normal partial pressure of water vapor, or containing vapor of an organic liquid such alcohol, acetone and acetonitrile, having a high dielectric constant. The physical properties of the charge transfer complex is improved by incorporating a polymeric matrix, which may or may not act also as the donor component of the complex.

Lithium–Organic Charge Transfer Complex

K.R. Hill; U.S. Patent 3,653,966; April 4, 1972; assigned to P.R. Mallory & Co., Incorporated describes a high energy organic electrolyte cell comprised of a light metal anode and a cathode, the active material of which is an organic charge transfer complex. The complex is essentially insoluble in the cell organic electrolyte and does not generate soluble discharge products. The cell is substantially nonself-discharging and has extended shelf-life.

There has been developed further in the process a cell which is assembled without the physical placement of a separator element between the cell cathode and anode members. In assembly of the cell, an anode of a light metal such as lithium is placed in direct contact with a cathode formed of an active material consisting of an organic charge transfer complex, there being formed in situ on the anode by electrochemical reaction between the anode and active cathode material an ionically conductive high resistance film comprised of a compound of the anode and active cathode materials and capable of functioning as the cell separator. The separator is chemically active and self-adjusting, the reaction forming the separator recurring upon reexposure of the anode surface to provide the cell with requisite internal resistance at all times.

The active cathode material of cells of the process comprises an organic charge transfer complex consisting of an organic acceptor and an organic donor. Acceptors include halogenides such as chloranil, bromanil or iodanil. Suitable donors may comprise p-phenylene diamine, 3,8-diamino pyrene, dimethyl aniline, tetramethyl-p-phenylene diamine, or 3,10-diamino pyrene. While various combinations of acceptor and donor are usable, a particularly preferred active cathode material is the complexation product comprising chloranil plus p-phenylene diamine:

A preferred ratio for this complexation product is 3:2. While the halogenide is soluble in organic solvents, the complexation product is essentially insoluble in organic solvents. The charge transfer complex may be prepared by any suitable method for the preparation of organic semiconductors. Exemplary methods employed in the formation of the active cathode material of practical cells are as follows, method B involving the preferred ratio of components.

Method A: Five parts chloranil and three parts p-phenylene diamine are dissolved separately in a heated organic solvent such as propylene carbonate, gamma-butyrol-acetone or methyl formate. The chloranil requires about 40 ml. solvent per gram and heating to 80°C. with stirring. The p-phenylene diamine requires about 10 ml. solvent per gram. When most of the chloranil is dissolved, the temperature of the two solutions are equalized. The solutions are then mixed and stirred until cool. The greenish-black complex thus formed is filtered with suction, washed on the filter with cold absolute ethanol and air dried or dried by evaporation of the solvent.

Method B: Method A is practiced using two parts chloranil and three parts p-phenyl-ene diamine. The resulting air dried greenish-black filter cake is then redispersed in a small amount of cold ethanol, refiltered and vacuum dried or dried by evaporation of the solvent.

Method C: In this process, method B is practiced using equal parts of chloranil and p-phenylene diamine.

A charge transfer complex prepared by any of these methods may either be employed alone as the cell cathode or it may be mixed with electrically conductive particulate matter which is chemically inert. For example, graphite may comprise 5% of the cathode by weight.

Prior to the final step of the above methods where the complex is dried in air, by vacuum or dried by evaporation of the solvent, the cell electrolyte is preferably mixed with the complex to form a composite cathode-electrolyte or catholyte. For example, the complex may be thoroughly mixed in a mortar with a 0.1 to 1.0 molar lithium fluoroborate-propylene carbonate solution. The resulting black powder has a dry appearance but flows plastically under pressure. Lithium perchlorate or lithium tetrachloroaluminate may be used in place of lithium fluoroborate as the electrolyte salt. Such mixture and intimate contact between complex and electrolyte is of course permissible by reason of the insolubility of the complex in the electrolyte.

In constructing practical cells in accordance with the process, the cathode or cath-ode-electrolyte mix may be pressed directly into a suitable cell can. In this case, the ethanol slurry or paste constituting the mix is air dried or vacuum dried prior to pressing. Alternatively, the paste may be rubbed onto a hot plate or heated cell container and the solvent removed by evaporation.

Cell assembly is completed by placing an anode member comprised of a light metal such as lithium upon and in direct contact with the cathode or cathode-electrolyte mix in the cell container and sealing same from atmospheric reactants. The electro-chemical reaction occurring in a cell comprising the chloranil/p-phenylene diamine complex and a lithium anode is presumed to be as shown on the following page.

$$(2m) Li + (m) \; [\text{chloranil}] \cdot (n) [\text{NH}_2\text{-ring-NH}_2] \rightleftharpoons (m) [\text{Li}_2\text{chloranil salt}] + (n) [\text{NH}_2\text{-ring-NH}_2] + (2m)e^-$$

In this reaction it will be observed that lithium is effective to dissociate the cathodic complex liberating 2m electrons and reacting with chloranil to form the salt $Li_2O_2C_6Cl_4$. Formation of the lithium chloranil salt on the anode at the anode-cathode interface is, in effect, the in situ formation of a chemically active cell separator constituting a compound of the anode and active cathode material. The thin film of this salt is present instantaneously upon cell assembly and by reformation throughout cell life.

The in situ separator is ionically conductive, permitting migration of lithium ions to the cell cathode and at the same time prevents short-circuiting of the cell by maintaining an electronic barrier between anode and cathode. By the film reformation capabilities of the cell components, the film is self-adjusting, healing imperfections as they occur, i.e., as the lithium anode becomes exposed or unfilmed. This separator characteristic is in direct contrast to that of the chemically inert cell separator which, upon the occurrence of pinhole failures or like imperfections, gives rise to internal shorting and cell failure.

Example 1: A method A cathode was made using 0.0552 gram of active material admixed with 5% graphite. This mixture absent any electrolyte, was air dried and pressed into a pellet at 5,000 lbs./in.^{2} and inserted in a cell can. An anode of lithium was placed upon the pellet and the cell was sealed. An open circuit voltage measurement of 3.25 volts was obtained. The cell was then discharged through a 330,000 ohm resistor until its output voltage dropped to 1.0 volt, discharge extending over a 140 hour period.

Example 2: A method A cathode was made using 0.0963 gram of active cathode material. The material was air dried without the addition of conductive particulate matter or electrolyte and pressed into a pellet at 3,000 lbs./in.^{2}. The pellet was inserted into a cell can and an anode of lithium placed thereupon. Upon sealing of the can an open circuit voltage of 3.20 volts was obtained. The cell was discharged through a 330,000 ohm resistor until its output voltage dropped to 1.0 volt, discharge extending over a 164 hour period.

Example 3: A method A cathode was made using 0.0115 gram of active material. The material was employed without the addition of conductive particulate matter or electrolyte. Instead of air drying the ethanol slurry of active cathode material, the material was deposited as a slurry and the solvent removed by evaporation. A lithium anode was placed upon the deposit and the cell sealed. An open circuit voltage of 3.17 volts was obtained. Discharge into a 270,000 ohm resistive load extended over a 38 hour period before cell output voltage dropped to 1.0 volt.

Example 4: A method B cathode was made using 0.0550 gram of active material admixed after vacuum filtering with 41% by weight of a 1.0 molar solution of lithium tetrafluoroborate in propylene carbonate. The mixture was hand pressed into a pellet and inserted into a cell can. A lithium anode was placed thereupon and the cell sealed. An open circuit voltage of 2.98 was obtained. Discharge of the cell into a 270,000 ohm resistive load until cell voltage dropped to 1.0 volt extended over a 108 hour period.

Example 5:   A cathode was made by method C above, using 0.0074 gram of active material admixed, after vacuum filtration, with 41% by weight of a 1.0 molar solution of lithium tetrafluoroborate in propylene carbonate. The material was hand pressed into a pellet and inserted into a cell can. A lithium anode was placed upon the cathode and the cell sealed. The cell provided an open circuit voltage of 3.14 volts. Discharge of the cell into a 270,000 ohm resistive load until the cell output voltage decreased to 1.0 volt extended over a period of 37 hours.

From the examples, it will be evident that cells employing a cathode comprised of the active material complex consisting of 2 parts chloranil and 3 parts p-phenylene diamine gives superior performance, providing utilization efficiencies in excess of 20% of the theoretical cell efficiency. The theoretical energy density of the cell of the process has been calculated to be 173 watt hours per pound. The examples also illustrate that improved performance in cells may be expected where the active cathode material was applied to the substrate and the solvent removed by evaporation and where an electrolyte solution is employed in the cathode mix.

It is shown further, in Examples 1, 2 and 3, that the organic charge transfer complex prepared with excess donor material may be employed without adding an electrolyte. In these cases, it is presumed that there is present some of the organic solvent employed in the preparation of the active cathode material and that the lithium anode forms salts therein providing cations rendering the cell ionically conductive to support ionic conduction in the cells.

Sprayed-in-Place Cell with Tellurium Tetraiodide Cathode

A process described by G. Ervin III; U.S. Patent 3,701,685; October 31, 1972; assigned to North American Rockwell Corporation provides a sprayed-in-place thick-film solid state electric cell on a substrate surface which is formed by successively spraying on this surface a silver-containing anode, a silver-ion-conducting solid electrolyte, and a nonmetal cathode containing TeI_4 as electron-acceptor material, using an inert organic solvent. Because of the negligible iodine vapor pressure of the cathode material, once the cell is sprayed in place and suitably encapsulated, it has an almost unlimited shelf-life.

The sprayed-in-place solid state electric cell may be formed on any suitable substrate surface, both conductive and nonconductive. Suitably conductive substrates include tantalum, titanium, and silver. Since silver serves as an electron-donor material, for certain applications using a silver substrate formation of a silver-containing film may be omitted. The solvent containing the silver-ion-conducting solid electrolyte is then directly sprayed onto the silver substrate. Suitable insulating substrate surfaces include alumina, glass, quartz, ceramics, and printed-circuit-board components.

It is considered an essential feature of this process that the cathode film that is formed contain TeI_4, alone or in admixed or complexed form, as electron-acceptor cathode material because of its essentially negligible iodine vapor pressure. These electron-acceptor cathode materials are described more fully in U.S. Patent

3,647,549. As pointed out in the above patent, while TeI$_4$ may be used alone as electron-acceptor material in the cathode composition, it is generally preferred to use it in complexed form. Substantially any material which does not interfere with the electrochemical cell reaction, such as by decomposing the solid electrolyte, may be used to form a mixture or complex with TeI$_4$. The electron-acceptor components of the cathode compositions characterized as complexes of TeI$_4$ may constitute simple mixtures, single-phase solid compounds, or multiphase mixtures of several such compounds. For example, a tetravalent tellurium heterohalide may be utilized, e.g., TeCl$_3$I, and this is regarded as a TeI$_4$ complex, TeI$_4 \cdot 3$TeCl$_4$, for the purpose of the process.

Tellurium tetraiodide itself may be readily prepared by reaction of tellurium metal with elemental iodine. In preparing the various preferred tellurium tetraiodide complexes, TeI$_4$ may be directly reacted in the solid state with the MI or QI component in desired selected proportions. Alternatively, a solid state reaction in a closed vessel may be performed wherein tellurium metal, elemental iodine and the desired MI or QI material are reacted in suitable proportions at an elevated temperature, suitably between 100° and 200°C. to form the desired cathode component. At the same time, carbon and electrolyte material may be optionally and preferably included with the electron-acceptor component so as to provide a final cathode composition consisting of a mixture of complexed tellurium tetraiodide component, carbon, and electrolyte material.

While silver alone may be used as anode material, preferably it is intermixed with solid electrolyte material and finely divided conductive carbon. Particularly preferred as silver-ion-conducting solid electrolyte because of their high conductivity are the ionically conductive compositions shown in U.S. Patent 3,443,997 and 3,476,606, i.e., the alkali metal silver iodide and organic ammonium silver iodide electrolytes, respectively.

The most advantageous results are obtained when both the anode and the cathode films are of composite structure and contain finely divided carbon dispersed therein. Generally a solid electolyte material is also dispersed in the anode, and optionally in the cathode. Thus for a typical preferred thick-film solid state cell, the anode film consists of an intimate mixture of silver, solid electrolyte material such as RbAg$_4$I$_5$, and finely divided conductive carbon. In U.S. Patent 3,503,810 is shown a method of preparing a suitable anode composition. The electrolyte film consists of RbAg$_4$I$_5$, and the cathode film consists of a mixture of Rb$_2$TeI$_6$, RbAg$_4$I$_5$, and carbon.

Example 1: The preparation of thin film of RbAg$_4$I$_5$ is as follows. RbAg$_4$I$_5$ (5 to 6 grams) was dissolved in 30 ml. warm pyridine containing 1 ml. of water. A glass slide was prewarmed to 40°C., and the solution was sprayed onto this substrate using an artist's air brush. The thickness of the deposited film was varied by repeated spray applications, the glass slide being warmed briefly after each application. X-ray analysis of the film showed it to be pure RbAg$_4$I$_5$.

The foregoing procedure was also used to deposit a thin film of the RbAg$_4$I$_5$ electrolyte on a silver substrate surface. Using similar procedures, thin films of KAg$_4$I$_5$, NH$_4$Ag$_4$I$_5$ and the quaternary ammonium electrolytes may be prepared.

Example 2: Film deposition from acetone solutions is described as follows. To

1 liter of predried acetone was added 124.3 grams RbI and 415.8 grams AgI corresponding to a molar ratio of RbI·3AgI, the solution being heated with constant stirring. Any precipitated material was allowed to settle, and the clear supernatant was decanted. The final solution contained 36.9 grams electrolyte per 100 ml. of acetone. To different portions of the solution was added about 1% by weight of methyl methacrylate resin. These solutions, both with and without the synthetic resin, were sprayed on various substrates including glass, titanium, tantalum, nickel alloy, pressed zirconium carbide pellet, pressed carbon pellet, silver, and a pressed standard anode pellet consisting of carbon, $RbAg_4I_5$, and silver. In all instances smooth adherent conductive coatings between 25 and 75 microns in thickness were obtained.

In another series of runs, about 10% by weight of carbon was added to the 3:1 AgI-RbI acetone solution and sprayed on glass and titanium substrates using an artist's air brush. Smooth, uniform adherent films were obtained on volatilization of the acetone. The dried carbon-electrolyte coating was then sprayed with the AgI-RbI electrolyte solution in acetone; a second adherent conductive uniform coating was obtained. After this second coating was dried, it was sprayed with the carbon-electrolyte solution in acetone to form a third adherent uniform coating.

Example 3:   The preparation of sprayed thick film cell is as follows. A finely divided intimate mixture containing 2 parts of silver to 1 part of carbon was prepared by the chemical reduction of silver oxalate in the presence of carbon black. An anode spray was prepared by dissolving 5 grams of this silver-carbon mixture in 50 ml. of a solution of electrolyte in acetone prepared as in Example 2. The cathode spray mixture consisted of 300 grams of $RbI·3AgI$ in 750 ml. of predried acetone to which was added 2 grams of predried carbon black and 2 grams of Rb_2TeI_6.

The electron-acceptor component had been prepared by reacting tellurium dioxide with rubidium iodide in a 1:2 molar ratio in a hydriodic acid solution. All the spray mixtures were made up in a nitrogen-purged dry box, and dried nitrogen gas was used with a commercial air brush. Squares of a high temperature nickel alloy were sprayed with the cathode mix, and copper squares were sprayed with the anode mix. Dried films were formed by removing the acetone by treatment under vacuum at a temperature below the $RbAg_4I_5$ electrolyte melting point (232°C.). The formed anode and cathode films on the squares were then sprayed with the RbI-AgI electrolyte solution in acetone, and cells were assembled consisting of sequential layers of copper base plate, anode film, electrolyte film, cathode film, and nickel alloy base plate.

A typical experimental cell thus contained as the anode film a mixture of silver, $RbAg_4I_5$, and carbon black; as the electrolyte film $RbAg_4I_5$; and as the cathode film a mixture of Rb_2TeI_6, carbon black, and $RbAg_4I_5$. The cell had a thickness of 635 microns, including the nickel alloy and copper back-up plates, and the electrode cross-sectional area was 3.6 cm.^{2}. This cell had an open circuit voltage of 0.53 volt at room temperature. When discharged at 0.6 ma. it yielded 15 minutes of operation to 80% of the open circuit voltage. This cell was then recharged and an identical discharge curve was obtained on the second cycle. Tests on other similarly prepared cells yielded flash currents as high as 300 ma.

In a similar manner, a solid state capacitor may be prepared by spraying successive layers of anode, electrolyte, and cathode, the cathode film consisting of only carbon and electrolyte and containing no active electron-acceptor cathode material. The thick-film cells are generally further processed by encapsulation in a suitable container or by spraying the edges of the cell with a suitable potting compound such as an epoxy resin to seal the cell against entrance of moisture. Thereby the cells have a substantially unlimited shelf-life because of the noniodine-yielding electron-acceptor component present in the cathode film. Also, where the active electron-acceptor cathode component is omitted from the cathode composition, thick-film solid state capacitors may be formed using the present sprayed-in-place techniques.

Furthermore, the cells are particularly advantageous for operation at high temperatures compared with cells using polyiodide cathodes in that high temperature operation with such latter cells is limited by the lower melting point of the polyiodide cathodes. For example, continuous operation at temperatures above 100°C. is not ordinarily feasible with many of these polyiodide materials. Furthermore, because of the iodine volatility and the resulting porosity of any formed film and the need to contain evolved iodine, such polyiodide cathodes cannot be utilized to make thick-film solid state cells. The thick-film solid state cells of the process containing the noniodine-yielding tellurium tetraiodide type electron-acceptor material in the cathode are operable to the melting point of $RbAg_4I_5$ (232°C.).

In related work, J.H. Christie and J.R. Humphrey; U.S. Patent 3,647,549; Mar. 7, 1972; assigned to North American Rockwell Corporation describe an electron-acceptor material selected from tellurium tetraiodide and its complexes as a cathode component for a solid state electric cell containing a silver anode and a silver-ion-conducting solid electrolyte. The tellurium tetraiodide compositions utilized as electron-acceptor cathode component consist of tellurium tetraiodide alone or as a complex of tellurium tetraiodide preferably with selected alkali metal iodides or organic ammonium iodides. The cathode materials of this process are particularly suited for the preparation of and use in thick-film and thin-film solid state batteries as well as in pellet-type cells because of their negligible iodine vapor pressure.

Elemental Iodine and Cyanides of Zinc or Cadmium

A process described by D.V. Louzos and G.W. Mellors; U.S. Patent 3,655,453; April 11, 1972; assigned to Union Carbide Corporation relates to cathodes for solid state batteries. The theoretical advantage of batteries composed only of dry solid materials offer such promise that considerable effort has been expended in recent years to provide materials which would make possible the practical attainment of such advantages.

For example, electrolytes having ionic conductivity superior to that of silver iodide used in the cell of U.S. Patent 2,894,053 have recently been developed such as binary silver iodide-silver sulfide compositions, described by Reuter and Hardel in Naturwissenschaften 48, 161; the binary silver iodide-potassium iodide materials described in U.S. Patent 3,443,997 and British Patent 1,140,938; the binary silver iodide-alkali metal cyanide compositions described in U.S. Patent 3,582,291. In U.S. Patent 3,653,968, cells utilizing such electrolytes are described in which the cathode is an iodine complex such as iodine polyhalides and charge transfer complexes containing iodine. This process comprises a material suitable for use as the cathode in solid electrolyte cells using any of the electrolytes described in the patents referred to, which material is the reaction product of elemental iodine and a metal cyanide, for instance of zinc or cadmium or alkali metals or mixtures of cyanides of zinc, cadmium and alkali metals. Specific examples of the materials are the reaction products of iodine and zinc cyanide [$Zn(CN)_2$], iodine and potassium zinc cyanide [$KZn(CN)_3$] and iodine and potassium cyanide (KCN) and iodine and cadmium cyanide [$Cd(CN)_2$]. These products may be obtained by heating substantially equal molar proportions of the element and the respective cyanide at 220°C. in a sealed vessel in the absence of water for several hours until reaction is complete. This procedure results in the production of a solid having a characteristic brown to black color and to a degree the characteristic appearance of a sintered mass. It has a distinctly lower vapor pressure of iodine at room temperature than elemental iodine.

Evidence that the reaction products so produced are true compounds is shown by infrared spectra. For instance, the reaction product of iodine and zinc cyanide so produced has an infrared spectrum with five distinct peaks at wave numbers 1,620, 1,340, 1,145, 1,095 and 800 cm.$^{-1}$. No evidence of CNI or ZnI_2, both of which compounds are known to form when iodine and metal cyanide are reacted in aqueous solution exist in the spectrum. When the reaction product of I_2 and $KZn(CN)_3$ produced as described is heated the characteristic color of iodine is absent even when the material is heated to 160°F. in a sealed glass tube. Of the specific examples given, this material appears to be the most stable at elevated temperatures. All of the materials are quite stable at room temperature and substantially no loss of iodine occurs on storage at room temperature.

A number of cells using the material of the process as cathode have been prepared. Thus, a cathode pellet was made of equal quantities of the iodine-cyanide product and the selected electrolyte with powdered graphite and acetylene black. The anode was made in a similar way, for example by preparing a pellet of powdered silver (0.8 to 1.2 micron fineness), electrolyte and a mixture of acetylene black and graphite. Between anode and cathode was placed a layer of the selected electrolyte. In the following table are set forth examples of cell systems produced together with test data showing their operability. For comparison a test made utilizing the compound CNI as cathode is included. This system was totally inoperative not even showing an open circuit voltage.

Cell system	Open circuit voltage	Polarization scans— Short term Operation established	
		mA./in.²	At volts
Ag/KAg₄I₄CN/Zn(CN)₂+I₂	0.67	50	0.50
Ag/KAg₄I₄CN/KZn(CN)₃+I₂	0.61	14	0.48
Ag/KAg₄I₄CN/KCN+I₂	0.67+	73	0.55
Ag/KAg₄I₄CN/CNI	0	0	0
Ag/4AgI+KI+Zn(Cn)₂/KZn(CN)₃+I₂	0.67	57	0.50
Ag/4AgI+KI+Zn(Cn)₂/KCN+I₂	0.68+	84	0.63
Ag/RbAg₄I₅/KZn(CN)₃+I₂	0.60	7	0.52
Ag/KAg₄I₄CN/KAg₄I4CN+I₂	0.58	2.2	0.51

It will be observed from the table that the cell Ag/4AgI–KI–Zn(CN)₂/KCNI₂ performed particularly well. In a further test of a cell of this system, the cell was continuously discharged at 72°F. across 4,000 ohms resistance, a current drain of 0.17 ma. Cell voltage of 0.68 volt was maintained substantially constant for 100 hours. At 120 hours the voltage had fallen to 0.65. Cell voltage dropped to 0.4 volt between 120 and 130 hours.

Polyhalides of Iodine

A process described by D.V. Louzos; U.S. Patent 3,653,968; April 4, 1972;

assigned to Union Carbide Corporation relates to solid state electrochemical cells, as well as solid state batteries comprising an assembly of such cells. Previous workers have described compounds and mixtures of compounds which are solid ionic conductors at room temperature and have specific conductances sufficiently high to permit their use in practical battery applications. In addition, many of these compounds are less sensitive to moisture than those which have previously been available.

It has been found that the combination of a silver or copper anode and specific solid iodine-containing compounds having at least a fraction of an iodine atom at a valence of zero as a cathode yield solid state cells which display good shelf-life stability and are powerful enough for practical battery applications.

Referring to Figure 6.10a, there is shown a solid state cell, generally designated (10), formed of three layers, an anode (12), an electrolyte layer (14) and a cathode (16), each shown in nonscaler, simplified form. The anode (12), is a layer of metallic silver which may be a thin sheet of silver foil or a thin layer of silver metal deposited on one side of electrolyte layer (14). The electrolyte layer (14) is a thin disc formed by compressing powdered electrolyte, e.g., monopotassium tetrasilver tetraiodide monocyanide (KAg_4I_4CN). The cathode (16) is also a pressed powder disc formed from an iodine-containing compound or complex, e.g., 2 perylene·3 iodine. The three layers of cell (10) are held together by intimate contact between the particles at the interfaces between the various layers which is achieved by compression at the time of manufacture. Thereafter, the cell will function without external pressure being applied.

To operate cell (10) is is merely necessary to make electrical contact to the anode (12) and the cathode (16). If desired, several cells can be formed into a stack in the conventional manner and electrical contact would be made to the anode and cathode at opposite ends of the stack.

Figure 6.10b shows a more refined cell construction employing additional parts, but the basic components of this cell, generally designated (20), will function in much the same manner as the three layers of the simplified cell of Figure 6.10a. Cell (20) comprises an electrolyte layer (22), which is identical to electrolyte layer (14) of cell (10), having an anode assembly on one side and a cathode assembly on the other side.

The anode assembly comprises a disc which is a pressed silver powder anode (24) adjacent the electrolyte layer (22) and an anode collector (26) on the outside surface of anode (24). The collector can be any compatible electrically conductive material, e.g., silver metal foil. The cathode assembly comprises a disc which is an iodine-containing pressed powder cathode (28), e.g., 2 perylene·3 iodine complex having a metallic cathode collector (30), e.g., a thin sheet of nickel, on its outer surface.

The entire cell assembly is compressed during manufacture to establish good electrical contact and a physical bond between the various layers and the cell is thereafter encapsulated in plastic (32) to form a sealed unit cell. The layers of plastic covering the anode collector (26) and cathode collector (20) may be punctured to allow external contact to be made with the cell, e.g., by filling the punctures with a conductive epoxy sealer. A preferred anode mixture can be prepared from silver powder mixed with powdered electrolyte, acetylene black and graphite. These components

FIGURE 6.10:   SOLID STATE ELECTROCHEMICAL CELLS

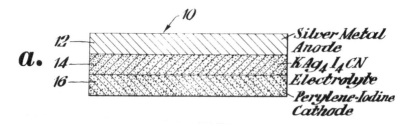

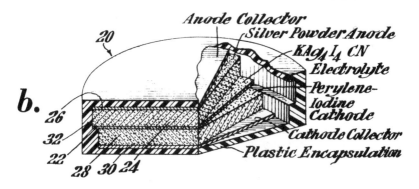

(a) Cross-Sectional View of Solid State Cell
(b) Cross-Sectional View of Packaged Solid State Cell

Source: D.V. Louzos; U.S. Patent 3,653,968; April 4, 1972

are first dried over phosphorus pentoxide in a nitrogen atmosphere and mixed to form a uniform composition. A typical preferred formulation would be 0.70 gram silver powder, 0.15 gram electrolyte, 0.05 gram graphite and 0.001 gram acetylene black.

It has been found that by using a fine silver powder the capacity of the cell is substantially increased. This improved performance is probably due to the resulting increased surface area of the anode–electrolyte interface. It has been found, for example, that by using silver powder of 0.9 to 1.4 microns in diameter the capacity of the cell is 2 1/2 times greater than that of a cell using 5 to 8 micron silver powder. The most preferred silver powder has a discrete particle size in the range of 0.9 to 1.4 microns.

In general, solid electrolytes useful in the cells of the process conform to the general formula $MCN \cdot 4AgI$ where M is potassium, rubidium or cesium or a mixture. Illustrative of these compounds are $KCN \cdot 4AgI$, $RbCN \cdot 4AgI$, $CsCN \cdot 4AgI$, etc. Cathode materials useful in the cells of the process are those compounds or complexes of iodine which have at least a fraction of an iodine atom at a valence of zero. Particularly useful are the so-called polyhalides of iodine and charge transfer complexes containing iodine. Particularly useful polyhalides are cesium triiodide, potassium triiodide, rubidium triiodide, cesium pentaiodide, cesium ennea-iodide, tetramethylammonium triiodide, tetramethylammonium pentaiodide and tetramethyl-ammonium ennea-iodide.

The second general class of materials suitable as cathode materials are the so-called charge transfer complexes. The perylene-iodine and phenothiazine-iodine complexes are particularly preferred and include such complexes as: perylene·I_2, 2 perylene·$3I_2$, perylene·$2I_2$, phenothiazine·I_2, 2 phenothiazine·$3I_2$, phenothiazine·$2I_2$ and perylene·$1/4I_2$.

Example 1:   A Ag/KAg$_4$I$_4$CN/2 perylene·$3I_2$ cell was constructed substantially similar to that shown in Figure 6.10b. The anode was a mixture of 0.70 gram of silver powder having a discrete particle size in the range of 0.9 to 1.4 microns, 0.15 gram powdered KAg$_4$I$_4$CN electrolyte, 0.05 gram graphite and 0.001 gram acetylene black. These components were individually dried over P_2O_5 in a dry box having a nitrogen atmosphere and mixed together in a mortar and pestle. The resulting mixture was then doctor-bladed in a cylindrical mold with a silver sheet substrate as a collector.

The cathode was fabricated in much the same manner as the anode and was formed from 1.0 gram 2 perylene·$3I_2$, 2.0 grams KAg$_4$I$_4$CN electrolyte, 0.6 gram graphite and 0.01 gram acetylene black. After being dried and mixed the powdered cathode mix was doctor-bladed in a cylindrical mold on the nickel sheet substrate. The electrolyte was 1.1 gram of powdered KAg$_4$I$_4$CN compressed into a disc 0.004 inch thick.

Anode, cathode and electrolyte subassemblies were individually compressed and were assembled and again compressed. The entire assembly was then encapsulated in polymerized methyl methacrylate. The completed cell measured 1.0 inch in diameter and 0.1 inch total thickness. Electrical contact with the cell was made by drilling through the methyl methacrylate layers covering the anode and cathode collectors to contact the metal foils. The cell had an open circuit voltage of 0.64 volt, a flash current of 0.75 amp. per square inch and discharge was possible for short periods at 50 milliamperes per square inch.

Example 2:   A Ag/KAg$_4$I$_4$CN/RbI$_3$ cell was fabricated in the manner described in Example 1 except that rubidium triiodide was substituted for the 2 perylene·$3I_2$. The cathode mix was 0.418 gram of rubidium triiodide, 0.827 gram of KAg$_4$I$_4$CN electrolyte, 0.310 gram of graphite, and 0.006 gram of acetylene black. The electrolyte layer was 0.55 gram of KAg$_4$I$_4$CN. The cathode mix was 0.645 gram of silver powder, 0.826 gram KAg$_4$I$_4$CN, 0.201 gram graphite and 0.006 gram acetylene black. The cell had an open circuit voltage of 0.67 volt, a flash current of 0.70 amp. per square inch and discharge was possible for short periods at 50 milliamperes per square inch.

Example 3:   A Cu/KAg$_4$I$_4$CN/2 perylene·$3I_2$ cell employing a copper anode was assembled in substantially the manner described in Example 1. The anode mix consisted of 0.645 gram of 100 mesh copper powder, 0.593 gram of KAg$_4$I$_4$CN electrolyte, 0.201 gram of graphite and 0.006 gram of acetylene black well mixed and compressed on a copper sheet anode collector. The electrolyte was a compressed layer of 1.1 gram of KAg$_4$I$_4$CN. The cathode mix was 0.323 gram of 2 perylene·$3I_2$,

0.827 gram of KAg_4I_4CN electrolyte, 0.310 gram of graphite and 0.006 gram of acetylene black compressed on a cathode collector of porous nickel. The open circuit voltage of this cell was 0.603 volt and the cell delivered for a short time 1.9 ma. at 0.57 volt on a 300 ohm load and 9.4 ma. at 0.47 volt on a 50 ohm load.

Lithium Iodide–Ammonium Iodide Separator

In a process described by C.C. Liang and J. Epstein; U.S. Patent 3,513,027; May 19, 1970; assigned to P.R. Mallory & Company, Incorporated a solid electrolyte battery is provided with a separator having the composition $4LI-NH_4I$, having a high conductivity. It is particularly useful with a cathode of composition MAg_4I_5, where M may be either NH_4 or K or Rb; and with an anode of active metal, such as lithium.

Cathode: The depolarizer is a complex of the composition MAg_4I_5 (M = NH_4, K or Rb). The conductivity of this complex is 10^{-2} ohm^{-1} cm.$^{-1}$ at room temperature. In addition to MAg_4I_5, any heavy metal or transition metal iodide such as PbI_2, HgI_2, CuI, AgI, can also be used as a depolarizer material. The cathode should comprise a mixture of the depolarizer and the $4LiI \cdot NH_4I$ electrolyte. The use of the electrolyte material in the cathode is adapted to decrease the polarization of the electrode and facilitate the ionic conduction.

Separator: The separator is a complex or solid solution of the composition $4LiI \cdot NH_4I$ which can be made by slowly drying an aqueous solution containing 80 mol percent of LiI and 20 mol percent of NH_4I. The conductivity of this complex or solid solution is 2×10^{-6} ohm^{-1} cm.$^{-1}$ at room temperature which is higher than that of either LiI (10^{-7} ohm^{-1} cm.$^{-1}$) or NH_4I (10^{-8} ohm^{-1} cm.$^{-1}$).

The complex MAg (M = NH_4, K or Rb) is known to have a high ionic conductivity and has been used as a separator of solid electrolyte cells. However, due to the presence of Ag^+ ions, the complex can be used with a Ag anode only. More active metals such as Zn, Mg, Li, etc. would be oxidized by the Ag^+ complex. Therefore, the conventional use of the complex MAg_4I_5 as a separator is very limited. In this process, the complex MAg_4I_5 is used as the cathode material together with the active anodes such as Li, and a cell voltage higher than 2.1 volts can be obtained. The complex or solid solution $4LiI \cdot NH_4I$ is suitable as a separator for high voltage solid electrolyte cells. It does not react with active metals such as Li yet its conductivity is higher than that of either LiI or NH_4I.

Operation: A test cell, $Li/NH_4I \cdot LiI/NH_4Ag_4I_5$ was made and its physical dimensions were: surface area, 1.2 cm.^{2}; thickness of the Li anode, 0.04 cm.; thickness of the $NH_4I \cdot 4LI$ separator, 0.065 cm.; and thickness of the $NH_4Ag_4I_5$ cathode, 0.065 cm.

The open circuit voltage of the test cell was 2.1 volts, as expected by thermodynamic calculation. The cell voltage under a 100 kilohm load was 1.5 volts at room temperature. No significant change in load voltage was observed after the cell has been discharged continuously for five days. In addition to the anode, other active metals such as Mg, La, Ba and Zn can also be used as an anode material. The current capability of the solid electrolyte system of this process represents a considerable improvement over the system using LiI as the separator. If a 0.065 cm. thick LiI separator were used in the cell, the cell voltage under the same 100 kilohms would be 0.29 volt. The significant aspect of this process is that the separator material NH₄I·4LiI, unlike the Ag complex, can be used in solid electrolyte systems with anodes more active than Ag. The conductivity of the complex is higher than any known separator in its class. The details of construction of a cell and the assembly of a battery unit, together with a graph of the voltage under certain test conditions, are shown in Figure 6.11.

FIGURE 6.11:   SOLID ELECTROLYTE BATTERY

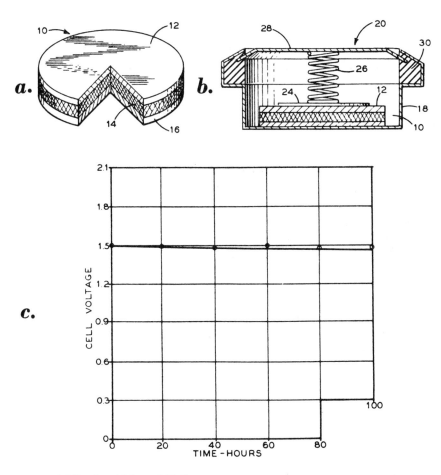

(a) Sectional View of Cell
(b) Sectional View of Battery Unit
(c) Cell Voltage Under a Load of 100 Kilohms as a Function of Time

Source: C.C. Liang and J. Epstein; U.S. Patent 3,513,027; May 19, 1970

As shown in Figure 6.11a, a typical cell structure (10), according to the process, is provided with an anode (12), a separator (14), and a cathode (16) having the chemical composition specified above for the test cell.

Figure 6.11b shows a cell (10) assembled in a can or metal container (18) to serve and be operable as a battery unit (20). The cell (10) is shown seated with its cathode (16) physically and electrically in contact with the metal floor (22) of the can (18), over the full area of the cathode for low resistance contact. A contact plate (24) engages and seats on the top exposed surface of the anode (12) to serve as a terminal for the anode. A compression spring (26) is disposed between contact plate (24) and a disc plate (28) that serves as a cover to close can (18), and that also serves as an accessible outer terminal for anode (12). The disc plate (28) also serves as a physical reaction element to hold the spring compressed to maintain any desired pressure on the contact plate (24). Both ends of contact spring may be soldered or welded to the two plates (24) and (28) for good electrical contact.

The outer anode terminal plate (28) is supported at its peripheral rim edge border on suitable insulating and sealing material (30) that seals the can closed and separates the anode terminal plate (28) from the metal of the can (18). The space within the can may be suitably filled with an inert setting plastic material, to hold the cell (10) and the related elements in place, and may be part of the insulating material of seal (30). Figure 6.11c shows a graph of the load voltage of the cell after discharge for five days through a load of 100 kilohms, and illustrates the effectiveness of the depolarizing action that prevents dropping of the voltage curve.

SODIUM-SULFUR WITH SOLID ELECTROLYTES

Sintered Polycrystalline Plate

J.T. Kummer; U.S. Patent 3,468,709; September 23, 1969; assigned to Ford Motor Company describes high energy density batteries which employ a molten alkali metal as the anodic reactant. In particular, this process is concerned with improvements in the cell or cells which employ a molten alkali metal anodic reactant and a sulfur-comprising cathodic reactant-electrolyte that is electrochemically reversibly reactive with cations of the alkali metal.

This type of cell is separated into an anodic reaction zone and a cathodic reaction zone by a solid electrolyte, preferably a crystalline object or membrane that is selectively conductive to cations of the alkali metal anode-reactant and essentially impermeable to other contents of these zones, e.g., the anodic reactant in elemental form, anions of the cathodic reactant and the cathodic reactant in elemental or compound form.

In the discharge half cycle of the cell, the alkali metal anodic reactant, e.g., molten sodium, releases electrons to the anode and thence to the external circuit with resultant formation of cations in the anodic reaction compartment. Sulfur atoms in the cathodic reaction zone accept electrons from the cathode and external circuit forming anions. The positively charged alkali metal ions are attracted to the cathodic reaction zone and pass through the cationically-conductive solid electrolyte to associate with the negatively charged sulfur ions. In the charging half cycle, the current flow is reversed by applying an extraneous source of electrical energy and the alkali metal ions are driven back through the solid electrolyte into the anodic reaction zone. The solid electrolyte employed may be a polycrystalline object formed by sintering crystals consisting essentially of a structural lattice and sodium ions which are mobile in relation to such lattice under influence of an electric field. The structural lattice consists essentially of a major proportion by weight of ions of aluminum and oxygen and a minor proportion by weight of a metal having a valence not greater than 2 in crystal lattice combination, e.g., lithium or magnesium. The solid electrolyte is prepared in plate form as follows.

(1) In powdered form Na_2CO_2, $LiNO_3$, and Al_2O_3 were added to a vessel and mechanically mixed for 30 minutes. The Al_2O_3 employed was in the form of 0.05 micron particles (Linde B). The weight percent of the individual oxides employed were Li_2O, 0.99; Na_2O, 9.99; and Al_2O_3, 39.02.
(2) The mix was heated at 1250°C. for one hour.
(3) The sample was mixed with a wax binder (Carbowax) and mechanically pressed into flat plates.
(4) The plates were then isostatically pressed at 90,000 psi.
(5) The wax binder was removed by slowly heating the plates to 550°C.
(6) The plates were sintered in an electric furnace at 1460°C. During sintering, the plates were kept in a covered crucible in the presence of packing powder of the same composition as the mix.

Several cell designs are described to minimize the internal impedance with increased power output capacity.

Alumina with Sodium Ion Conduction

A process described by A. Iwai, M. Harata, Y. Ogawa, and S. Hasegawa; U.S. Patent 3,671,324; June 20, 1972; assigned to Tokyo Shibaura Electric Company, Limited, Japan involves a β-alumina solid electrolyte having excellent electrical conductivity and sintering property, which is suited for use, for example, in Na-S batteries.

β-Alumina is known to be a solid electrolyte having a good conductivity of Na^+ ion. The crystal system of β-alumina is in a hexagonal form, and no electrical conductivity due to the movement of electrons is observed. However, in the basal plane, the mobility of Na^+ ion is large, and electrical conductivity due to the movement of Na^+ ion is 0.03 ohm^{-1} cm.$^{-1}$ at room temperature and 0.3 ohm^{-1} cm.$^{-1}$ at 300°C. On the other hand, Na^+ ions do not at all move in the direction of the basal axis. Since β-alumina has a high conductivity of Na^+ ion, a battery can be constructed by positioning β-alumina between an anode of molten metal sodium and a cathode of sulfur.

Further, β-alumina functions as a secondary battery or a storage battery by being sandwiched between plates of alkali ferrite, and a sintered mass of β-alumina is used for this purpose. The orientation of fine crystals forming the mass is not generally constant. Thus the electrical conductivity of the sintered mass of β-alumina is less than the value obtained in the basal plane of a single crystal, and is of the order of 0.01 to 0.1 ohm^{-1} cm.$^{-1}$ at 300°C. although it is more or less dependent upon the state of sintering. It is generally desired that the conductivity of β-alumina is high when the conduction of Na^+ ion is to be utilized. It is known that the conductivity of Na^+ ion can be promoted by the addition of magnesium oxide to β-alumina. More particularly, the electrical conductivity of the sintered mass without any additive ranges 0.001 to 0.002 ohm^{-1} cm.$^{-1}$, while when MgO is added the same can be increased to 0.005 ohm^{-1} cm.$^{-1}$ at room temperature by properly selecting the amount to be added. Addition of Mg to β-alumina may create an advantage in respect of electrical conductivity, but produces the disadvantages that sintering property is inferior, physical strength is poor and that considerably high sintering temperature is required, all due to a low degree of shrinkage.

This process provides a β-alumina solid electrolyte utilizing the phenomenon of conduction of Na^+ ion, where 0.9 to 20 mol percent of CuO is added and baked, the amount being calculated in terms of the amount of a mixture of NaO, Al_2O_3 and CuO.

Example 1 through 10: To a β-alumina powder of a particle size of 1.2μ are added an additive to be described later and sodium carbonate. The powdery mixture is press molded into a pressed mass. The powdery mass is then subjected to preliminary sintering by placing it in a sealed envelope and heating it for one hour at a temperature of 1400°C. The sintered mass is pulverized and press molded again to form a pressed mass. The mass thus obtained is introduced into a sealed envelope and baked for 4 hours at 1650°C., avoiding a loss of sodium due to evaporation. The ratio of mixed ingredients of the baked mass and the electrical conductivity by Na^+ ion at 350°C. was measured using molten sodium as electrodes. The results are shown below.

Example	β-Alumina (mol percent)	Additives (mol percent)	Sodium carbonate (mol percent)	Electrical conductivity at 350° C. ($\Omega^{-1}cm.^{-1}$)
1	87.6(98.8)		12.4(1.2)	4.5×10^{-2}
2	75.7(75.7)	CuO: 9.1(0.96)	15.(15.2)	5.3×10^{-2}
3	69.4(69.6)	NiO: 13.9(1.6)	15.7(1.8)	5.6×10^{-2}
4	69.4(69.6)	CuO: 13.9(1.6)	16.7(1.8)	6.5×10^{-2}
5	57.5(94.4)	CuO: 23.0(3.1)	19.5(2.5)	8.6×10^{-1}
6	42.7(90.2)	CuO: 34.2(5.9)	23.1(3.9)	1.00×10^{-2}
7	37.9(88.2)	CuO: 37.9(7.2)	24.2(4.6)	1.01×10^{-1}
8	21.1(76.7)	CuO: 50.6(15.0)	28.3(8.3)	5.9×10^{-2}
9	16.8(71.2)	CuO: 54.0(18.6)	29.2(10.2)	5.1×10^{-2}
10	5.1(39.8)	CuO: 62.6(40.0)	32.3(20.2)	(1)

[1] Immeasurable due to Na attack on sample.

The values in the parentheses represent those in mol percent in the case where the mixture should consist of Na_2O, Al_2O_3 and NiO or CuO. In the table, Example 1 shows a case in which no additive is used, Example 3 a case in which NiO is added as the additive, and Example 10 a case where the additive is used in an excess amount lying outside the scope of the process. In the composition of each of the species given in the table, sodium was introduced in order to compensate the lack of valence due to the introduction of Cu(divalent) to the position of Al(trivalent) in the crystalline lattice of β-alumina. The ratio of incorporation is as follows.

$$1.21Na_2O/11Al_2O_3/(0.14+x/2)Na_2CO_3/xCuO$$

It has been found through experiments that the ratio of Na_2O and Al_2O_3 is preferably from 1.41:11 to 3.11:11, respectively as the amount of incorporation of CuO increases from 0.9 to 20 mol percent. In addition to sodium carbonate as described above, sodium salts such as sodium nitrate, sodium sulfate, sodium oxalate, sodium chloride or the like may be used as the source of supply of sodium. x represents the amount of CuO added; $1.21Na_2O \cdot 11Al_2O_3$ indicates the chemical composition of the β-alumina used. The amount of sodium carbonate to be incorporated consists of the amount $(x/2)$ of Na to be introduced into the crystalline lattice of the β-alumina and the amount (0.14) to be decreased due to evaporation during the baking process. When the amount of Na is less than the ratio given above, $CuAl_2O_4$ is observed in the phase of formation, while an excess amount results in the production of $NaAlO_2$. In either case, the substance produced will act to increase the resistivity of the electrolyte.

An increase in the electrical conductivity of β-alumina is entirely dependent upon Na^+ ion. In order to ascertain that the contribution of electronic conduction can be ignored, variation of current on a time basis was measured by applying a DC electric field to electrodes vapor-deposited with gold at both ends of a rod-shaped specimen piece.

It was observed that the intensity of the current decayed as the time lapsed and was less than 0.3% of the initial current after 3 hours in respect of all of the specimen pieces. In other words, since Na^+ ion was only concerned with the current, the external electrical field was gradually canceled by the polarization due to movement of the Na^+ ion, thus resulting in sharp decay in current. This shows that electronic conduction can be ignored in the examples described and that an increase in electrical conductivity can be taken as an increase in ion conduction.

X-ray diffraction analysis of each specimen indicates that only a so-called β-alumina phase (equivalent to $Na_2O \cdot 11Al_2O_3$) is observed when the amount of the additive which is calculated as being mixed with Na_2O, Al_2O_3 or the like is 6 mol percent and that a phase equivalent to $Na_2O \cdot 5Al_2O_3$ is observed with the amount in excess of that range.

If the amount exceeds 20 mol percent, spinel and sodium aluminate begin to be produced to coexist and will degrade electrical conductivity. Further, when the additive, i.e., CuO, amounts to more than 40 mol percent, the substances formed will consist of spinel, sodium aluminate and the additive itself, so that they exhibit no anticorrosive power to metal sodium and hence can not be applied to Na-S batteries. No sufficient advantage can be expected when the amount of incorporation of CuO is less than 0.9 mol percent.

In order to explain the sintering property of the β-alumina, the rates of shrinkage with the addition of Ni, Mg and Cu, respectively, will be compared in Figure 6.12. Figure 6.12 shows the rates of shrinkage of β-alumina having a particle size of 1.2μ in respect of compositions of baked material added with the above described impurities together with sodium. Baking was effected for 4 hours at 1700°C. in respect of β-alumina added with Ni or Mg, and for 4 hours at 1650°C. when copper was added.

Figure 6.12 clearly shows that shrinkage is larger when copper is added than when either of magnesium or nickel is employed. Such very large shrinkage is measured in spite of the low baking temperature. The degree of shrinkage will thus be furthered if comparison is made at a temperature of 1700°C.

Shrinkage was also measured on a pressed mass sintered for 2 hours at 1600°C. and found to be 16.7%, the mass being formed of 42.7 mol percent β-alumina, 34.2 mol percent copper oxide, and 23.1 mol percent sodium carbonate. The shrinkage was superior to that of a pressed mass baked at 1700°C. for 4 hours, and porcelain

FIGURE 6.12:   ALUMINA SOLID ELECTROLYTE — COMPARATIVE DATA FOR
LINEAR SHRINKAGE

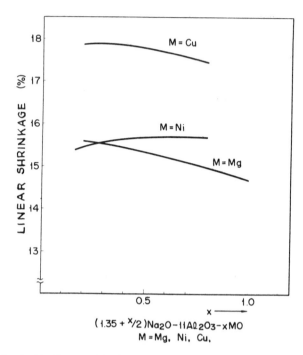

$$(1.35 + {}^{x}/2)Na_2O - 11Al_2O_3 - xMO$$
$$M = Mg, Ni, Cu,$$

Source: A. Iwai, M. Harata, Y. Ogawa, and S. Hasegawa; U.S. Patent
3,671,324; June 20, 1972

having sufficiently low porosity could be obtained. The electrical conductivity
was measured at 350°C. to 8.1 x 10^{-2} ohm^{-1} cm.$^{-1}$. β-Alumina is required to
have sufficiently low internal resistivity and a large physical rigidity so as to be
suited for a solid electrolyte. In order to satisfy these conditions, a sintering temperature above 1700°C. is necessary when MgO is to be added. When CuO is used,
however, a temperature of 1600°C. for sintering is sufficient to achieve the purposes, and hence the use of this material is particularly advantageous to manufacture.

Tubular Hollow Glass Fibers and Impregnated Glass Separator

W.E. Brown, R.G. Heitz, and C.A. Levine; U.S. Patent 3,679,480; July 25,
1972; assigned to The Dow Chemical Company describe a battery assembly which
comprises a sealed container, which is divided or positioned into mutually exclusive
anode and cathode chambers. Partitioning of the anode and cathode chambers is
achieved by means of a solid electrolyte-membrane, sometimes also referred to as
the electrolyte or separator, in the form of a plurality of hollow fine fibers held
together as a bundle by means of a tube sheet or common header. The fibers may
be of glass or ceramic materials. They extend into the cathode chamber but are
sealed at this end. They are in open communication with the anode chamber. For

example, single strands of hollow fibers are sealed at the end extending into the cathode chamber and are open to the anode chamber at the other end. Continuous hollow fibers may be looped into the cathode chamber so that both ends communicate with only the anode chamber.

The electrolyte (separator) is usually made of a material such as a polycrystalline ceramic, amorphous glass or an impregnated matrix, the latter having embedded an essentially nonmigrating salt or liquid which is permeable only to the anode metal ion. The electrolyte (separator) is further characterized as transmitting ions of the alkali metal anode between the anode and cathode chambers but as being substantially nonconducting to electrons. Within the anode chamber is contained a liquid alkali metal, such as sodium or potassium. The cathode chamber contains a reducible material, such as sulfur, selenium, tellurium, tetracyanoethylene, etc.

A preferred anode-cathode-electrolyte system consists of a liquid sodium anode, a liquid sulfur-sodium sulfide mixture as cathode system, e.g., a sodium polysulfide, and a sodium ion conductive glass or ceramic electrolyte. This system is capable of producing energy densities of over 300 watt-hours per pound at operating temperatures as low as 330°C.

A preferred form for the electrolyte is fine, hollow fibers where the individual fibers have an outside diameter/wall thickness ratio of at least 3, ordinarily from 3 to 20 and preferably from 4 to 10. Usually within these ratios, fibers having an outside diameter from 20 to 1,000 microns and a wall thickness of from 5 to 100 microns are used. Such hollow fibers provide a high strength, thin walled membrane and give a high ion conductivity. They also provide a very large surface area to volume ratio. Although less advantageous in the latter respect, fibers as large as 5,000 microns outside diameter and having walls as thick as 1,000 microns can be employed when fabricated from more highly ion conductive materials, e.g., certain porcelains.

A battery cell of the process employing tubular hollow fibers as electrolyte as shown in Figure 6.13a illustrates one form of the process. In this example, a multiplicity of hollow glass fibers (10) fabricated from a conductive glass within the size range set forth above and having their lower ends sealed off are positioned in parallel substantially uniformly spaced apart relationship and sealed into a common header (12).

A molten alkali metal (13), for example sodium, substantially fills the hollow fibers and header. An anode lead (14) is positioned in the header (12) contacting the molten anode (13) and the assembly sealed. The anode and electrolyte-separator assembly is placed in a container (16) which serves as a reservoir for the molten cathode (18) (e.g., sulfur containing sodium sulfide). A cathode lead assembly (20) is positioned within the vessel (16) in contact with the molten cathode material (18) and the entire battery assembly sealed with top assembled (22) so as to be vapor and liquid tight.

To assure that both the anode and cathode are maintained in the molten state ordinarily the vessel (16) is jacketed with an insulating cover (24). Alternatively, if desired, this cover (24) in turn can be fitted with an electrical resistance heater (26) adjacent the outer wall (28) of the vessel (16). In a second form of the battery Figure 6.13b, the electrolyte is in the form of a large number of hollow fine glass

FIGURE 6.13:   ELECTRIC CELL ASSEMBLY USING HOLLOW GLASS FIBERS AS
THE ELECTROLYTE SEPARATOR

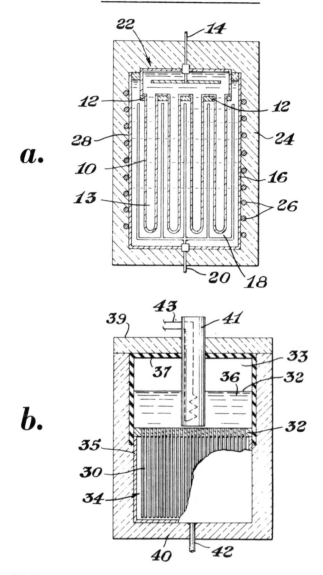

Source: W.E. Brown, R.G. Heitz, and C.A. Levine; U.S. Patent 3,679,480;
July 25, 1972

fibers (30) sealed within a tube sheet (32) prepared from a potassium and sulfur resistant potting resin. The assembled tube bundle is sealed within an electrically
insulating anode can (37), of glass, which in turn is sealed to an electrically conductive cathode metal can (35). Thus, is defined an anode compartment (33)

containing potassium (36) and a cathode chamber (34) containing potassium sulfide (not shown) in exterior relationship to the fibers (30). Extending through, and sealed within, the glass anode can (37) and penetrating substantially into the anode compartment to contact potassium is a graphite anode lead (41) in which there is embedded an electrical, resistance-heating circuit (43) for melting the potassium. The cathode lead (42) is electrically attached to the cathode can (35). The resulting assembly is placed within a thermal insulator comprising a top (39) and bottom (40) of a thermal insulating material.

In the operation of the battery cell, as current is drawn from the battery, the molten alkali metal anode gives up electrons and forms the corresponding metal ions. The electrons go through an external circuit doing work while the resultant alkali metal ions diffuse or otherwise are transported through the thin wall electrolyte and migrate toward the cathode.

At the molten cathode, electrons are fed into the cathode chamber through the cathode lead from the external circuit forming anions with the molten cathode material, for example sulfur. These anions are, in effect, neutralized by reaction with the alkali metal ions migrating through the electrolyte thereby forming the alkali metal salt. This reaction continues through the discharge cycle of the battery.

To recharge the battery, a source of current is attached to the leads so as to feed electrons through the anode lead (41) to the molten sodium or potassium anode and the positive lead from the power source is attached to the cathode lead (42) of the battery. As the voltage of the power source is increased over the battery voltage, the exact reverse of the electrode reactions presented for the discharge cycle takes place. Alkali metal ions pass through the separator; alkali metal is regenerated and the reduced cathode material is oxidized to its original state.

An unexpected advantage of the system particularly when utilizing the sodium-sodium sulfide-sulfur electrodes is that much more rapid recharging of the battery can be carred out without any adverse effects. In conventional lead-acid storage batteries, permanent damage occurs unless a slow trickle charge is applied during the recharge cycle.

In a specific example, a 2 volt cell is prepared using liquid sodium as the anode and sodium polysulfide as the liquid cathode. This cell is designed to operate at a temperature of 300°C. For this cell, 54,000 ion conductive glass fibers each having an inside diameter of 48.2 microns and a wall thickness of 12.05 microns are placed in parallel relationship one to another to provide a center to center fiber spacing of 192 microns. This occupies a cross-sectional area of 20 cm.^{2}. The overall appearance of the parallel shaped fibers is a cylinder.

The fibers are held in place by cementing with a low melting adhesive glass. The fibers are attached through one open end to a common header of a porcelain insulating flange by means of a low melting glass adhesive. The other end of each of the fibers is cut to a length of 8 inches and sealed off by heating. The so-fabricated cell bundle, designed to provide an effective area of glass in the battery cell of 61 cm.^{2}/cm.^{3} of the cell, can be placed in a metal container which also serves as a cathode reservoir. The internal IR loss (voltage drop) from the glass is 0.13 volt at 1 ma./cm.^{2}. Lead wires are attached to the anode assembly and to the metal case holding the cathode. The case can be filled under a reduced pressure with a predetermined amount of a liquid sulfur–sodium sulfide mixture to provide a sulfur–sodium sulfide composition corresponding to Na_2S_{22}. The volume of the liquid cathode is 0.89 cm.^{3}/cm.^{3} of the cell. The assembly of the glass tubes and insulating flange can be fastened to a header compartment and this assembly filled with liquid sodium also under a reduced pressure. Connections for the lead wires are affixed to the anode and cathode compartments. The container seals for the latter unit are constructed so as to be vapor and liquid tight.

Multiples of these cells can be connected in series to give batteries of predetermined voltage for a variety of uses. To illustrate, 120 of these cells can be connected in a combination of parallel and series arrangement to provide a power source capable of delivering 6 kilowatts for 7 hours at 60 volts.

In order to assure maintenance of the operating temperature, utilization of 2 inches of glass wool insulation or its equivalent around the cell prevents complete discharge of the battery even if idled for as long as a 24 hour period. The battery is reversibly rechargeable and shows no degradation after many hundreds of discharge–charge cycles.

In related work, W.E. Brown, R.G. Heitz, and C.A. Levine; U.S. Patent 3,672,995; June 27, 1972; assigned to The Dow Chemical Company describe an improved electrolyte-separator for liquid anode/liquid cathode cells where its function is to conduct cations of the anode material without transmitting electrons. The electrolyte-separator is an impregnated glass or ceramic, open matrix where the discontinuous or disperse phase is an essentially nonmigrating, cation-conducting salt. Such impregnated matrixes are particularly useful in batteries having molten alkali metal anodes and alkali metal sulfur mixtures for cathodes.

Example 1:   Twenty closed end tubes of porous $2CaO \cdot SiO_2$, each 25 cm. long and having 40% porosity, have their pores impregnated with $Na_2O \cdot 2SiO_2$ by immersing them in molten $Na_2O \cdot 2SiO_2$ at 950°C. The open ends of the tubes are not immersed to avoid filling the central cores of the tubes with $Na_2O \cdot SiO_2$. The $2CaO \cdot SiO_2$ tubes are 1,000μ o.d. and 600μ i.d.

Formation of a continuous layer of $Na_2O \cdot 2SiO_2$ around the $2CaO \cdot SiO_2$ is avoided by keeping the tubes immersed until they reach the temperature of the molten $Na_2O \cdot 2SiO_2$. By impregnating at 950°C., the temperature is high relative to the melting point of the $Na_2O \cdot 2SiO_2$ so that it is highly fluid and will run off readily as the tubes are removed from the melt. Radiant heat is applied to the emerging coated tubes to keep them from cooling too rapidly. Retention of the $Na_2O \cdot 2SiO_2$ in the pores of the $2CaO \cdot SiO_2$ prior to cooling is by capillary attraction.

The impregnated tubes are assembled in a parallel fashion and the open ends held in a header assembly which is then sealed to an anode cup which will act as a reservoir for the anode material. The tubes and the anode cup are then filled with molten sodium metal. A wire (the anode lead), which contacts the sodium metal in the anode cup, is brought through and sealed outside the anode cup. This assembly is immersed in a melt of Na_2S_4. A metal cathode lead is also immersed in the Na_2S_4 and one end brought out as the cathode lead from the cell. When this cell assembly is held at 300°C., an open circuit voltage of 1.98 volts is obtained. Drawing 0.1 ampere from the cell reduces the terminal voltage to 1.90 volts. On the charging cycle, 2.065 volts impressed across the leads produces a charging current of 0.1 ampere.

Example 2:   Another example of a matrix membrane suitable for the preparation of an impregnated separator is a phase separated glass. An example of such glass is a silicate based glass containing approximately 40 mol percent B_2O_3. This glass composition is drawn into fine fibers which are held at an elevated temperature for sufficient time to allow phase separation to take place. The phase separation gives a matrix of SiO_2 encompassing channels of sodium borate, i.e., the nonmigrating salt. These fine fibers are assembled into bundles and the bundles assembled into battery cells as described in Example 1.

Alkali Metal Polysulfide Catholyte

C.A. Levine and G.S. Fujioka; U.S. Patent 3,663,294; May 16, 1972; assigned to The Dow Chemical Company describe a battery cell having a molten alkali metal anode and a molten alkali metal polysulfide catholyte separated by a silicate glass as an alkali metal ion permeable membrane. The improvement involves the use of an alkali metal polysulfide having a low hydroxide content as the catholyte. Use of such a material substantially increases the membrane life.

Recent battery technology, such as that described in U.S. Patent 3,476,602, teaches the use of a molten alkali metal anode and a molten alkali metal polysulfide catholyte separated by an alkali metal ion permeable membrane. In one example of the battery, the membrane is made of a silicate glass. As the battery discharges, alkali metal ions migrate through the membrane into the alkali metal polysulfide catholyte.

The battery must be operated at a temperature greater than 300°C. in order to maintain the catholyte in the molten state. A problem develops when contacting silicate glass with an alkali metal polysulfide at these elevated temperatures due to cracking and pitting of the glass. Since the glass membrane is preferably quite thin, any such cracking and pitting will greatly reduce the operating life of the cell.

It is believed that the cracking and pitting of the glass is primarily caused by the presence of alkali metal hydroxide impurities in the polysulfide. Alkali metal polysulfides normally contain significant amounts of alkali metal hydroxide. This is the case since any water present in the alkali metal sulfide from which the polysulfide is prepared will react with the sulfide at elevated temperatures to form the corresponding hydroxide which is highly detrimental to silicate glass at temperatures above 300°C.

In this process, the alkali metal polysulfide is preferably a sodium polysulfide. Normally, a sodium tetrasulfide is used as the catholyte when the battery is assembled. Since sodium tetrasulfide is the preferred alkali metal polysulfide, the following discussion will be concerned with its use.

Since the hydroxide ion is a major factor causing degradation of the glass membrane in the battery, any reduction in its concentration is beneficial. Reduction of the hydroxide level to below 0.15 weight percent of the alkali metal polysulfide is desirable and reduction to below 0.10 weight percent is preferred. Sodium tetrasulfide having a hydroxide ion content below this level has been shown to be less detrimental to glass by a factor of greater than 2 than is the case where the hydroxide ion content is greater than 0.9 weight percent. A problem is presented in preparing sodium tetrasulfide which is free of hydroxide impurity. Anhydrous sodium tetrasulfide is not available commercially and must be prepared by the reaction between sodium sulfide and sulfur. Since anhydrous sodium sulfide is also not available commerically, a composition containing as little water as possible must be prepared and then reacted with sulfur to form the tetrasulfide. Careful drying of the sodium sulfide will remove most of the water of hydration; however, no practical method is available for producing commercial quantities of sodium sulfide containing small enough amounts of water to keep the hydroxide content below the level necessary for extended use in the battery. Thus, it is necessary to remove hydroxide ions from the sodium tetrasulfide after its preparation.

One method of removing hydroxide ion from the tetrasulfide is to heat it to at least its melting point but below its decomposition point, whichever is lower, and contact it with particulate aluminum under an inert atmosphere. In order to accomplish substantially complete removal of the hydroxide impurities, the amount of aluminum added should be 1 weight percent of the polysulfide being treated. Quantities of aluminum ranging from 2 to 5% of the material are preferred. Small particles, i.e., those having dimensions of less than 100 mesh on the U.S. standard sieve series are preferred due to their greater surface area. The aluminum particles are usually left in contact with the molten polysulfide for from 15 to 30 minutes to cause effective purification. Longer reaction periods may be employed to remove substantially all of the hydroxide impurities.

Example 1: Sodium tetrasulfide suitable for subsequent purification was prepared as follows. Commercial reagent grade hydrated sodium monosulfide was dehydrated by boiling out the water with gradual heating to 300°C. The resulting solid material was heated to 300° to 400°C. under a blanket of nitrogen. Commercial sulfur (99.8%) was vaporized and passed through an 800°C. zone before being collected in a nitrogen filled receiver. The essentially anhydrous sodium sulfide was mixed with a stoichiometric amount of the sulfur to provide a ratio of sodium to sulfur of 2:4. This composition was digested at 500°C. for 16 hours and filtered to yield the Na_2S_4 starting material.

About 25 grams of the Na_2S_4 prepared by the above procedure was heated with several small immersed strips of aluminum foil for 66 hours at a temperature range of from 400° to 500°C. The total exposed aluminum surface was between 6 and 8 cm.[2]. The purified Na_2S_4 (20 grams) was recovered by filtration of the hot melt through a fritted glass filter.

The purified Na_2S_4 was tested by exposing hollow glass fibers (composition $2Na_2O \cdot 4SiO_2 \cdot B_2O_3$) in an imposed stressed configuration to the polysulfide at a temperature of 300°C. for 112.5 hours. The fibers, which had 100μ o.d. and 50μ i.d. were bent into U-shapes having a radius of 8 mm.

The curved portion was immersed in the test sodium tetrasulfide under a nitrogen atmosphere. Attack on the glass was evidenced by weakening of the fiber as detected in Instron tensile strength measurements. Instron testing of seven replicate test fibers which had been immersed in the purified sodium tetrasulfide revealed failures occurring between 105 and 335 grams load with a medium of 200 grams. Unstressed control fibers not exposed to sulfide melt failed at a medium load of approximately 250 grams. Identical stressed fibers exposed to sodium tetrasulfide prepared as described herein for 90 hours but not reated with aluminum failed below 100 grams load.

Example 2: Samples of glass of composition $Na_2O \cdot 2SiO_2 \cdot 1/2B_2O_3$ were placed in glass ampoules with sodium tetrasulfide prepared as described in Example 1. Sample A was purified by being heated with 1 weight percent glass wool having a composition based on weight of SiO_2 (80.5%), B_2O_3 (12.9%), Na_2O (3.8%), K_2O (0.4%) and Al_2O_3 (2.2%), for 20 hours at 350°C. and filtered through a glass frit. Sample B was not purified. The ampoules were sealed and heated to 300°C. At the end of the indicated time the ampoules were opened and the glass examined. The results of these runs are presented in the following table.

Na_2S_4 Sample	Exposure Time	Appearance of Glass
A	4 days	No evidence of corrosion
	42 days	No pitting or surface cracks, some strain cracks
B	63 hours	Surface cracks and beginning of crazed surface

Additional alkali metal polysulfides which can be purified by the method and subsequently used in the battery cell with greater efficiency are Li_2S_2, Na_2S_3, Na_2S_5, K_2S_3, Rb_2S_2, Rb_2S_5, Cs_2S_2, Cs_2S_3 and Cs_2S_6.

THERMAL AND ZINC-AIR CELLS

THERMAL CELLS

Ceramic Sandwich Construction

K.O. Hever and J.T. Kummer; U.S. Patent 3,499,796; March 10, 1970; assigned to Ford Motor Company describe an energy storage device comprising a ceramic sandwich where a pair of electronically and cationically conductive crystalline objects are in cation-exchange relationship with and separated by a cationically conductive, electronically insulative, crystalline object.

The outer members of the ceramic sandwich comprise an electronically conductive structural lattice and cations which migrate in relation to the lattice under influence of an electric field, the lattice consisting essentially of ions of a metal electrically reversible between two valence states and ions of oxygen in crystal lattice combination. These are exemplified by polycrystalline objects prepared by sintering crystals formed by heating together at crystal-forming temperature oxides of iron and potassium and made electronically conductive by reduction of some of the ferric ions to ferrous ions by doping or other conventional means.

The inner member of the ceramic sandwich, termed "separator", comprises a structural lattice that is electronically insulating and cations which migrate in relation to the lattice under influence of an electric field. The separator is exemplified by a polycrystalline object prepared by sintering crystals formed by heating together at crystal-forming temperature oxides of aluminum and potassium. A common cation is employed as the conductive cation in both the other members and the separator.

As a solid state capacitor, the device has many advantages relative to conventional capacitors. It provides a high capacitance per unit volume, i.e., typically of the order of 10 to 30 farads per cubic centimeter or equivalent to permittivity of the order of 10^{14}. As a consequence of this, the device will also find application in filter circuits and as a DC block where the very high values of capacitance involved allow a low impedance at very low frequencies while still providing a very high impedance for direct current, e.g., typically an impedance of about 41 ohms at frequencies equal to or greater than about 0.3 cps with an impedance of the order of 1 megohm for DC or greater. The device can be operated at high temperatures, i.e., at least as high as about 500°C. The capacitance of the device varies with DC bias, i.e., the device is nonlinear and therefore tunable. The device behaves symmetrically to DC bias, i.e., there is no inherent polarity.

On application of a difference of electrical potential across the sandwich, the following processes take place: at the positive electrode alkali metal ions pass into the separator and an equivalent number of electrons are given up to the external circuit; at the negative electrode alkali metal ions enter from the separator and an equivalent number of electrons are accepted from the external circuit. As a rechargeable solid state battery, the device has the advantages of durability, small size, long shelf life, and the ability to function in a gravity-free environment or under a wide range of operating conditions.

The separator may be a polycrystalline slab or wafer comprising crystals formed from aluminum oxide and sodium oxide. The separator may also be a polycrystalline slab or wafer comprising crystals formed from a major component of aluminum oxide and a remainder where the major proportion is sodium oxide and the minor proportion consists essentially of the oxide of a metal having a valence not greater than 2, preferably lithium and/or magnesium. The following example illustrates the process.

Example 1:   Powders of Na_2CO_3, Fe_2O_3, TiO_2, and Al_2O_3 were mixed in relative concentrations to provide a molar composition equal to $Na_2O \cdot 5$ $(Fe_{0.95}Ti_{0.05}Al_2O_3)$. This mixture was heated at 1000°C. for one hour. The resultant crystals were mixed with a wax binder and cylindrical discs were isostatically pressed at about 20,000 psi. These discs measured about 0.5" in diameter and had an average weight of about 0.4 g. Two such discs were placed on opposite sides of a 5/8" square plate of fusion cast beta-alumina $(Na_2O \cdot 11Al_2O_3)$–alpha alumina (Al_2O_3) eutectic. The sandwich was wrapped in 0.0005" thick platinum foil. The sandwich in foil was then heated at 1400°C. for one hour to sinter the discs and plate into a unitary object. The foil was cut away except for those portions covering the outer flat faces of the aforementioned discs. The foil was bonded to these faces during the sintering process.

Figure 7.1a shows the steady state charge-voltage curve for this sample device at 300°C. Figure 7.1b shows the discharge curves from a starting voltage of 1 volt under 1,000 and 5,000 ohm loads at 300°C.

FIGURE 7.1:   CERAMIC SANDWICH TYPE SOLID STATE STORAGE DEVICE

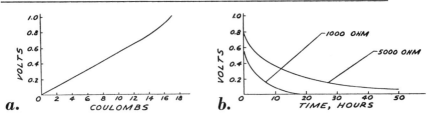

(a) Potential Discharge Voltage Relative to Charge in Coulombs at 300°C.
(b) Discharge Voltage at 300°C. Under 1,000 and 5,000 Ohm Loads

Source: K.O. Hever and J.T. Kummer; U.S. Patent 3,499,796; March 10, 1970

Example 2:   The procedure of Example 1 is repeated with a separator which was prepared in the following manner.

1. In powdered form Na_2O — 10.02 wt. percent (introduced as Na_2CO_3), Li_2O — 0.66 wt. percent (introduced as $LiNO_3$) and Al_2O_3 — 89.32 wt. percent were added to a vessel and mechanically mixed for 30 minutes.
2. The mixture was heated at 1250°C. for one hour to form crystals.
3. The sample was mixed with a wax binder and mechanically pressed into pellets.
4. The pellets were then isostatically pressed at 90,000 psi.
5. The wax binder was removed by gradually heating the pellets to about 550°C.
6. The pellets were sintered for 16 hours at 1520°C. in an electric furnace in a covered crucible in the presence of packing powder of the same composition as the powders from which the crystals were prepared.

Example 3:   The procedure of Example 2 is repeated with the outer members of the sandwich formed of crystals prepared as in Example 1. These crystals are mixed with wax, compressed by being isostatically pressed at 90,000 psi. The resulting pellets are gradually heated to about 500°C. to remove the binder and then sintered for 3 hours at 1450°C.

Cerium Oxide-Sintered Zirconium and Yttrium Oxides Disc

A process described by J.T. Kummer and T.Y. Tien; U.S. Patent 3,607,415; September 21, 1971; assigned to Ford Motor Company provides a solid state energy storage device in which an electrical field produces anions in one side member that migrate through an electronically insulative intermediate member to the other side member. The device comprises a ceramic sandwich having two side members in anion exchange relationship with and separated by an intermediate member. Each of the side members consists essentially of a polycrystalline material comprising an electronically conductive structural lattice containing anions that migrate relative to the lattice. Ions of a metal in two valence states and ions of oxygen preferably make up the lattice. The intermediate member consists essentially of an electronically insulative structural lattice containing oxygen anions that migrate in relation to the lattice under the influence of an electric field.

Heating the device to a temperature of at least about 500°C. and applying an electrical field moves oxygen anions from one side member to the other side member through the intermediate member. Good operations can be obtained at temperatures ranging up to about 1500°C., although the device is most economical in the 500° to 1000°C. range. Electrons surrendered by the anions move through an external electrical circuit connecting the side members. When the field is removed, the anions return to the original side member through the intermediate member and the electrons return to the original side member through the external circuit, thereby producing an electrical current in the external circuit. Thus the device is capable of acting as a solid state capacitor or energy storage device.

Example: Powders of chemically pure zirconium oxide and yttrium oxide are mixed together thoroughly in acetone, dried and sintered at about 1350°C. for 24 hours in a platinum container. The reaction product is ground, pressed into a disc at about 40,000 psi and sintered at 1800° to 2000°C. in an inert atmosphere. End members are prepared by mixing powdered ceric oxide in proportions selected to provide an overall composition having a molar ratio of oxygen to cerium between 1.90 and 1.95. After thorough mixing, the powders are pressed at about 40,000 psi into discs and sintered at 1800° to 2000°C. in an inert atmosphere.

Two of the cerium oxide discs are placed on opposite sides of the zirconium oxide-yttrium oxide disc and a sheet of 0.0005" thick platinum foil is placed on each exterior surface of the cerium oxide discs. The resulting sandwich is hot pressed at 1300°C. and 5,000 psi for about 1 hour to produce a unitary structure. Hot pressing produces an excellent anion exchange relationship between the end members and the intermediate member and between the platinum foils and the end members.

When the device is assembled to an external electrical circuit that applies an electrical potential thereto and its temperature is raised to about 500° to 1500°C., anions of oxygen migrate through the intermediate member while the corresponding electrons migrate through the external circuit. When the external circuit is interrupted and the electrical field is removed, the unbalanced situation existing within the device produces an electrical potential across its terminals. When the external circuit is reconnected, anions return to the original side member through the intermediate member and electrons flow through the external circuit to link up with the returning anions.

Thus the device serves as an energy storage device such as a capacitor or battery. The device is particularly useful in conjunction with fuel cells that operate at elevated temperatures where the device can be used to store energy to meet temporary demands in excess of the rated capacity of the fuel cell.

Tellurium-Containing Carbon Electrode

R.A. Rightmire, J.L. Benak and J.E. Metcalfe III; U.S. Patent 3,567,516; Mar. 2, 1971; assigned to The Standard Oil Company have found that the capacity of an electrical energy storage device containing a high surface area carbon cathode, a metallic anode, and a fused salt electrolyte in contact with the electrodes, can be greatly enhanced by the addition of a tellurium compound directly to the electrolyte or to the carbon cathode. On passage of an electric current through the system, the tellurium becomes permanently bonded to the carbon of the cathode thereby forming an active tellurium species that manifests itself as a characteristic plateau in the discharge profile of the cell.

The tellurium may be added to the system as any tellurium compound that is soluble in the electrolyte, stable in the environment of the cell and is compatible with the ions of the system so that metals foreign to the system will neither contaminate nor plate out on the surface of the metallic anode. Those compounds suitable for the purpose of addition include tellurium metal, tellurous and telluric halides, oxides, and acids, the tellurides and the tellurate and telluric salts of the alkali and alkaline earth metals. Examples of these compounds include $TeCl_2$, $TeCl_4$, $TeBr_2$, $TeBr_4$, TeI_2, TeI_4, TeF_2, TeF_4, TeO, TeO_2, TeO_3, K_2TeO_3, K_2TeO_4, Na_2TeO_3, and Na_2TeO_4. Those particularly suitable are the tellurium halides and the tellurates, tellurites and tellurides of lithium and potassium. One of the main advantages associated with the tellurium additives in this system is that tellurium can exist in either a positive or a negative valence state. The system is therefore reversible and there is little chance for loss of tellurium by being transformed into inactive, insoluble, tellurium metal during the operation of the cell. Any tellurium metal that is lost from the carbon cathode is reduced at the anode to the negative valence state, $Te^=$. The anion thus formed is soluble in the electrolyte and will again migrate from the electrolyte to the cathode where it is bonded to the carbon and can form the active species.

The amount of tellurium required in the system to bring about a discernible enhancement in energy storage capacity is more dependent upon the design of the cell and the rate of reaction in forming the active tellurium-carbon complex, than by the rate of diffusion of tellurium into the cathode. Sufficient amounts of tellurium metal should be present in the system to drive the reaction forming the active tellurium species to completion. However, not so great an excess of tellurium should be present so as to cause excessive leakage of current. Tellurium may be added to the electrolyte or to the carbon electrode in amounts such that the condition cathode contains from 5 to 40% by weight, and preferably from 10 to 35% by weight, of tellurium metal, based on the weight of carbon.

In Figure 7.2a a schematic test cell (10) is shown. The tellurium-containing carbon electrode (12) and metallic anode (11) are positioned one from another, in spaced relationship, immersed in an electrolyte (17) held in a heat-resistant glass tube or stainless steel tube (18). Carbon electrode (12) is fixed rigidly to a graphite current carrier (13) and the metallic anode (11) is fixed rigidly to a steel current carrier (14). The container comprising the electrolyte and electrodes is purged of atmospheric air and an inert gas is introduced into the container. The open end of the container is then sealed with a cap (15) of inert material, such as lava or ceramic. The following examples illustrate the process.

Example 1:   A cathode was obtained from a charcoal prepared by Pure Carbon Co. The commercial grade carbon had the following physical properties: a total pore volume of 0.566 cm.^{3}/g., a surface area of 400 m.^{2}/g., an average density of 0.90 g./cm.^{3}, a Scleroscope hardness of 35 to 45, and an ash content of 10%. The carbon electrode contained 0.092 inch^3 of active carbon.

The solid aluminum-lithium anode, approximately 1.0 x 0.5 x 0.15 inch, initially contained 13% by weight of lithium. Both electrodes were immersed in an electrolyte containing 160 g. of a eutectic salt mixture of lithium chloride and potassium chloride. The eutectic mixture had a composition of 59 mol percent lithium chloride and 41 mol percent potassium chloride and had a melting point of 352°C. The cell assembly was contained in a stainless steel test tube with a total inside volume of 13.4 inches^3. An argon atmosphere was established within the cell and the cell was operated at a temperature between 450° and 500°C. The electrodes were conditioned in the electrolyte by cycling the cell at a constant voltage of 3.34 volts for 30 minutes and then discharging at a constant current to 1.0 volt. The cell was cycled for 20 cycles. At a constant current discharge of 200 milliamperes, the cell delivered 185 watt-minutes per inch^3 of carbon.

Example 2:   The experimental conditions of Example 1 were repeated except that 1.6 g. of potassium telluride were added to the electrolyte in the discharge state and the cell was then cycled in the same manner as described. At a constant discharge of 200 milliamperes, the cell delivered 315 watt-minutes per inch^3 of carbon.

FIGURE 7.2: ADDITION OF TELLURIUM COMPOUND TO ELECTRICAL DEVICE

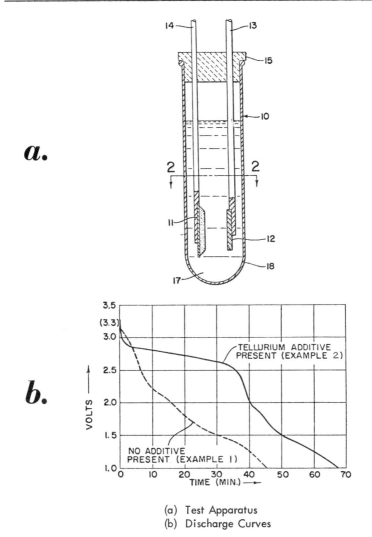

(a) Test Apparatus
(b) Discharge Curves

Source: R.A. Rightmire, J.L. Benak and J.E. Metcalfe III; U.S. Patent 3,567,516;
March 2, 1971

A comparison of the discharge curves obtained for the cells of Examples 1 and 2 is
shown in Figure 7.2b. A sloping discharge profile is observed for the cell in Example 1 without the tellurium additive, whereas with the tellurium additive, in
Example 2, a characteristic plateau in the discharge curve appears at a reaction
potential of about 2.5 to 3.34 volts. As a result of a shift in the reaction to a
higher voltage, a significant increase in watt-minutes is observed for the cell containing the tellurium additive.

Example 3:   The cell and operating conditions of Example 2 were repeated except that 1.5 grams of potassium telluride (K_2Te) were added to the electrolyte and the carbon electrode had a volume of 0.124 inch^3. At a constant discharge of 400 milliamperes, the cell delivered 416 watt-minutes per inch^3.

Example 4:   The experimental conditions of Example 2 were repeated with the exception that the cathode was preconditioned by cycling in a prototype energy storage cell, and the preconditioned electrode was then removed from the prototype cell and placed in a cell containing fresh electrolyte. The cathode contained 19.92 inches^3 of active carbon and had the same carbon composition as that employed in Example 1. The prototype cell contained 35 pounds of the eutectic salt mixture of lithium chloride-potassium chloride to which had been added an excess of potassium telluride. The solid aluminum-lithium anode of the auxiliary cell had the dimensions of 6 3/32 x 5 27/32 x 0.050 inches and the cell volume was equivalent to 44.6 inches^3 (total inside volume). The auxiliary cell contained 200 g. of the lithium chloride-potassium chloride eutectic salt mixture, but did not contain any potassium telluride. At a constant current discharge of 30 amperes, the auxiliary cell delivered 297 watt minutes per inch^3 of carbon cathode.

Different Electrolyte Concentration in Two Compartments

A process described by L.A. King and D.W. Seegmiller; U.S. Patent 3,671,322; June 20, 1972 involves a thermal cell employing aluminum electrodes and a nonaqueous, fusible electrolyte.

The interest in the operation of high altitude aircraft, missiles, and space vehicles, as well as various military operations, has created a need for lightweight, maintenance free, high energy density power supplies characterized by a long shelf life. The use of lightweight power supplies in weight-sensitive space vehicles, the use of backpack batteries for downed pilots and infantrymen, as well as the use of field recharging units for smaller hand carried conventional secondary batteries, are examples of various applications that require a lightweight power supply of the type described in this process.

A particularly attractive candidate for use as an electrode material in lightweight power supplies is the metal aluminum. The metal is readily available, easily purified, inexpensive and capable of being fabricated into any reasonable shape. Consequently, a number of attempts have been made to utilize this metal as an electrode material. However, the problems of self discharge, aluminum passivation in acidic aqueous electrolytes, or aluminum corrosion in basic aqueous electrolytes have not been completely solved by these prior art attempts at using aluminum.

With this process, however, it has been found that the problems of discharge, corrosion and passivation, which were encountered in previous attempts at using aluminum electrodes, have been solved by a thermal cell which comprises a container having two interconnected compartments; a fusible electrolyte mixture of aluminum chloride and sodium chloride contained in each compartment; and aluminum electrodes immersed in the electrolyte mixture.

A nonaqueous, fusible electrolyte comprising a mixture of about 50 to 75 mol percent aluminum chloride and about 50 to 25 mol percent sodium chloride is contained in each compartment in contact with the anode and cathode. The electrolyte in each compartment is similar in component content but differs in the relative concentration of aluminum chloride contained in each mixture. If each compartment contained the same relative concentration of electrolyte ingredients, then the resultant voltage of the cell would be zero. However, if the respective concentrations of aluminum chloride in each compartment is different, then the cell is capable of generating electrical power. The magnitude of power or voltage produced by the cell is a direct correlation of the difference between the aluminum chloride concentrations in the electrolytes in each of the compartments.

At normal temperatures the electrolyte is solid and the cell remains inactive. Heating the cell, however, to a temperature within the range of about 125° to 180°C. renders the electrolyte molten and activates the cell. The cell is capable of producing voltages in excess of 0.2 volt with energy densities nearly three times greater than those available from conventional lead-acid and nickel-cadmium batteries.

The solid electrolyte of the thermal cell comprises a mixture of about 50 to 75 mol percent aluminum chloride and 50 to 25 mol percent sodium chloride. The following table describes two examples of specific electrolyte compositions together with an indication of their operating temperatures and the magnitude of voltage produced by using the specific electrolyte in combination with the aluminum electrodes.

<table>
<tr><td colspan="7" align="center">Electrolyte Concentration in Mol Percent</td></tr>
<tr><td></td><td colspan="2">Compartment A</td><td colspan="2">Compartment B</td><td></td><td></td></tr>
<tr><td>Cell No.</td><td>$AlCl_3$</td><td>NaCl</td><td>$AlCl_3$</td><td>NaCl</td><td>Temp., °C.</td><td>Voltage</td></tr>
<tr><td>1</td><td>50.9</td><td>49.1</td><td>70.7</td><td>29.3</td><td>179</td><td>0.230</td></tr>
<tr><td>2</td><td>68.8</td><td>31.2</td><td>70.7</td><td>29.3</td><td>179</td><td>0.007</td></tr>
</table>

The solid electrolytes are rendered molten at temperatures ranging from about 125° to 180°C. A 60–40 ratio of $AlCl_3$ to NaCl melts at the lower temperature. Heating the cell to the appropriate temperature converts it from an inactive state to an active state with the resulting generation of electrical power.

Both the anode and cathode of the cell are fabricated from metallic aluminum. Each is separately positioned within one of the two compartments containing the electrolyte mixtures. The two compartments are interconnected such that the two electrolyte mixtures come in physical contact. Generally, a porous or wick-like material, such as asbestos fiber, is placed within the interconnecting passageway to form a porous bridge between two electrolyte mixtures thereby providing sufficient physical contact while simultaneously preventing any undue mixing of the two electrolyte mixture concentrations.

Referring to Figure 7.3, there is shown an electro-chemical cell comprising a container (10) having two separate compartments (12) and (14). A capillary-sized passageway (16) connects compartments (12) and (14) for the purpose of providing physical contact between a first, solid $AlCl_3$-NaCl electrolyte mixture (18) and a second, solid $AlCl_3$-NaCl electrolyte mixture (20). Aluminum electrodes (22) and (24) are immersed in their respective electrolyte mixtures and function as the anode and cathode components of the cell. The electrodes, in turn, are connected to suitable tungsten or platinum leads (26) and (28). The leads (26) and (28), in turn, are connected to an external load, not shown, which consumes the electrical power

FIGURE 7.3:   CELL WITH ALUMINUM ELECTRODES AND TWO COMPARTMENTS

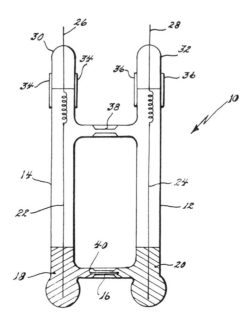

Source: L.A. King and D.W. Seegmiller; U.S. Patent 3,671,322; June 20, 1972

generated by the cell. The removable top portions (30) and (31) of the compartments (12) and (14) are sealed, for example, by means of Teflon sleeve seals (34)
and (36). A pressure equalizing capillary (38) interconnects the top portions of the
compartments (12) and (14). The passageway (16) contains a suitable porous or wick-
like fibrous material (40), such as asbestos fibers, to prevent undue mixing between
the contacting electrolyte mixtures (18) and (20).

The container (10) can be made of Pyrex glass or some other suitable inert material.
The cell is provided with suitable heating means, not shown, for activating the
cell. For example, electric coils may be used or the cell could be coated with a
paste of water, filler material, and a pyrotechnic agent. A mixture of zirconium
and barium chromate could serve as the pyrotechnic agent. The paste would be
applied wet to the surface of the cell and allowed to dry. An electric match, connected to a source of electric power, would be embedded in the pyrotechnic paste
and activated by closing a suitable switching arrangement. The match would ignite
the pyrotechnic material which, in turn, would generate sufficient heat to fuse
the two electrolytic salt mixtures, thereby activating the battery.

In the operation of the cell, using the electrolyte mixtures of the table shown on
the previous page, it was found that the capillary (38) eliminated any detectable
mass transport of melt through the asbestos fiber (40). The electrodes (22) and (24)
were made from "Baker Analyzed" reagent grade 0.032" diameter aluminum wire.
The melt in each compartment or arm of the cell was stirred with a Teflon covered

magnetic stirring bar. The cell was immersed in a conventional, well-stirred, constant temperature, silicone oil bath in order to provide adequate operating temperatures. Melt concentration ratios were changed by adding known weights of either $AlCl_3$ or NaCl to one of the arms of the cell. A new, steady cell voltage was reached once the added salt was dissolved and well mixed.

Only purified salt mixtures were utilized for the electrolyte compositions. The purification of $AlCl_3$ was accomplished by repetitive sublimation of Baker Analyzed reagent grade $AlCl_3$ in dry argon along the length of a 60 cm. long, 5.4 cm. o.d. Pyrex tube and finally into a receptacle which was capped with a standard taper plug. Baker Analyzed reagent grade NaCl was dried by fusion in air. The two chlorides were stored, handled and mixed in a glove box filled with dry air. Electrolyte compositions were fixed by weight. Water-clear, colorless, molten mixtures were obtained after being subjected to the appropriate operating temperatures. The resulting cell voltages were measured with a Leeds and Northrup K-3 potentiometer and a Leeds and Northrup Electronic DC Null Detector.

Alkali Salt Electrolyte

J. Greenberg, P. Pike and T.E. Seitz; U.S. Patent 3,554,806; January 12, 1971; assigned to the U.S. Administrator of the National Aeronautics and Space Administration describe a heat activated cell having an anode made of one or more alkali metals and a cathode made of an oxidizing material which will react with the anode to form a stable salt. The anode and cathode are separated by an alkali salt electrolyte deposited in a thin layer on a divider made of a porous electrically conductive material. The salt is heated to a temperature sufficiently below its melting temperature that it does not disintegrate.

As shown in Figure 7.4, an EMF cell constructed in accordance with the process may include a suitable container (10). The container (10) may be made of an electrical insulating material such as ceramic or may even be a metal. In order to divide the cell into sections for containing an anode material (11) and a cathode material (12), respectively, a cup (13) made of a porous electrically conductive material is disposed in the container (10). The cup (13) forms a chamber (14) for the anode material (11) and a chamber (15) for the cathode material (12) and thus serves as a divider. Porous carbon, metal or ceramic may be used as material for the cup (13).

The anode material (11) may be either a single alkali metal or a mixture of any two or more of the alkali metals in any proportion. The alkali metals include lithium, sodium, potassium, rubidium, and cesium. These are all monovalent metals in the first group of the periodic system. The cathode material (12) may be any oxidizing material which will react with the anode to form a stable salt. By way of example, the cathode material (12) may include one or more of the following: halogen gases, halogen compounds, nitrates, ferrous compounds, cobaltous compounds, sulfur and selenium. In general, any of the well-known multivalent cathode materials can be used satisfactorily.

To provide a solid electrolyte which also serves as a separator between the anode and cathode materials, a layer comprising one or more anhydrous alkali salts is coated onto the anode side or exterior of the cup (13). The cup (13) serves as support means for the alkali salt electrolyte. The electrolyte (16) may be applied by

FIGURE 7.4:   HEAT ACTIVATED CELL WITH ALKALI SALT ELECTROLYTE

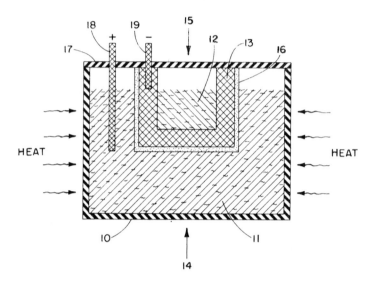

Source: J. Greenberg, P. Pike and T.E. Seitz; U.S. Patent 3,554,806;
January 12, 1971

dipping or by vacuum deposition. To the end that the alkali salt layer will be
highly conductive to ions, i.e., its resistance is desired to be less than 100 ohms,
its thickness is one millimeter or less, and it is heated by heat applied to the cell.
However, the temperature of the alkali salt electrolyte (16) must not exceed a maximum which is sufficiently below its melting point that it maintains its structural
integrity as a layer on the cup (13). In general, the temperature of the alkali salt
electrolyte should be less than about 50% of its melting temperature.

The EMF cell is completed by a top cover plate (17) and electrodes (18) and (19).
The cover plate (17) seals the chambers (14) and (15) and may be made of any suitable electrical insulating material. The electrodes (18) and (19) may be carbon or
any metal which does not react with the anode, cathode, or electrolyte materials.
The electrode (18) extends through the cover plate (17) into the anode material (11)
while the electrode (19) extends through the cover plate (17) into the cup (13). It
will be understood that, if desired, the electrode (19) may be positioned to extend
through the cover plate (17) into the cathode material (12) without contacting the
cup (13).

If the anode and cathode materials (11) and (12) do not completely fill the respective
chambers (14) and (15), a nonoxidizing, inert gas may be used to fill the unoccupied
space. Examples of some suitable gases include the so-called noble gases, nitrogen
and helium. The following examples illustrate the process.

Example 1:   The anode material (11) is sodium. Because the melting point of
sodium is about 100°C., sulfur which has a melting point of about 119°C. may be

advantageously used as the cathode material (12). To minimize vaporization of the sodium and the sulfur, the cell must operate at a relatively low temperature. The electrolyte should have a relatively low melting point. In this case, NaOH, which has a melting point of 318.4°C., is used as the electrolyte (16) and is coated on the exterior surface of the cup (13) which is made of porous carbon. The cell is heated to maintain an operating temperature in the range of from about 100° to 250°C. preferred.

Example 2:   The anode material (11) is sodium and the cathode material (12) is iodine. The electrolyte (16) is a eutectic mixture of 59 molar percent LiCl and 41 molar percent KCl whose melting point is about 380°C. and the cup is porous carbon. The cell normally operates at a temperature of about 200°C. but will perform satisfactorily in a range of 150° to 250°C.

Example 3:   The anode material (11) is sodium, the cathode material (12) is iodine, and the electrolyte (16) is NaCl. The cell normally operates at a temperature of about 300°C. but may be operated in a range of 200° to 400°C.

Example 4:   The anode material (11) is sodium and the cathode material (12) is iodine. The electrolyte (16) is a eutectic mixture of 10 molar percent NaCl, 40 molar percent LiCl and 50 molar percent KCl. The melting point of the electrolyte is about 400°C. and the cell is normally operated at a temperature of about 200°C. The operating temperature range of the cell may vary from 150° to 250°C.

The cell of Example 3 is the simplest form of a cell of the process. That is, each of the anode and cathode materials is one of the constituents of the alkali salt electrolyte. With an alkali metal anode and an alkali salt electrolyte, the nonmetallic constituent of the salt may be used as a cathode material in a cell built in accordance with the process. However, as indicated previously and as described in Example 1, other suitable oxidizing materials, such as sulfur, may be used as cathodes.

In general, the cell is operated at a temperature high enough to melt the alkali metal anode material but not high enough to melt the alkali salt electrolyte. With such a temperature, the cathode may be either a liquid or a gas depending on the material used. It will be seen from Examples 2 and 4 that the electrolyte may be comprised of a plurality of alkali salts. However, the objective of providing a low temperature cell should be kept in mind and the alkali salt or alkali salt mixture should be chosen accordingly.

Aluminum Salt Electrolyte

J. Greenberg; U.S. Patent 3,635,765; January 18, 1972; assigned to the U.S. Administrator of the National Aeronautics and Space Administration describes a heat activated electromotive force cell having an anode formed of aluminum and a cathode comprising an oxidizing material such as sulfur. The cathode material is supported in a container or in a matrix such as porous carbon.

An aluminum salt layer electrolyte such as $AlCl_3$ separates the anode from the cathode. To minimize vaporization of the aluminum salt, an alkali halide salt may be used with the aluminum salt. The cell may be operated at temperatures up to a point where either the cathode material or the electrolyte is molten.

ZINC-AIR CELLS

Addition of Gallium or Tin to Zinc Electrode

A process described by H.G. Oswin; U.S. Patent 3,623,911; November 30, 1971; assigned to Leesona Corporation provides zinc electrodes containing a minor amount of a metal which is capable of creating cation vacancies in a zinc oxide lattice such as gallium or tin. The gallium or tin additive helps to prevent passivation of the zinc during rapid discharge of the electrodes in an electrochemical cell and reduces the amount of cell electrolyte needed for efficient performance. The gallium or tin is preferably introduced into the zinc by means of alloying, e.g., by electrolytic co-reduction of zinc and gallium and/or tin compounds.

Example 1:   A silver screen provided with a terminal strip was pasted with an aqueous slurry containing 4.049 g. of zinc oxide and 0.075 g. of potassium stannate trihydrate. The pasted screen was wrapped in Kraft paper and dried in a hot air oven at 50°C. for one hour. The assembly was then placed in a reducing tank containing an aqueous 5 weight percent KOH solution. The reduction was carried out at 4 volts and 0.16 amp. per square inch for 20 hours to insure complete reduction of zinc oxide.

The assembly, now containing a spongy zinc alloy deposit, was removed from the KOH electrolyte and carefully washed and dried under vacuum to remove all traces of electrolyte and trapped gases. The screen/zinc alloy assembly was then subjected to 75 psi of pressure and heated to 250°C. in an inert atmosphere. The resulting zinc/tin alloy sheet had a composition of approximately 99.5 atomic percent zinc and 0.5 atomic percent tin.

The porous zinc/tin alloy sheet containing the silver current collector was utilized as the consumable metal anode of a zinc-air bi-cell. The design of the cell utilizes an envelope cathode comprising a hydrophobic polymer member having the inner surface coated with a uniform admixture of a catalytic material. In this case the polymer member is PTFE with the catalytic layer containing particles of platinum black with polytetrafluoroethylene.

The air cathode had a platinum loading of approximately 7 mg./cm.^{2}. The electrolyte was a 30% aqueous KOH solution retained in a hydrophilic matrix. The zinc-air cell was completely discharged at a current density of 75 ma./cm.^{2}. The cell was charged at a current density of 35 ma./cm.^{2} geometric surface area of zinc, the current being applied by the pulsed technique, 10 milliseconds on and 20 milliseconds off. During the course of 20 charge/discharge cycles, no evidence of passivation was observed.

Example 2:   The procedure of Example 1 was followed with the exception that 4,065 grams of zinc oxide and 0.006 gram of gallium hydroxide were used. The anode alloy metal contained approximately 99.9 atomic percent zinc and 0.1 atomic percent gallium. The performance of the zinc/gallium electrode was comparable to the zinc/tin electrode.

Sulfide Ion Crystallizing Agents

A process described by E.V. Steffensen, J. Orshich and J. Steffensen; U.S.

Patent 3,649,362; March 14, 1972; assigned to McGraw-Edison Company involves
providing in an alkaline-zinc anode primary cell an agent which regenerates the
electrolyte by causing the zinc reaction end products to crystallize out as zinc-
hydroxide when the electrolyte becomes saturated with zincate during discharge of
the cell. Further, the process provides a fibrous mass below the electrodes having
a large surface area onto which the crystals of zinc hydroxide grow in preference to
forming onto the electrodes. In this way the zinc reaction end products are kept
from coating or clogging the electrodes and from interfering with the continuing
operation of the cell.

The crystallizing agent may be any compound or compounds which function to
release sulfide or silicate ions to the electrolyte and may be selected from the
group consisting of the silicates of potassium and sodium, sulfur in its precipitated
or sublimed forms, thiourea (NH_2CSNH_2) and the sulfides of calcium, zinc (precipitated form), potassium, sodium, strontium, antimony, mercury, barium, aluminum, and phosphorus. The action of the sulfide ion crystallizing agents is improved
by adding a small amount of zinc oxide especially as to air depolarized cells which
are to have good life on open circuit. The fibrous mass may be of glass wool,
cotton linters, shredded or loosely folded Kraft paper, polyethylene fibers, asbestos wool, steel wool or nylon fibers.

The primary battery shown in Figure 7.5 is of the air depolarizing type comprising
a container (10) as of hard rubber having a centrally located air depolarizing cathode (11) of a porous carbon material preferably of a petroleum coke base, which
may be cylindrically shaped and extended from the top of the cell downwardly

FIGURE 7.5: AIR DEPOLARIZING TYPE PRIMARY BATTERY

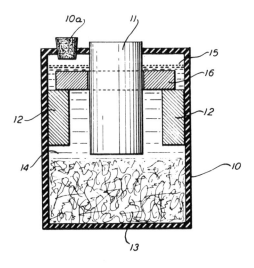

Source: E.V. Steffensen, J. Orshich and J. Steffensen; U.S. Patent 3,649,362;
           March 14, 1972

through a major portion of the height of the cell. Opposite to the carbon cathode are one or more zinc anodes (12) of which two are shown each of a block form. The space in the cell below the electrodes contains a fibrous mass (13) preferably of glass wool or cotton linters, and the entire free space in the cell is filled with an alkaline electrolyte solution (14) preferably of potassium hydroxide to a level (15) above the zinc anodes.

Previously the deleterious action of zincates in zinc anode alkaline electrolyte primary cells has been attacked by adding lime to the electrolyte as in the manner described in U.S. Patent 2,450,472. The action of the lime is to combine with the alkali zincate and form regenerated electrolyte and highly insoluble calcium zincate. Although this method of attacking the zincate problem has been successful, it is accompanied by certain disadvantages, one of which is the expense of producing the lime in the special form necessary for the purpose intended and a second of which is the low volume efficiency of the lime.

For example, the electrolyte solution may be 5 to 6 N potassium hydroxide, the silicate agent may be present in the amount of 4 to 12 grams per liter of electrolyte solution calculated as silicon dioxide, and the sulfide agent may be present in the amount of 0.5 to 2 grams per liter of electrolyte solution calculated as sulfur. Two or more of these silicates and sulfides may be used in the same cell in which case the amounts of the respective agents are proportionately reduced. The zinc oxide is preferably added as an initial ingredient in an amount which is at least the chemical equivalent of the sulfur-containing agent.

The operation in a preferred cell using sodium sulfide, zinc oxide and potassium hydroxide is as follows. When the user adds water to the cell, the sodium sulfide dissolves giving sulfide ions and the zinc oxide reacts therewith to form zinc sulfide. As the battery is discharged, more zincate goes into solution as the zinc metal anode is consumed and more zinc sulfide dissolves until finally a saturation point is reached causing crystallization of zinc hydroxide upon any suitable substrate of large surface area such as is provided by the fibrous mass (13).

The crystallization of zinc hydroxide as against the deposition of zinc oxide is carried out at the expense of water from the electrolyte solution, and has the effect therefore of causing the electrolyte solution to become slightly more concentrated as the cell is discharged. The prime advantage of crystallizing zinc hydroxide instead of precipitating zinc oxide or reacting the zinc electrolyte reaction product with lime is that the hydroxide crystallizes out onto the fibrous mass leaving the electrodes clean and at their utmost efficiency, and the crystallizing agent requires very little space compared to the large volume required by lime permitting therefore much more reaction end products to be contained per unit volume of the battery.

These advantages of the process are shown by the following battery discharge tests. Two different batteries of equal size made with aqueous potassium hydroxide and with similar air depolarizing carbon cathodes and zinc anodes, but with one having the greatest possible quantity of highly efficient lime to react with the zinc electrode reaction products and the other having a sodium sulfide crystallizing agent and a glass wool fibrous mass, were discharged at 250 ma. The battery with lime delivered 1,200 ampere hours at useful voltage before voltage fell sharply; at the end of discharge, the cell was filled with reacted lime and bluish zinc oxide that had continued to form after the lime was fully reacted. On the other hand, the battery with sulfide and glass wool in accordance with the process delivered 1,500 ampere hours at useful voltage before voltage fell sharply; at the end of discharge the glass wool mass was converted into a highly crystalline mass of zinc hydroxide and there was still much room in the battery beneath the electrodes in which more crystals of zinc hydroxide could have grown had even larger zinc electrodes been provided. This example shows that the process makes possible significant and practically important increases in capacity per unit volume in zinc batteries of the air breathing type by compacting or sequestering the zinc electrode reaction products in less space than has previously been possible using lime.

Polyterpene Resin Treated Carbon Electrode

A process described by E.G. Munck and R.F. Hauser; U.S. Patent 3,562,018; February 9, 1971; assigned to Union Carbide Corporation involves a gas permeable battery electrode of carbon having on at least its electrochemically active surface portions a quantity of polyterpene resin sufficient to render the electrode repellent to battery electrolyte. The resin may be applied only to the surface portions of the electrode or may be impregnated throughout the body of the electrode.

Polyterpene resins suitable for use in the process are polymers of α or β-pinene or mixture which are natural materials obtained from pine tree rosin. Their molecular weights may be as high as 2,000. They are available commercially under trade names Piccolyte and Nypene and their properties and preparation are discussed in the book "Polymers and Resins" by Brage Golding, published by D. Van Nostrand Company.

The application of the polyterpene resin to an electrode to be treated may be accomplished simply by preparing a solution (preferably warmed) of the resin in a solvent and immersing the electrode in the solution for the time sufficient to permit the solution to penetrate the pores of the electrode in at least its surface portions. Suitable solvents are those derived from petroleum, such as toluene, xylene, benzene or kerosene. Other volatile solvents may be used, but since they will be removed from the treated electrode by heating, they must have boiling points below a temperature (about 170°C.) at which the electrodes would be damaged during heating.

Similarly, the solvent used should of course not react with the carbon electrode or its binder. The solvent preferably should have a saponification number near zero. Preferred solvents are toluene, kerosene and ethylene dichloride. After immersion and removal of the electrode from the solution, the electrode is heated to volatilize and remove the solvent.

Typically, a solution of as little as 1.5% resin in toluene may be used. The solution preferably is warmed to below the boiling point of the solvent, but it may be at room temperature. The quantity of resin in the solution may be as little as 0.5% by weight or as high as 25% or more. Immersion time may be as short as 5 seconds to as long as one hour, but it is desirable that the time be shorter rather than longer, and the more time is dependent in part on the quantity of resin in solution. For the composite electrodes used in fuel cells, five minutes immersion in a 1.5% solution of polyterpene in toluene is satisfactory; good results have been obtained on electrodes for air depolarized cells in 5 seconds immersion in a solution of 0.5%

polyterpene in toluene at room temperature. It is important, of course, that the quantity of resin incorporated in the electrode not destroy its gas permeability. Accordingly, the use of dilute solution is preferred. Removal of solvent, as stated, is accomplished by heating. A simple guide for completion of solvent removal is odor detection, heating being continued until no solvent odor emitting from the treated electrode can be detected. This may require heating for several hours. Usually heating for five to eight hours has been found to achieve this end.

A number of electrodes of the types used in air depolarized cells and in fuel cells have been prepared and have been tested in comparison with conventional electrodes used commercially. In such tests the electrodes have been placed in cells having conventional aqueous alkaline electrolyte. The cells were placed on normal drain for varying periods of time, and the voltages under load were determined at fixed intervals. The open circuit voltages of the electrodes were measured before and after the tests. The extent of electrolyte penetration was also determined in electrodes which were subjected to the same tests.

The electrodes of the process showed the minimum penetration of electrolyte and while their electrical discharge characteristics were at least as good as, if not always superior to those of the conventional electrodes, they recovered their initial open circuit voltages much more quickly than did the conventional electrodes.

One of the advantages of the electrodes of the process is that they are capable of use under heavier drains than the conventional electrodes. Under normal conditions of use of conventional air depolarized cells, a continuous drain of about 60 milliamperes is encountered, with peaks up to about 700 milliamperes. Under such conditions maximum output of these cells is about 1,500 to 3,000 ampere-hours depending upon the construction of the cells. Generally it has been found that if heavier drains than about 600 milliamperes are placed on the conventional cells for long periods of time, they become inoperative due principally to electrolyte penetration of the carbon electrode. As the following data show, electrodes of the process withstand continuous drains of 900 milliamperes with peaks of 1.6 amperes.

Electrodes of 4 x 5 x 1.25 inches were prepared. One group was treated by immersion in an 0.5% solution of polyterpene resin in toluene for different times, a second in a 1.5% solution and a third in a 25% solution, after which they were heated in air to 150°C. for five hours. These electrodes were then placed in cells and tested with a 900 milliampere drain imposed for 23 hours. Then a drain of 1.6 amperes was imposed for 1 hour. This cycle was repeated in each case for three days. Similar conditions were imposed on cells containing conventionally wetproofed carbon electrodes.

Treatment of Electrodes

Electrode No.	Solution	Immersion Time, sec.
A-1	0.5% in toluene	5
A-2	0.5% in toluene	30
A-3	0.5% in toluene	3,600
B-1	1.5% in toluene	5
B-2	1.5% in toluene	30
B-3	1.5% in toluene	3,600
C-1	25% in toluene	6
C-2	25% in toluene	30
C-3	25% in toluene	3,600
D	Conventional	--

In the following table, OCV stands for open circuit voltage in volts, and CCV stands for closed circuit voltage in volts.

Tests of Electrodes

Electrode No.	Initial OCV	Average CCV, 900 ma.	Average CCV, 1.6 amp.	Final OCV	Electrolyte Pickup, g.
A-1	1.34	1.20	0.95	1.34	30
A-2	1.34	1.20	0.95	1.34	30
A-3	1.34	1.19	0.95	1.34	28
B-1	1.35	1.19	0.97	1.34	27
B-2	1.35	1.19	0.97	1.34	26
B-3	1.35	1.14	0.82	1.33	26
C-1	1.33	0.82	0.40	1.24	26
C-2	1.31	0.60	0.40	0.80	23
C-3	1.31	0.35	0.15	0.75	26
D	1.38	*	**	0.19	175

*Actual data 900 ma. drain = 0.90; 0.47; 0.10 volt.
**Actual data 1.6 amp. drain = 0.37; 0.10; 0.00 volt.

The data above indicate that best results are obtained with dilute solutions of resin and short immersion time. They also show the failure of conventional commercial electrodes to withstand the conditions of test.

A number of similar tests made with solutions of polyterpene in other solvents heated similarly were made. Other tests made on electrodes treated only with toluene, kerosene, or other solvents showed that the solvent alone does not produce significant wetproofing.

Gas Diffusion Cathode Containing Manganese Oxide Catalyst

A process described by J. Paynter and J.R. Morgan; U.S. Patent 3,649,361; March 14, 1973; assigned to General Electric Company relates to electrochemical cells which have at least one gas diffusion electrode containing a non-noble or base metal catalyst, an aqueous alkaline electrolyte, and room temperature operation.

An electrochemical cell with a gas diffusion cathode is a device capable of generating electricity by electrochemically combining an oxidizable reactant, termed a fuel, and a reducible reactant, termed an oxidant. Such cells, each of which is comprised of spaced electrodes ionically connected by an electrolyte, include fuel cells, and metal-air cells. When fuel and oxidant are concurrently and separately supplied to the electrodes of the cell, an electrical potential will develop across the electrodes. When an electrical load is provided across the electrodes an electrical current flows between, the electrical energy thus represented being generated by the electrocatalytic oxidation of fuel at one electrode and simultaneous electrocatalytic reduction of oxidant at the other.

While various non-noble or base metal materials are known to be active for the electroreduction of oxygen in an aqueous alkaline electrolyte cell, such materials can be employed only at cell operating temperatures usually greater than 100°C. to produce a performance similar to a noble metal catalyst. When such an electrochemical cell employs an alkaline electrolyte and is operated at room temperature, it is generally considered necessary to employ a noble metal as the catalyst for the gas diffusion cathode. A non-noble or base metal material employed as the catalyst shows a serious decline in performance as opposed to a noble metal, such as platinum, which maintains its high level of activity. Thus, while the substitution of a non-noble or base metal catalyst for a noble metal catalyst could be employed in such an aqueous alkaline cell at room temperature, a large loss in performance is encountered, especially if high current drains are required.

This process is directed to an improved electrochemical cell, which has an aqueous alkaline electrolyte, operates at room temperature, and has a special gas diffusion electrode containing a non-noble or base metal catalyst whereby the cell exhibits performance comparable to such a cell employing a gas diffusion cathode with a noble metal.

This process is directed to an electrochemical cell with at least one gas diffusion electrode, at least one anode spaced from the cathode, an aqueous alkaline electrolyte in contact with both the cathode and the anode, and the gas diffusion cathode comprises a mixture of a finely divided, electrically conductive carbon powder, and a catalyst material selected from the class consisting of a manganese oxide, and a manganese oxide and at least one other base metal catalyst.

In Figure 7.6a there is shown generally an electrode (10) which has a terminal grid in the form of a metal wire screen (11) which serves the functions of transmitting electrical current and providing reinforcement for the electrode. An electrical connection in the form of an electrical lead (12) is connected directly to screen (11). Electrode (10) has at (13) a mixture of a finely divided, electrically conductive powder, and a catalyst material selected from the class consisting of a manganese oxide, and a manganese oxide and at least one other non-noble or base metal catalyst held together by a binder and in electronically conductive relation with wire screen (11).

For example, the catalyst is held together and bonded to screen (11) by a binder material of polytetrafluoroethylene (PTFE). Electrode material (13) surrounds both the screen (11) and a portion of electrical lead (12). The ratio of such a binder to the catalyst material may be from about 5 to 50% by weight, with the preferred range being from about 10 to 30% by weight. If desired, a hydrophobic film (14) is shown bonded to one surface of electrode material (13) to prevent electrolyte drowning of electrode (10). This film is desirable if the electrode is to be used with a free aqueous electrolyte.

In Figure 7.6b there is shown generally an electrode (20) which has an electrically conductive, porous substrate (21), such as carbon, with a catalyst material (22) of a manganese oxide, or a manganese oxide with one or more non-noble or base metal catalysts impregnated into and coating substrate (21). An electrical lead (23) is affixed to electrode (21) in any suitable manner.

In Figure 7.6c there is shown generally at (30) an electrochemical cell in the form of a fuel cell which comprises gas diffusion cathode (10) from Figure 7.6a and an anode (31), separated by an annular electrolyte gasket (32). Electrolyte inlet conduit (33) and electrolyte outlet conduit (34) are sealingly related to the electrolyte gasket to circulate a free aqueous electrolyte to and from electrolyte chamber

FIGURE 7.6:   GAS DIFFUSION CATHODE

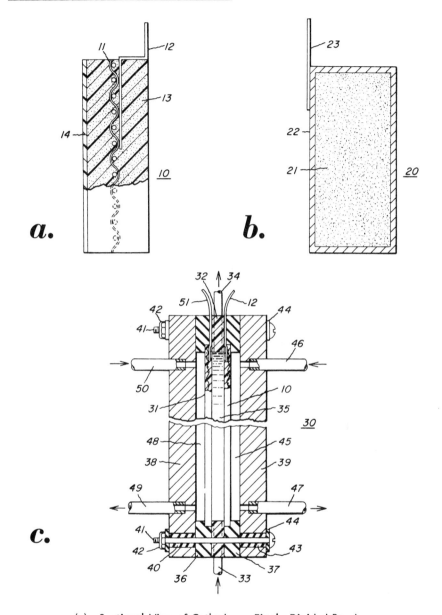

(a)   Sectional View of Cathode — Finely Divided Powder
(b)   Sectional View of Cathode — Porous Substrate
(c)   Sectional View of Fuel Cell

Source: J. Paynter and J.R. Morgan; U.S. Patent 3,649,361; March 14, 1972

(35) formed by the anode, cathode and gasket. An anode gasket (36) and a cathode gasket (37) are positioned adjacent opposite faces of the electrolyte gasket to seal and hold the anode and cathode in assembled relation. Identical end plates (38) and (39) are associated with the anode and cathode gaskets, respectively, in sealing relation therewith. To hold the gaskets and end plates in assembled relation a plurality of tie bolts (40) are provided, each having a threaded end (41) and a nut (42) mounted thereon. To insure against any possibility of internal short circuiting of the fuel cell electrodes, the tie bolts are provided with insulative bushings (43) within each end plate and with an insulative washer (44) adjacent each terminus.

An oxidant chamber (45) is formed by the cathode gasket, cathode, and end plate (39). An oxidant inlet conduit (46) is sealingly associated with the end plate to allow oxidant to be fed to the oxidant chamber while an oxidant outlet conduit (47) is similarly associated with the end plate to allow the purge of oxidant. Where the fuel cell is to be operated on ambient air, no end plate (39) is required. The anode, anode gasket, and end plate (38) similarly cooperate to form a fuel chamber (48). A fuel outlet conduit (49), similar to oxidant outlet conduit (47) is provided. A fuel inlet conduit (50) is shown for providing a fuel to the fuel chamber. An electrical lead (51) is provided for anode (31). Such a fuel cell operates generally on a gaseous fuel.

Example 1:   A supported catalyst material was prepared in accordance with the process by diluting 7.19 g. of 50 weight percent manganous nitrate aqueous solution with water to a volume of 25 milliliters. Carbon powder in the amount of 9.93 g. was added to the solution which was then stirred to obtain a creamy paste. The solvent was evaporated from the mixture by heating the mixture on a hot plate with a surface temperature of 150°C., while the mixture was agitated constantly. The resulting dried powder was ground and then transferred to a 5.5 inch diameter petri dish.

The dried powder was then heated further on a hot plate which had a surface temperature of 150°C. for 2 1/2 hours. The temperature was raised to 200°C. for one hour and was then raised to 250°C. for one hour. The heating was continued at 300°C. for one hour, which was then followed by heating at 330°C. for one-half hour. The resulting material was ground and sieved through a 325-mesh screen.

The above heating process decomposed thermally the metallic salt on the carbon powder thereby producing a supported catalyst material containing approximately 10 weight percent manganese. This supported catalyst material was then made into a PTFE bonded fuel cell electrode by the method described in U.S. Patent 3,297,484. The resulting electrode contained approximately 4.6 milligrams of manganese per square centimeter.

Example 2:   A supported catalyst material was prepared by the thermal decomposition of manganese acetate on carbon powder. 2.62 g. of manganese acetate was dissolved at a temperature of 85°C. in 13.0 ml. of pyridine. Carbon powder in the amount of 5.29 g. was added to the solution which was then stirred to obtain a creamy paste. The pyridine solvent was evaporated from the mixture by heating the mixture on a hot plate with a surface temperature of 150°C. while the mixture was agitated constantly. The resulting dried powder was ground and then transferred to a 5.5 inch diameter petri dish. The dried powder was then heated further on a

hot plate which had a surface temperature of 200°C. for one hour. The temperature was then raised to 250°C. and the heating continued for 1.5 hours. The heating was continued at 300°C. for two hours, which was then followed by heating at
320°C. for one hour. The resulting material was ground and sieved through a 325-
mesh screen. The above heating process decomposed thermally the metallic salt on
the carbon powder thereby producing a supported catalyst material containing approximately 10 weight percent manganese. This supported catalyst material was then
made into a PTFE-bonded fuel cell electrode by the method described in U.S. Patent
3,297,484. The resulting electrode contained approximately 3.8 mg. of manganese
per square centimeter.

Example 3:   An unsupported catalyst material was prepared by the thermal decomposition of manganous acetate and silver acetate. 33.46 g. of manganous acetate
and 11.60 g. of silver acetate were mixed together and placed on a 5.5 inch diameter petri dish. The powders were heated on a hot plate which had a surface temperature of 150°C. for one and one-half hours. The resulting material was ground
and sieved through a 325-mesh screen. The temperature was then raised to 200°C.
and the heating continued for one-half hour. The heating was continued at 250°C.
for one-half hour which was then followed by heating at 275°C. until the powder
ignited. The final heating was at 325°C. for one and one-half hours. The resulting
material was ground and sieved through a 325-mesh screen.

The above heating process decomposed thermally the mixture of metallic salts
thereby producing an unsupported catalyst material which contained approximately
50 weight percent manganese and 50 weight percent silver. 50 weight percent of
carbon powder was added to and mixed with the decomposed salts, and was then
made into a PTFE-bonded fuel cell electrode by the method described in U.S.
Patent 3,297,484. The resulting electrode contained approximately 11 mg. of
manganese and approximately 11 mg. of silver per square centimeter.

Example 4:   A supported catalyst material containing 10 weight percent platinum
on carbon was purchased commercially. This supported catalyst was then made into
a PTFE-bonded fuel cell electrode by the method described in U.S. Patent 3,297,484.
The electrode contained 4.6 mg. of platinum per square centimeter

Example 5:   Carbon powder was formed into a PTFE-bonded fuel cell electrode by
the method described in U.S. Patent 3,297,484. This electrode contained 58 mg.
of carbon per square centimeter.

Example 6:   Five electrochemical cells were assembled as shown in Figure 5.6c.
Each cell employed a platinum anode, contained an electrolyte of 27 weight percent potassium hydroxide, and was operated at a temperature of 25°C. Oxygen
was supplied to each cathode as the oxidant while hydrogen was supplied to each
anode as the fuel.

These cells, which are set forth in the table on the following page under cell
numbers 1 through 5, which correspond to Examples 1 through 5, contain the
respective cathode of each of the above examples. In this table there is shown
the voltage in volts with internal resistance loss removed between the cathode
and a hydrogen reference electrode in the same electrolyte, and the current density
in milliamperes per square centimeter.

Current density, ma./cm.²	Cell 1, Cathode 1 to reference volts	Cell 2, Cathode 2 to reference volts	Cell 3, Cathode 3 to reference volts	Cell 4, Cathode 4 to reference volts	Cell 5, Cathode 5 to reference volts
1	0.93	0.93	0.95	0.95	0.85
2	0.92	0.91	0.94	0.94	0.84
5	0.90	0.89	0.93	0.92	0.82
10	0.89	0.87	0.90	0.91	0.80
20	0.87	0.85	0.88	0.89	0.78
60	0.82	0.79	0.83	0.87	0.71
80	0.81	0.77	0.81	0.86	0.67
120	0.78	0.74	0.78	0.84	0.57
160	0.76	0.72	0.76	0.82	0.48

It will be seen from the above table that cells 1, 2 and 3, which include a cathode having at least one non-noble metal made in accordance with the process provided high performance similar to cell 4 whose cathode includes a platinum catalyst. Cell 5 has a cathode which employs only carbon.

Disc Shaped Carbon Electrode

E. Fangradt and W. Wenzlaff; U.S. Patent 3,489,616; January 13, 1970; assigned to VEB Berliner Akkumulatoren- and Elementefabrik, Germany describe galvanic atmospheric-oxygen cell having a negative zinc electrode, a positive carbon electrode and an alkaline electrolyte. The subassembly is formed in a thermoplastic, preferably polyethylene, sealing ring by superimposed sequential arrangement of the carbon electrode, a suction layer preferably made from absorbent paper, a metal cap preferably having at least one aeration aperture, and a thermoplastic, preferably polyethylene, cover foil.

The sealing ring has offset portions, a double metal cup being fitted partly outside and partly inside the sealing ring, with an absorbent cuff, preferably made of a gas-permeable cellulosic material, disposed between the sealing ring and the outer portion of the metal cup.

The process thus involves a heavily chargeable galvanic atmospheric-oxygen cell or electrolytic cell with alkaline electrolyte which permits keeping the harmful influence of the carbon dioxide of the atmosphere away from the electrolyte to a large extent, and protecting the carbon electrode from diffusion through of the electrolyte by a rapidly responding valve action, as well as preventing the leakage of the electrolyte entrained by the gas escaping to the outside at excess pressure, and finally allowing the access of atmospheric air only upon taking the cell into operation.

This problem is solved in that a disc shaped carbon electrode is used in the electrolytic cell having an alkaline electrolyte, the end face of which is covered with a perforated metal cap which carries on the outside a cover foil and encloses by the inner side a suction layer lying on the carbon electrode and made of absorbent paper. The electrode is so extrusion-coated with a packing ring that an intimate electrolyte-proof connection exists between the surface area of the carbon electrode and the packing ring, and therefore after the cell is set in operation by removal of the cover foil and clearance of the access of atmospheric air, the cell becomes active only from the face covered with the metal cap, and the alkaline electrolyte contained in the electrolyte-absorbing separator cannot penetrate between the carbon electrode and the packing ring.

GENERAL BATTERY DESIGN AND CONSTRUCTION

Flat Primary Cells with Two Half Cells

A process described by R.L. Glover; U.S. Patent 3,607,430; September 21, 1971; assigned to Union Carbide Corporation relates to flat primary galvanic cells and refers more particularly to an improved method for manufacturing such cells.

With the growth of portable electrically powered devices such as radios, tape recorders and the like, and particularly those of thin compact design, has come an increasing demand for thin flat batteries for powering them. Although many types of flat cells have been developed in the past for such use, few if any, have proved to have all of the characteristics required. Cells of the type described, for instance, in U.S. Patents 2,870,235 and 2,995,614, although providing excellent service characteristics, have found only limited use partly because of bulk and partly because they have proved difficult and costly to manufacture.

This process is based on the concept of producing a cathode laminate or subassembly and an anode laminate or subassembly in separate operations and combining the two laminates to complete the cell. The method includes the steps of providing an electrically conductive film of thermoplastic resin, covering one entire surface of the film with cathode mix and applying a separator to the cathode mix. A layer of immobilized electrolyte is supplied for the cell and may be applied to the separator to complete the cathode laminate or may be applied simultaneously with the anode laminate or before the laminates are brought together.

In either case, the cathode laminate is placed upon a web of electrically nonconductive heat sealable thermoplastic film material which is provided with a substantially centrally located aperture for making electric contact to the electrically conductive film which serves as the cathode collector of the completed cell. The cathode subassembly is placed upon the web in such fashion as to have a marginal portion about the entire cathode collector. The anode laminate is prepared from a metal sheet which will serve as the anode of the completed cell.

Onto one surface of it is secured a second web of electrically nonconductive thermoplastic film similarly provided with an aperture for making contact to the metal and
similarly arranged with marginal portions extending outwardly of the periphery of the
metal sheet. The opposite surface of the metal sheet is provided with a layer of viscous electrolyte paste backed by a separator of microporous paper. The anode laminate should be larger than the cathode laminate.

The anode and cathode laminates or subassemblies so prepared are placed in juxta-
position with the two separator layers in contact with each other and in such position
that a margin of anode laminate surrounds the edges of the cathode laminate. The
marginal portions of each plastic web are in overlapping relationship.

After application of pressure to eliminate entrapped air between the cell components
and to insure intimate contact therebetween, the marginal portions of the two webs
are sealed together, preferably by heat sealing. To complete the cell a sheet of
metal foil is secured to the web on which the cathode laminate was placed, and the
foil is deformed through the aperture in the web to make physical and electrical contact with the cathode collector film. Referring to Figure 8.1, a cathode laminate
is prepared by providing a film (10) of electrically conductive thermoplastic material
such as a vinyl film loaded with carbon (Condulon).

FIGURE 8.1:   MANUFACTURING PROCESS FOR FLAT PRIMARY CELLS WITH TWO
HALF CELLS

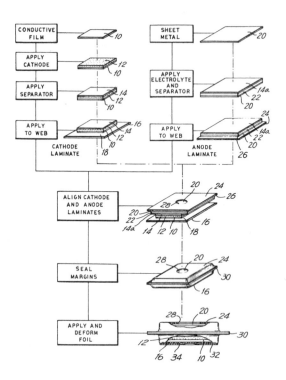

Source: R.L. Glover; U.S. Patent 3,607,430; September 21, 1971

Onto the film (10) is placed a cathode (12) by extrusion, printing, stencilling or doctoring. The cathode (10) may be the conventional Leclanche cathode mix of finely divided manganese dioxide and carbon particles. The cathode (12) is commensurate in size with the film (10). A separator (14) of conventional separator material such as microporous paper is placed on top of the cathode (12). In the example shown the cathode laminate is now placed on a web (16) of electrically nonconductive heat sealable thermoplastic film having a margin (18) extending outwardly of the periphery of the film (10). A central aperture in the web (16) is not visible.

To prepare the anode laminate a sheet (20) of anode metal such as zinc is provided. It is larger than the cathode laminate. As shown, a layer (22) of electrolyte is applied to one surface of the sheet (20) and above it is a separator (14a). In the Leclanche system the electrolyte is an aqueous solution of ammonium chloride usually containing zinc chloride which is immobilized to form a paste. The layer (22) may be applied by printing or stencilling technique. The electrolyte coated anode sheet (20) is now placed on a web (24) of the same material as the web (16). The web (24) has a marginal portion (26) extending outwardly of the sheet (20) about its entire periphery. This completes the anode laminate.

The cathode laminate and anode laminate are now aligned, as shown with the cathode laminate being bottommost and the anode laminate being turned over to place the separator (14a) in contact with the separator (14) and the marginal portions (18), (26) of the webs (16), (24) overlapping. The marginal portions of the anode laminate surround the edges of the cathode laminate. As shown an aperture (28) in the anode web (24) provides access to the anode (20) for electrical connection. The marginal portions of the webs (16), (24) are sealed together, preferably by conventional heat sealing, forming a seal (30) entirely surrounding the edges of the cell.

To strengthen the cell as well as to improve contact with the cathode a metal foil (32), for example, tin coated steel is adhesively secured to the outside of the web (16) and is deformed through the aperture (34) of the web (16) into contact with the cathode collector film (10). The cell is now complete.

Internally Bonded Cells

C.A. Grulke and T.A. Reilly; U.S. Patent 3,617,387; November 2, 1971; assigned to Union Carbide Corporation describe an electric current producing cell having a metal anode, a body of cathode-depolarizer mix and a cathode collector where the anode is physically separated from the body of cathode-depolarizer mix but is adhesively secured to one surface of the body of cathode-depolarizer mix by a polymeric adhesive providing an ionically conductive path between the anode and the body of cathode-depolarizer mix, and the cathode collector is adhesively secured to the opposite surface of the body of cathode-depolarizer mix by an electronically conductive polymeric adhesive.

Typical of adhesives which are useful in the process are polymeric materials which are adhesive, cohesive, conductive and stable. In addition, the polymeric material must be conductive. This conductivity can be ionic or electronic or both depending upon the particular function for which the adhesive is used. For example, when used between components of a single electrode, e.g., cathode and cathode collector, primarily to hold the components together, the adhesive must be electronically conductive and can be ionically conductive. The same is true when the adhesive is used between cells in a stack. However, when the adhesive functions as the electrolyte for the cell and separates the anode and cathode, it must be ionically conductive only. The adhesive can itself be conductive or can be modified as will be described in greater detail to provide a conductive path through the layer of adhesive. An additional requirement is that the adhesive be stable both with the materials of construction of the cell and with the by-products of the cell reaction. This stability is necessary during the shelf life of the cell and during discharge.

Suitable adhesive materials can be made from natural and synthetic condensation polymers such as wax, starch, methylcellulose, carboxymethylcellulose, hydroxyethylcellulose, locust bean gum, gum arabic, gum karaya and gum tragacanth. These polymers are preferably at least partially insolubilized, for example, through a cross-linking reaction with formaldehyde, to control electrolyte absorption by the adhesive.

Vinyl polymers are particularly useful for forming adhesives since they can be made to be stable in the presence of strong electrolytes and oxidants such as used in the cathode and generally can be made sufficiently permeable to electrolyte. These polymers are particularly well suited for use in preparing adhesive electrolytes since they may be ionically conductive by themselves or may be used in combination with liquid electrolytes containing added salts for increased conductivity. For example, the usual Leclanche electrolyte consisting of an aqueous solution of ammonium chloride and zinc chloride may be dispersed in the polymeric composition. The manner in which several specific materials of cell construction may be combined into complete unit cells is exemplified in Table 1.

TABLE 1: TYPICAL INTERNALLY BONDED CELL CONSTRUCTIONS

Anode	Electrolyte	Separator	Cathode-depolarizer [1]	Adhesive [1]	Cathode collector
Zinc	Water plus reaction product of poly acrylamide, urea and acetamide.	Tissue paper	Manganese dioxide	Reaction product of polyacrylamide, urea and acetamide.	Grafoil.
Do	Hydroxyethyl cellulose plus aqueous solution of magnesium chloride and ammonium chloride.	do	do	Hydroxyethyl cellulose	Do.
Manganese	Reaction product of polyacrylamide, urea and acetamide.	do	do	Reaction product of polyacrylamide, urea and acetamide.	Do.
Zinc	Polyvinyl chloride plus aqueous solution of magnesium chloride and ammonium chloride.	do	do	Poylvinyl chloride	Do.
Do	Hydroxyethyl cellulose plus aqueous solution of potassium hydroxide.	α-cellulose	do	Hydroxyethyl cellulose	Do.
Do	Aqueous solution of potassium hydroxide plus reaction product of poly acrylamide, urea and acetamide.	Nylon cloth	do	Reaction product of polyacryl amide,	Do.
Magnesium	Water plus reaction product of poly-acrylic acid and glyoxal.	α-cellulose	do	Reaction product of polyacrylic acid and glyoxal.	Graphite disk.
Do	Aqueous solution of potassium hydroxide plus reaction product of poly-acrylic acid and glyoxal.	do	do	Reaction product of polyacrylic acid and glyoxal.	Do.
Aluminum	Aqueous solution of potassium hydroxide plus reaction product of poly-acrylic acid and glyoxal.	do	do	do	Do.
Do	Water plus reaction product of poly-acrylic acid and glyoxal.	do	do	do	Do.
Lead	Polyvinyl chloride plus aqueous solution of fluoroboric acid.	do	Lead dioxide	Vinyl choride	Carbon coated zinc.
Zinc	Epoxy-amine plus aqueous solution of fluoroboric acid.	do	Manganese dioxide	Epoxy amine	Do.
Do	Melamine formaldehyde plus aqueous solution of zinc chloride and ammonium chloride.	do	do	Melamine formaldehyde	Do.
Do	Polyvinyl formaldehyde plus aqueous solution of zinc chloride and ammonium chloride.	do	do	Polyvinyl formaldehyde	Do.
Do	Hydroxyethyl cellulose plus aqueous solution of potassium hydroxide.	do	Mercuric oxide	Hydroxyethyl cellulose	Graphite.
Cadmium	Hydroxyethyl cellulose plus aqueous solution of potassium hydroxide.	Polypropylene.	Nickel oxide [NiO(OH)].	do	Nickel.
Zinc	Reaction product of polyacrylamide, urea and acetamide.	Tissue paper	Ceric oxide	do	Graphite.
Copper	Reaction product of polyacrylamide, urea and acetamide.	α-cellulose	Cupric oxide	Reaction product of polyacrylamide, urea, and acetamide.	Do.

[1] A conductive particulate material, e.g. carbon, is added to both cathode-depolarizer and adhesive for electronic conductivity.

Example 1: A 0.008" thick zinc disk measuring 0.5" in diameter was coated around its periphery with a vinyl adhesive and a circle of vinyl tubing was heat sealed to the adhesive and allowed to extend upward from the surface of the zinc to form a basket shape. Into the basket was placed one drop of electrolyte of the reaction product of polyacrylamide, urea and acetamide. A 0.0005" thick tissue paper separator was added and another drop of electrolyte. This assembly was allowed to air dry.

A cathode-depolarizer mix cake 0.5" diameter by 0.068" thick was prepared from finely divided manganese dioxide, acetylene black and insolubilized hydroxyethylcellulose in the approximate ratio of 50:6:1, respectively. The mix cake was added to the basket and a few drops of a conductive adhesive formed by adding carbon particles to a hydroxyethylcellulose solution was placed on the surface of the mix cake. A thin sheet of Grafoil coated with the carbon-hydroxyethylcellulose adhesive was added and the entire assembly was heat cured at 135°C. The voltage of the resulting cell was 1.68 v. and at a 22 to 30% water content in the electrolyte layer, displays a flat discharge curve at 1.20 bolts at 15 ma./sq. in. of electrode area.

Example 2: A thin sheet of zinc was coated on one side with a layer of carbon particles dispersed in vinyl resin. A 0.008" thick cathode-depolarizer mix layer of manganese dioxide, graphite and vinyl binder in a ratio of about 10:9:1, respectively, was added to the coated side of the zinc. The vinyl binder was a mixture of diisobutyl ketone, vinyl chloride-vinyl acetate copolymer and acrylate in a ratio of about 20:4:1, respectively. The assembly was air dried and cured at 150°C. and 4,000 psi of pressure.

Next the separator, a sheet of condenser paper, 0.004" in thickness, was impregnated with 10 mg./in.^{2} (based on the dry weight of the binder material) of the binder solution given above diluted 1:2 with diisobutyl ketone. The anode, a 0.016" thick sheet of zinc, was degreased and coated on one side with the same binder solution given above to form a thin layer (1.0 mg./in.^{2}, dry resin weight based on zinc area). The coating was then air dried. The vinyl treated separator was placed on top of the mix and the anode on top of the separator. The entire assembly was then heated to 65°C. at 15 psi pressure to seal all the layers to each other.

When the cell was cool, aqueous electrolyte was introduced to the interior of the cell through two small lateral holes previously drilled in the vinyl gasket. The electrolyte was a 10% by weight aqueous solution of fluoroboric acid. After the filling procedure was completed, the holes were sealed with hot wax containing a plasticizer. The resulting cell was 3" in diameter and was discharged at 8 ohms resistance in series and performed as set forth in Table 2.

TABLE 2

Time (minutes)	Voltage	Amperage
0	1.61	0.200
5	1.57	0.128
10	1.52	0.132
15	1.44	1.27
20	1.37	1.20

(continued)

TABLE 2:   (continued)

Time (minutes)	Voltage	Amperage
25	1.31	0.116
30	1.27	0.112
40	1.20	0.110
100	1.00	0.078

Example 3:   A cell assembly was made using a zinc anode, lead dioxide cathode and a paper separator. A modified epoxy thermosetting resin was used to bond the components together to yield a cell possessing good adhesion even after storage at high temperatures.

A carbon coated zinc electrode, normally referred to as a duplex electrode was selected as the cathode collector. Onto it was applied a thin layer of a conductive adhesive consisting of approximately 50% graphite particles, 30% of an epoxy resin having an epoxide equivalent of 180 to 195, an average molecular weight of 350 to 400 and a viscosity of 11,000 to 13,500 cp. at 25°C. and 20% of a polyamine curing agent having an amine value of 290 to 320 and a viscosity of 40,000 to 60,000 centipoise at 25°C. in methyl acetate.

A mixture of equal parts lead dioxide and graphite containing 1 part of epoxy-amine binder for each 50 parts of lead dioxide-graphite mixture was spread onto the adhesive coated carbon surface, air dried and pressed at 2,500 psi into a 0.008" thick cake. A 0.004" thick condenser paper separator was impregnated with 10% by weight epoxy-amine binder and air dried. The separator was applied to the mix cake and the anode layer of zinc was applied to the separator.

The entire assembly was held under a pressure of 25 psi at 75°C. for 15 minutes to thermoset the epoxy-amine adhesive. At the time of testing a 10% by weight aqueous solution of fluoroboric acid was added. The resulting cell had a 1 in.^{2} electrode surface and, when tested under 8 ohms resistance in series, performed as follows.

TABLE 3

Time (minutes)	Voltage	Amperage
0	2.15	0.160
5	1.80	0.120
10	1.82	0.122
15	1.82	0.122
20	1.75	0.118
25	1.70	0.116
30	1.65	0.111
35	1.61	0.107
40	1.57	0.104
200	1.80	0.105

The adhesives of the process are effective in greatly reducing the internal resistance within the cell thereby making possible construction of thin cell batteries having relatively high current capacities.

Mass Production of Batteries Using Stacked Flat Components

In a process described by J.F. Jammet; U.S. Patent 3,679,489; July 25, 1972; assigned to Ste. des Accumulateurs Fixes et de Traction SA, France batteries are mass produced by initial formation from plastic material of trays of substantially inverted truncated pyramidic form. These formed trays are perforated at their bottoms and have adhesive applied around the openings on internal bottom faces of the trays.

Then, the perforated trays have mounted therein successively a negative duplex electrode, for example, of zinc whose lower face is covered with a conductive coating, the superposed separator impregnated with electrolyte lying over the upper face of the electrode and a depolarizer positive active material is superposed over each separator. The filled trays are severed and nested successively. Then, respective positive and negative contact terminals are applied at the opposite ends of the battery assembly. Thereafter, the so-formed battery assembly is mounted in stretched outer sheath of extensible plastic material.

The process thus relates to a battery of electrochemical generators comprising flat constituents wherein the flat constituents are respectively piled up in successive trays which are themselves stacked and nested in each other and then compressed, characterized by the fact that the battery thus assembled is additionally contained in an outer plastic extensible sheath which has been contracted after insertion into it of the battery and is submitted by the contraction to axial as well as radial compression, this compression being practically uniform from one cell to the other in the entire height of the battery, the dimensions of the sheath, prior to stretching and mounting, being inferior to that of the battery. The advantage of such an external sheath is that the piled or stacked generators are each submitted to individual compression.

Referring to Figure 8.2a, the tray forming apparatus comprises the following. A fixed hollow body (2) bordered by a frame (3), wherein a plate (4) provided with bosses such as (5), whose external dimensions correspond to the internal dimensions of the trays to be formed is movable reciprocally. A fixed screen (6), having apertures through which the bosses (5) can pass spans the body (2), being appropriately supported by the frame (3).

The bosses (5) are shaped in conformity with the desired taper of the corresponding trays to be formed. This taper has been intentionally exaggerated in the figures to facilitate understanding of the manufacturing process.

A chamber (7) is provided in body (2) beneath the plate (4) into which either vacuum can be created or else compressed air can be injected, for example, as illustrated by the hollow pistons (8) extending from the plate (4) and which may serve for moving the plate (4) carrying the bosses (5). These hollow pistons may have either vacuum or compressed air directed therethrough from suitable sources and communicate with chamber (7) via passage (8'). A movable hollow body (9) bordered by a frame (9') having the dimensions of the frame edge (3) corresponding shape, in which heating elements such as infrared lamps (10), for example, are fixed is mounted movably

FIGURE 8.2:   MASS PRODUCTION OF BATTERIES USING STACKED FLAT
COMPONENTS

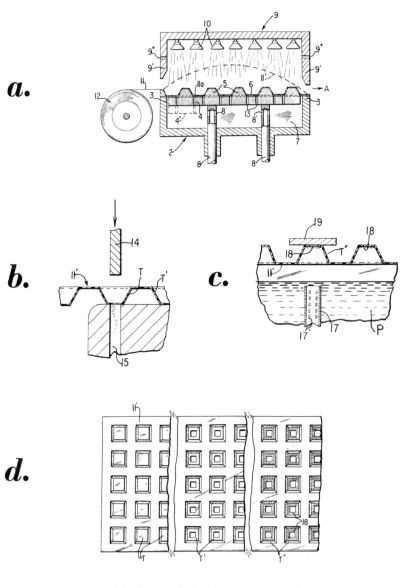

a.

b.

c.

d.

(a)   Detail of Manufacturing Apparatus

(b)   Detail of Perforation Apparatus

(c)   Detail of Coating Apparatus

(d)   Top View of Strip

(continued)

FIGURE 8.2: (continued)

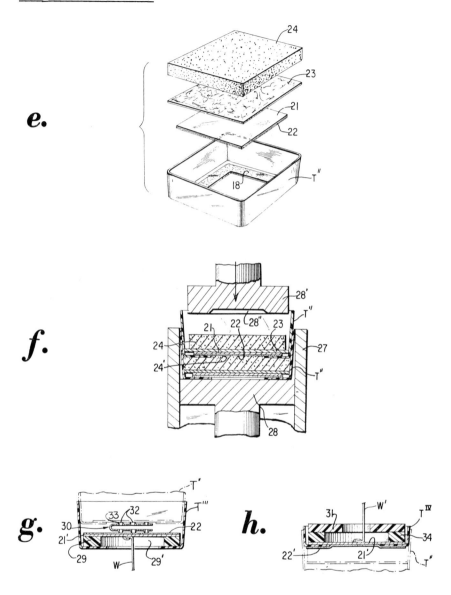

e.

f.

g.

h.

(e) Exploded View of One Cell

(f) Stacking and Compression Step

(g)(h) Cross-Sectional Views of Opposite Ends
        of Stacked Cells with Terminals

(continued)

FIGURE 8.2:   (continued)

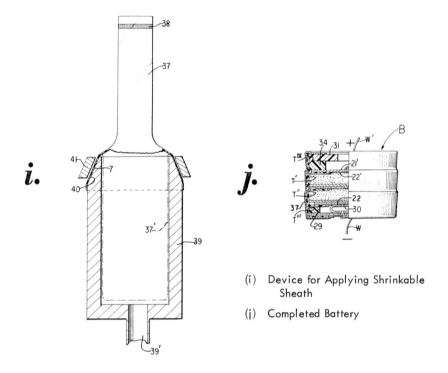

i.

j.

(i)   Device for Applying Shrinkable
       Sheath

(j)   Completed Battery

Source: J.F. Jammet; U.S. Patent 3,679,489; July 25, 1972

over body (2). It is provided in its frame edge (9') with vent passages (9"). Tray forming with the described apparatus proceeds as follows. A length (11) of the plastic strip is unwound intermittently from a reel (12) of such strip and is moved into place between the two bodies (2) and (9) so that when the body (9) is lowered downwardly, a stretched length (11) of the strip is clamped between upper edges of the lower bordering frame (3) and the lower edges of the upper frame (9') in practically airtight locking relationship.

Advantageously, as seen in the figure, the upper edge of the frame (3) at the outlet side for the strip (11), right side in Figure 8.2a, is lower than the other portions of the upper edge of the frame (3) so that the strip length can be moved readily into position.

According to the French Patent 1,252,174, describing a process for forming, vacuum is created in the chamber (7) and it is sufficient to press onto the upper surfaces of the bosses (5) the square portion of the length (11) of the plastic sheet that is locked between the frames (3) and (9). It is softened now by heating supplied by lamp (10) to its upper side. This technique does not yield satisfactory results with all plastic materials, being, for example, good with polystyrene but not with polyvinyl chloride plastic. According to the process, when the length (11) of the sheet which has been locked between the frames (3) and (9) has been heated by lamps (10), the sheet is stretched by injecting a fluid, such as air for example, with the necessary pressure into chamber (7) via passage (8') to blow up the softened length (11) of the plastic sheet to the position shown by the dotted line (11'). Afterwards, the heated stretched plastic sheet length (11) will be able to conform perfectly well to the shape of the bosses, for example, by creating vacuum in the chamber (7) via passages (8').

During this last vacuum step, the plastic sheet length (11) is no longer heated. Such a vacuum formation of this length (11') to initial tray conforming shape on bosses (5) is of very good quality, providing a practically constant wall thickness in the formed trays of plastic sheet length (11') and smooth surfaces. It should be noted that the screen (6) is very favorable to this formation because it permits localizing the action of vacuum transmitted through channels (13) defined in plate (4) defined about the peripheries of the bosses (5), i.e., at the zones where the last part of the formation of the plastic material occurs.

When tray formation has been finished the body (9) is elevated and at the same time air can be blown over the upper surface of the tray formed sheet length for cooling and hardening or setting. The tray formed sheet (11') is then automatically disengaged from the bosses (5) by lowering the plate (4) downwardly to the position shown by dotted outline (4').

It is to be noted that thereby since the right edge of the frame (2) is at the level of the screen (6), the tray formed strip length (11) is readily moved in the direction of arrow (A) at a determined pace by applying only a slight pulling force, for example.

The flat portion (11a) of the formed length (11') then arrives at the right end of the screen (6) and the body (9) is again lowered onto body (2) so that tray forming operation can be repeated on the next length (11) of strip, the device being designed to operate intermittently so that the tray formation can succeed longitudinally at regular intermittent intervals on successive lengths of strip drawn from reel (12) as is shown in Figure 8.2d where the formed trays (T) in strip length (11') (openings towards the observer) each have the shape of regular truncated pyramids.

The strip (11) is moved intermittently through the forming apparatus thus far described and in the depicted example each intermittent movement provided for the simultaneous production of 25 trays (T) at a time. At the outlet side of the forming device so far described, the strip lengths (11') bearing the formed trays (T) therein opening originally downward, is inverted or turned upside down opening upwardly so as to cut or punch out and remove the bottom center portions of each tray (T) to provide trays (T').

A punching out mechanism for such purposes is shown in Figure 8.2b which is a partial diagrammatic view of the perforating or punching out device. It comprises a plurality of punches such as (14) of a substantially square cross-section and corresponding underlying dies (15). This device is preferably designed to perforate the trays at the tray formation speed so that production can proceed substantially continuously. The result appears in Figure 8.2d where reference number (T') designates perforated trays whose bottoms have been perforated by punches (14) and dies (15). Beyond discharge end of the perforating or punching out device, the strip length bearing perforated trays (T') is again turned upside down and trays move forwards intermittently to another successive position illustrated in Figure 8.2c where an approximately square shaped hollow plug (17) with openings such as (17'), is dipped into an adhesive such as pitch (P), dissolved in a solvent as petrol, for example.

This plug is elevated and upon engaging the bottoms of the trays internally coats them around the respective perforations with an adhesive layer such as (18) seen in Figure 8.2d. During this coating step, the strip tray bearing (11') is held in place by appropriate holders and a counterplate (19) as seen in Figure 8.2c. This coating of the perforated trays (T') with adhesive is preferably effected at the speed of the conveyed strip to avoid accumulation of strip lengths between two consecutive stations.

Such a coating requires the use of a relatively fluid adhesive and, therefore, a strip after having been coated cannot be used immediately. It must be allowed to rest for a few hours so that the solvent can evaporate at least partially. Preferably, this is done by winding to strip bearing coated trays upon a roller or reel and temporary storage thereon.

When the adhesive has sufficiently solidified, the roller on which the strip has been wound and stored moves to the next stage of treatment where it is unrolled, openings of trays being directed upwardly. Then coated and perforated trays (T") are filled with the constituents of the cells. Successively, as shown in the exploded view of Figure 8.2e for a tray (T") number one, a duplex electrode with dimensions inferior to that of the tray into which it is to be centered by mechanical means is mounted in the tray (T"). This electrode is constituted by a zinc plate (21) whose lower face is coated with a conductive layer (22) on which the adhesive (18) previously applied over the bottom of the tray (T") adheres preventing electrolyte from intercell leakage, and so ensuring intercell sealing.

Then, a separator (23) of a square sheet of paper previously impregnated with electrolyte and then dried and having dimensions corresponding to those of the tray above the duplex electrodes and thus larger than the zinc plate (21) is mounted over the uncoated face of the plate (21).

The duplex electrodes and the separator are preferably cutoff successively from very long strips all of a suitable width so that no waste is possible. After severance, they are transferred and deposited inside the tray (T") as by pneumatic means. Since zinc plate (21) is of smaller dimensions than the bottom of the tray (T"), the separator can be constituted by a flat piece instead of being shaped as a hollow part with turned up edges because no direct contact can occur even accidentally between the zinc plate and the depolarizer mix owing to the fact that the zinc plate is centered and fixed to the bottom of the tray (T") by the adhesive (18).

Then, a depolarizer mix (24) impregnated with electrolyte and severed in an appropriate shape from a strip of compressed mix is deposited in the tray (T") over the separator (23) as by using a kind of core catcher ensuring its transfer into the tray (T").

The strip (11') further conveys the trays now filled with the cell constituents and in a following step (not illustrated) the filled trays (T") are inserted into corresponding apertures arranged in an endless belt, wherein they are maintained and transferred to a severing device such as a saw located just beneath the belt in order to separate individual trays (T") from the strip. The severed individual trays drop on a plate beneath the saw and are transferred thereby to the next operational position which is stacking of filled trays (T"). This stacking is effected in such a way that the individual cells formed by filled trays are series connected as is seen in Figure 8.2f where a tubular form (27) having dimensions corresponding to those of the filled trays (T") is provided and in which a lower piston (28) can move.

When no downward pressure is exerted on the piston (28) it is kept in balance in any position inside the form (27) in which it can move by friction. Such friction provides a braking action and an inertia effect against its displacements subsequent to the introduction of trays (T") filled with their constituents during the stacking operation.

In Figure 8.2f, two filled cell trays (T") are shown nested in each other. Before introducing the second tray, the lower cell tray has already been automatically placed in the form (27) and has had its constituents compressed by a downward movement of the upper piston (28') so that the piston (28) also is pushed down allowing the introduction of the second tray into the one already encased in the form (27). The second cell tray is then compressed in turn by the piston (28') as shown upon its moving downwards.

The compression exerted by the piston (28') on each successive cell tray introduced into the form (27) effectively brings the individual constituents of each cell tray nearer together, so promoting satisfactory electrochemical operation and in addition securing efficacious series connection of the cells. In fact, during this compression the upper piston (28') whose bottom face has an appropriate hollow shape (28") provides the mix with a central boss (24') which will lodge in the perforation of the bottom of the next following tray superposed on it.

In this way, the conductive coating (22) on the bottom face of each duplex electrode of each cell is effectively in contact with the depolarizer mix (24) of the next preceding cell, and this ensures good electrical series connection between adjacent cells.

When a specified number of filled trays has been piled up to form a stack successively compressed during piling, the piston (28) is elevated to permit the removal of the stack from the form (27) so that a terminal applying operation can follow and a special tray provided with the positive terminal is mounted in the top tray of the stack. Another tray fitted with the negative terminal is mounted at the bottom of the lowermost filled tray (T") of the stack. These negative and positive terminals are shown with their special trays (T''') and (T^{iv}) in Figures 8.2g and 8.2h, respectively.

As seen in Figure 8.2g, the negative or lower terminal is constituted by a tray (T''') in which is positioned a plastic frame (29) defining a space (29') between the bottom of the tray (T''') and the zinc sheet (21) of a duplex electrode, permitting completion of correct stacking notwithstanding the presence of a U-shaped metal part (30) provided with small slots such as (32) on both legs.

The rims of these slots, on the external sides of the U-legs, bear burrs (33) capable of penetrating the conductive layer (22) of the duplex electrode (21) in the tray (T") just above tray (T'''). In this manner, the negative terminal can simply be completed by welding a wire (W) to the zinc sheet (21) which appears in tray (T''') through the open frame (29) and then by pouring wax or other sealing compound in the empty volume (29). The positive terminal shown in Figure 8.2h comprises a perforated tray (T^{IV}) having a height inferior to that of the trays (T"), in which is placed a zinc plate (21') held in place by means of a sheet of conductive coating (22') applied to the external side of the bottom of the tray (T^{IV}).

In this tray (T^{IV}) above, the zinc plate (21') a square piece (31) of insulating material such as cardboard, perforated in its middle is positioned and adhered to a frame part (34). The thickness of this assembly is such that it projects substantially beyond the upper edges of the tray (T^{IV}), i.e., above the upper edge of the tray (T") containing cell constituents located directly below tray (T^{IV}), since these two edges are nearly at the same level. This arrangement permits final compression efficiently of the assembly of piled trays (T"), (T''') and (T^{IV}).

After the stack assembly has been fitted with both terminals, it is then submitted to a global compression of a few kg./cm.2 approximately. The side walls of all trays and external parts of end surfaces are then coated with several layers of wax or paraffin in order to insure sealing between individual cells in superposed trays (T"); to confer more rigidity to the stack; and to protect all the trays against the attack of a solvent during further manufacturing procedure, such protection being absolutely necessary when the trays are made of polystyrene.

Thereafter, as a next step, the coated battery assembly is submitted to a final compression which serves to bring its dimensions to definitive values. The process for effecting this compression comprises as a first stage utilization of the elasticity of a sheath made of a plastic material of smaller dimensions than the assembly which is overstretched to permit introduction of a battery assembly and, a second step, to deplastify the sheath already tightened and contracted around the stack, so that the diameter and length of the sheath further contract and hardens or sets to maintain the stack assembly and all its components in a definitive state of permanent compression.

The advantage of such a sheathing procedure is to provide an individual compression of those cells in the stack and may be effected, for example, by the following manner as is illustrated in Figure 8.2i. A flexible plastic tube (37), for example, of plasticized polyvinyl chloride is cut to desired length and closed at one end (38) as by welding. The diameter of this tube (37) is initially smaller than that of a battery assembly so that the latter cannot be introduced directly into it until it has been stretched.

This operation is effected by means of a tubular form (39) provided with an upper opening whose external walls (40) are cone-shaped and over which a clamping ring (41) can be slipped. The plastic tube (37) is first dipped in a solvent such as trichloroethylene and its open end is then forced over the conical wall (40) and then clamped in place by the ring (41). Any other suitable plastic that is stretchable and subject to additional contraction by treatment with a suitable deplastifying agent may be used for tube (37).

Then vacuum is created inside the tubular form (39) via its bottom opening (39') so that the part of the tube (37) outside the form (39) inverts or reverses itself and enters into close stretched contact with the inner walls of the form (39), which has dimensions that are slightly larger than those of the battery assembly stack [this reversed position is shown by dotted line (37')]. As a result of the inversion and stretching of tube (37) to condition (37'), the battery assembly can easily be introduced in the stretched tube (37') in tubular form. Then, when vacuum is broken at the lower opening (39') of tubular form (39) and ring (39) removed, the sheath (37') contracts and its sheath closes around the inserted battery assembly and compresses it and its constituents (41).

This operation can be very quickly performed before the deplastification resulting from the aforesaid dipping in solvent of tube (37) occurs. Then the two ends of the tube can be welded as by high frequency electric fields to shut the sheath at both ends of the confined assembly. When the plastic sheath or tube (37) contracts and shrinks back, the portions of the tray walls projecting higher, the mix in each respective nested tray are pressed against the external tray sides adjacent the bottom of the next upper tray.

This compression effect of sheath (37) in itself could be sufficient to ensure intercell sealing in the assembly; however, it is advantageous to combine this effect with the previous coating of wax or paraffin applied to the assembly to improve sealing reliability.

On the other hand, when the tube or sheath (37) shrinks initially as a result of stretch release, its ends fold over the upper and lower ends of the assembly so that longitudinal contraction of the sheath also causes a compression of the assembly which proceeds from cell to cell since the sheath engages also the flanged rims of the trays.

The second contraction of the sheath resulting from solvent evaporation emphasizes this effect. Thus, the overall global contraction effected by the sheath is the result of the sum of its partial contractions and engagement at the rims of piled trays effected by sheath (37). A completed battery B comprising a stack of trays and cells with shrunk-on or contracted outer sheath (37) applied as herein described is illustrated diagramatically on an exaggerated scale in Figure 8.2j.

More complete details of the use of the shrinkable plastic sheath technique for encasing batteries comprised of a stack of flat cells are given by J.F. Jammet; U.S. Patent 3,597,276; August 3, 1971; assigned to Ste. des Accumulateurs Fixes et de Traction SA, France.

Immobilized Adhesive Electrolyte

A battery described by W.H. Deierhoi, Jr.; U.S. Patent 3,563,805; Feb. 16, 1971; assigned to Union Carbide Corporation comprises a thin, flat primary galvanic cell having a sheet metal anode and a cathode of oxidic depolarizer mix with an electrolyte permeable separator, the cathode being within and bounded by a frame of moisture-impervious sealing material which is marginally sealed to the anode and to a cathode collector which is in intimate contact with the cathode.

An immobilized tacky, viscous electrolyte in contact with anode and cathode adhesively secures the two together. All components of the cell are adhesively secured in intimate contact, and no external sealing or pressure-exerting means is employed. The frame of sealing material surrounding the cathode aids to rigidify the cell and in maintaining low electrical resistance contacts between the cell components in cooperation with the sheet metal anode and the cathode collector to which it is marginally sealed. The process also comprises a battery of such cells arranged in electrical series relation and adhesively secured together without external pressure-applying means. In a preferred example when a battery is to be prepared the cathode collector of one cell is provided as a coating carried on and bonded to a surface of a sheet metal anode for a neighbor cell. The electrolyte for the cell is confined within the space defined by anode, cathode and frame.

Referring to Figure 8.3a, a number of component parts for assembly to produce the cell is shown. An anode subassembly comprises an anode (10) of sheet metal; a marginal layer (12a) of an adhesive sealing material to be placed on the anode (10); a layer (14a) of immobilized electrolyte; a separator (16) with marginal layers (12b), (12c) of adhesive, one on each side of the separator (16); and a second layer (14b) of electrolyte.

A cathode subassembly comprises a layer (14c) of electrolyte to be placed adjacent one surface of a cathode (18). A layer (20) of adhesive is provided between the other surface of the cathode (18) and a cathode collector (22). A fourth layer (12d) of sealing material is provided to be in contact with the cathode collector (22). A flash coating (24) of highly conductive metal such as gold or silver may be provided for the cathode collector (22).

Reference to Figure 8.3b will show that when the various components are assembled the layers (12a), (12b), (12c) and (12d) of adhesive are joined to form a frame (12) which marginally bounds the cathode (18) and is secured to the anode (10) and cathode collector (22) of the cell. The layers (14a), (14b), (14c) of electrolyte provide the cell electrolyte (14) which is in ionic or electrolytic contact with the anode (10) and cathode (18). The separator (16) performs its conventional function of preventing physical contact between anode (10) and cathode (18) and is permeable to the electrolyte.

To provide a battery of cells like that shown in Figure 8.3b, it is necessary merely to place the anode of one cell in intimate electronic contact with the cathode collector of another thus effecting a series connection. A more desirable construction for a series stack battery is one having a duplex electrode in which one surface of a sheet metal anode is provided within a coating of conductive carbonaceous material such as has been used for many years in primary batteries. Such a construction is shown in Figure 8.3c.

As seen in Figure 8.3c two cells are arranged one above the other and adhesively secured together. In the lower cell a sheet metal member (30) has an electronically conductive coating (32) of carbonaceous material bonded to one surface thereof onto which is secured a cathode (34) by a layer (33) of adhesive. The cathode (34) is bounded by a frame (36) of plastic adhesive material marginally secured to the coating (32).

A layer (38) of electrolyte surmounts the cathode (34) and is in ionically conductive contact therewith and with a separator (40) above it. Also in ionically conductive contact with the separator (40) is a sheet metal anode (42) the marginal portions of which are secured to the frame (36) of adhesive material. On the upper surface of the anode (42) is a layer (44) of conductive carbonaceous material which serves as the cathode collector for the upper cell. The upper cell includes a cathode (46) bounded by a frame (48) of adhesive sealing material and secured to the cathode

FIGURE 8.3:   THIN, FLAT PRIMARY CELLS AND BATTERIES

a.

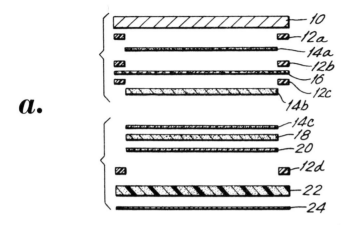

b.

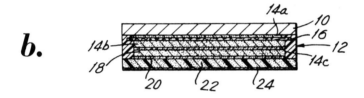

c.

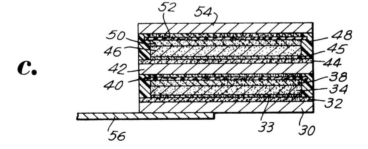

(a)   Exploded View of Parts

(b)   Assembled Cell

(c)   Assembled Battery

Source: W.H. Deierhoi, Jr.; U.S. Patent 3,563,805; February 16, 1971

collector (44) by a layer (45) of adhesive. A layer (50) of electrolyte is provided in contact with a separator (52) to which is adhesively affixed a sheet metal anode (54) the marginal portions of which are bonded to the frame (48). The stack of cells is thus connected in series relationship. It is secured to a conductor (56) which may, if desired, be connected to a second stack of cells.

An important feature of construction of cells and batteries is that it permits the use of electrolytes which are aqueous solutions and thus to maintain wet contact between appropriate cell elements. The electrolyte is immobilized by a thickening agent to produce a viscous tacky or sticky mass which can be held within the bounds of the adhesive frame without leakage from the cell. In the thin, flat cells this mass in addition to providing the electrolyte serves to hold the cell components together and the internal frame in cooperation with the anode member and cathode collector serve to maintain good, low resistance contact throughout the cell and to rigidify it.

A large number of cells and batteries embodying the process have been made and tested with good results. For example, in one series of tests cells of the general configuration and construction shown in Figures 8.3a and 8.3b were made utilizing a zinc anode 2" by 3" and 0.008" thick. The zinc was of high purity (99.99%) and one surface was amalgamated in conventional manner. A marginal layer of a vinyl adhesive was provided on the amalgamated surface by stencil. In some cases the margin was 1/4" wide, in others 5/32".

A thin layer of polyacrylamide thickened electrolyte was brushed onto the amalgamated surface of the anodes. The electrolyte contained 24% ammonium chloride, 6.5% zinc chloride and 3.3% polyacrylamide by weight in aqueous solution. Its viscosity was about 2,300 cp.

To both sides of a rag paper sheet 0.001" in thickness and 2" by 3" were applied marginal layers of the vinyl adhesive which had been applied to the zinc anode. The adhesive layers were dried by heating the paper in an oven in air at about 135°C., and the separator was then placed on the electrolyte coated anode with the adhesive margins congruent to complete the anode assembly, care being taken to press the separator firmly against the wet electrolyte to eliminate air bubbles. A layer of the same electrolyte was then applied to the upper surface of the separator.

The cathode subassembly was prepared by providing as a cathode collector a sheet of electrically conductive vinyl film (Condulon), the sheet being of the same dimensions as the anode and 0.007" in thickness. Onto one side of this film was vacuum deposited a flash coating of gold. Onto the marginal portion of the other side (uppermost) was deposited a layer of the vinyl adhesive used in the preparation of the anode assembly and of the same dimensions.

Within the boundaries of this deposit was placed a layer of an aqueous solution of polyacrylamide containing finely divided graphite, the layer serving as an adhesive for securing the cathode to the collector. It contained 5.5% polyacrylamide and 4 parts by weight of graphite with 1 part by weight of acetylene black. This layer was dried in air at 70°C. Onto this adhesive layer was then placed the cathode mix. The cathode mix was conventional containing finely divided manganese dioxide and about 18% finely divided carbonaceous material to improve conductivity and about 1% binder. It was wet with electrolyte. In some cases, the mix was applied in successive layers and in others in one layer.

In any event care was taken to deposit only within the boundaries of the vinyl adhesive margins. After application, the cathode was dried by heating in air at about 135°C., and a layer of electrolyte of the same composition above described was applied to its upper surface to complete the cathode subassembly.

To complete the cell the anode and cathode subassemblies were pressed together while the electrolyte layers on the separator and the cathode mix were wet and tacky. Care was taken to remove residual air and to align all of the layers of vinyl adhesive. When this had been accomplished the assembled cell was maintained in position and heat and pressure were applied to the marginal portions to seal all of the marginal layers of adhesive together. The overall thickness of cells produced as described ranged from about 0.035" to 0.045", variation being due to the use of thicker cathodes in some cases.

Performance tests of cells constructed in the manner described having a thickness of 0.045", and a cathode weight of 1.2 g., an anode weight of 5.8 g. and an active area of 4.0 in.^{2} show the cells to have a nominal open circuit voltage of 1.6 v., a theoretical zinc capacity of 4.82 ah. Typical voltage and current density characteristics at various loads are tabulated.

Load (ohms)	Closed Circuit Voltage	Current Density ma./in.^{2}
10	1.56	43
5	1.53	85
2	1.46	185
1	1.36	386
0.5	1.19	699

In other tests, cells of 4.0 in.^{2} active area have delivered 1.5 amp. for pulses of 1.25 seconds duration with 5 seconds recycling. The current density under load was 0.375 amp./in.^{2}. In some cells twenty such pulses were delivered before voltage fell to 1.0 v. For a specific requirement, pulses of 1.2 amp. for 1.1 seconds are called for with a 3 to 5 seconds recycle, and the cells are required to deliver ten pulses of this character over a temperature range of 30° to 120°F. Cells of the process more than met this test as will be seen. In some tests as many as fifty pulses have been obtained.

Other tests have shown that cells constructed by this process are capable of delivering required voltage at lower drains for long periods of time. For extra protection against loss of moisture from the cells utilizing the electrolyte described herein, a conventional hot melt adhesive may be used instead of the vinyl solution adhesive employed in the example.

Seal for Wafer Cells

In a process described by H.J. Strauss; U.S. Patent 3,525,647; August 25, 1970; assigned to Clevite Corporation an electrical battery is provided comprising a stack of wafer or flat cells, particularly cells utilizing an alkaline electrolyte. An improved seal is provided between the electrodes or terminals of each cell and the adjacent integument covering at the periphery of apertures provided in the covering to permit electrical contact to be made between adjacent cells. The seal comprises a complete ring of an adhesive bonding the film or integument covering to the electrode or terminal about the aperture. Additionally, an electrolyte absorbent material, as for example, carboxymethylcellulose, in the form of a dry film, completely surrounds the adhesive seal.

Any electrolyte which leaks around the edges of the electrode or terminal and which might otherwise penetrate an area where the seal between the adhesive and electrode surface is imperfect is instead absorbed and retained by the absorbent material, thereby preventing any attack of the leaking electrolyte upon the electrical connection to the cell.

MULTICELL CONSTRUCTION

Duplex Electrode Construction Using Conductive Plastic Carrier Strip

In a process described by B.C. Bergum, J.M. Bilhorn, K.H. Kenyon, W.R. Macaulay and J.A. Youngquist; U.S. Patent 3,694,266; September 26, 1972; assigned to ESB Incorporated duplex electrodes are constructed by placing positive and negative electrodes in contact with opposite sides of a continuous, electrically conductive plastic carrier strip. Use of the carrier strip as a substrate permits the positive and negative electrodes to be made from compositions which, during the construction of the duplex electrode, are unable or poorly suited to function as a substrate. Use of an electrically conductive carrier strip permits electrical current to be conducted between the electrodes without additional components or assembly steps.

The positive and negative electrodes are applied in intermittent deposits along the carrier strip with a deposit of positive electrode being centered opposite a deposit of negative electrode. During this construction process the resulting duplex electrodes are structurally and electrically connected together. The structural connection is desirable because high speed production machinery is better able to process flexible continuous strips than individual pieces.

The duplex electrodes are then assembled into multicell batteries. The assembly preferably occurs while the duplex electrodes are structurally and electrically connected by the continuous plastic carrier strip after which the carrier strip is subsequently cut between duplex electrodes to obtain structurally and electrically unconnected batteries. Alternatively, the carrier strip may be cut between duplex electrodes before those electrodes are assembled into multicell batteries.

Figure 8.4a is a schematic diagram showing a continuous, electrically conductive plastic carrier strip (50) from a roll or some other source of supply (250) being passed to the positive and negative electrode applicators (220) and (230), respectively, where the applicators place intermittent patch deposits of positive and negative electrodes (20) and (30), respectively, on opposite sides of the carrier strip from each other. Each patch deposit of negative electrode is substantially opposite a patch deposit of positive electrode. The applicators (220) and (230) may be spaced opposite one another so that they make their opposing patch deposits simultaneously, or they may be spaced apart so that one applicator first makes its patch deposit and later the

FIGURE 8.4:   MULTICELL BATTERY CONSTRUCTION USING DUPLEX ELECTRODES

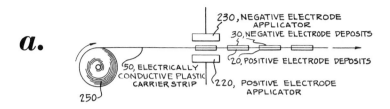

a.

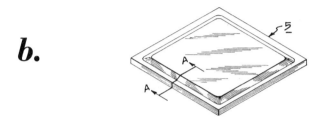

b.

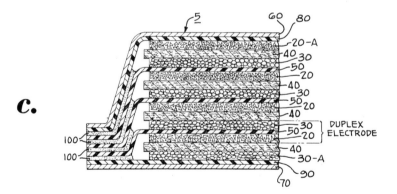

c.

(a)   Continuous Conductive Plastic Strip Carrier for Electrode Application

(b)   Oblique View of Multicell Battery

(c)   Cross-Section Along Line (A—A) of Figure 8.4b

(continued)

FIGURE 8.4:   (continued)

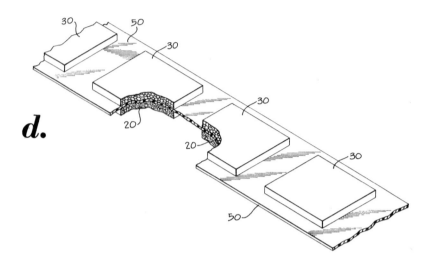

d.

Portion of Conductive Strip with Electrodes in Place on Both Sides

Source: B.C. Bergum, J.M. Bilhorn, K.H. Kenyon, W.R. Macaulay and J.A.
             Youngquist; U.S. Patent 3,694,266; September 26, 1972

other applicator makes the opposing patch deposit.   Figure 8.4d illustrates a portion
of the electrically conductive carrier strip with patches of positive and negative
electrodes placed on opposite sides.   It will be noted that a segment of the carrier
strip having a positive and negative electrode on its opposite sides defines a duplex
electrode, also known as a bipolar electrode.   As can be seen from Figures 8.4a and
8.4d, the duplex electrodes are structurally and electrically connected together
immediately after the carrier strip passes by the second of the two applicators; these
physical and electrical connections between duplex electrodes subsequently must be
broken but this can be done either before or after the duplex electrodes ãre assembled
into multicell batteries.

The eventual multicell battery product is the same whether the duplex electrodes constructed as shown in Figure 8.4a are assembled into multicell batteries before or
after being structurally and electrically unconnected from each other.   Figure 8.4b
shows such a multicell battery (5) in an oblique view.   Figure 8.4c shows a portion
of the multicell battery (5) in magnified cross-section and illustrates members of the
battery.

As Figure 8.4c shows, the battery (5) comprises the combination of an outer positive
electrode (20-A), an outer negative electrode (30-A) and at least one duplex electrode between electrodes (20-A) and (30-A), each duplex electrode being of the type
constructed by the method shown in Figure 8.4a.

As shown in Figure 8.4c, a duplex electrode comprises the combination of a segment of electrically conductive plastic carrier strip which functions as the intercell connector of the duplex electrode, together with deposits of positive and negative electrodes (20) and (30) respectively on the opposite sides of the segment. The multicell battery (5) also includes an electrolyte impregnated separator (40) between each positive electrode (20) or (20-A) and each negative electrode (30) or (30-A).

The multicell battery thus includes at least two cells, each cell comprising the combination of a positive electrode (20) or (20-A), a negative electrode (30) or (30-A) and an electrolyte impregnated separator (40) between the electrodes. Liquid impervious layers (80) and (90) which also function as current conducting means are in contact with the outer electrodes (20-A) and (30-A) respectively. Metal current collectors (60) and (70) which also function as vapor barriers are on the outside of layers (80) and (90). Electrolyte impervious sealing means and electrical insulating means around the electrolyte impregnated separators (40) are designated by the numeral (100).

Numerous advantages result from the construction illustrated schematically in Figures 8.4a and 8.4d and described above. Use of the carrier strip as a substrate permits the electrodes to be made from compositions which, during the construction of the duplex electrodes, are unable or poorly suited to be produced as continuous strips. Examples of electrodes which are unable or poorly suited to be produced as continuous strips include electrodes comprising particles of active material, contained in and dispersed throughout a porous matrix; flame spray deposits; and vapor deposits.

The conductive carrier strip also permits current to be conducted between the positive and negative electrodes in a duplex electrode without the use of any other electrically conductive members. The electrically conductive plastic also has the advantage of being electrochemically inert in the battery.

While continuous carrier strips made from metals such as zinc, aluminum, or steel could be used to achieve some of the advantages attained with this process, these metals tend to produce unwanted electrochemical reactions with the electrodes on one or both sides of the carrier strip unless the strips are coated on one or both sides with an electrolyte impervious, electrically conductive, electrochemically non-reactive material. Metal carrier strips which by themselves are electrochemically nonreactive, such as titanium, tantalum or gold are excessively expensive.

The use of the continuous carrier strip as a substrate along which intermittent deposits of electrodes are applied is also advantageous from the view point of manufacturing techniques. Modern high speed production machinery is better able to handle flexible, individual pieces.

Maximum advantage of this principle may be attained in conjunction with this process by using the continuous strip as a processing implement throughout the construction of duplex electrodes and the subsequent assembly of those electrodes into multicell batteries, leaving the step of cutting the carrier strip into segments until all other assembly steps required to assemble the multicell batteries have been taken. In this regard, it is preferred to assemble a plurality of structurally and electrically unconnected multicell batteries by beginning with the construction of duplex electrodes which are structurally and electrically connected together by the continuous

electrically conductive plastic carrier strip. This consists of placing intermittent
deposits of positive and negative electrodes on the carrier strip as shown in Figure
8.4a so that each patch deposit of negative electrode is on the other side of the
strip from and substantially centered opposite a patch deposit of positive electrode.
The next step in the process consists of assembling multicell batteries which are structurally and electrically connected together by at least one of the carrier strips having positive and negative electrodes deposited thereon.

This step comprises the acts of placing in alignment at least one such carrier strip
between outer positive and outer negative electrodes so that a duplex electrode is
between an outer positive electrode and an outer negative electrode; placing an
electrolyte impregnated separator between each positive and negative electrode,
and sealing around the perimeter of each electrolyte impregnated separator to produce a liquid impervious seal.

The process continues by sealing a liquid impervious layer around the electrodes
and electrolyte impregnated separators; and, connecting to each outer positive electrode electrically conductive means which extend to the exterior of the liquid impervious layer and connecting to each outer negative electrode additional electrically conductive means which extend to the exterior of the liquid impervious layer.

After the multicell batteries have been so assembled, each carrier strip is then cut
between duplex electrodes to obtain structurally and electrically unconnected multicell batteries; the carrier strip may be cut between each successive pair of duplex
electrodes, or it may be cut into increments each of which contains two or more
duplex electrodes so that the resultant batteries structurally connected by the incre-
ment are electrically connected in parallel.

During the assembly of the multicell batteries additional components may be processed in the form of continuous strips; alternatively, each of these additional components may also comprise a succession of structurally unconnected components placed
along the continuous conductive plastic carrier strip.

Figure 8.4c is helpful in illustrating these concepts. The multicell battery (5) shown
in Figure 8.4c may be made by using three of the electrically conductive plastic carrier strips (50) with positive and negative electrodes (20) and (30) applied intermittently on the opposite sides of each as shown in Figures 8.4a and 8.4d. The electrolyte impregnated separators (40) shown in Figure 8.4c were assembled into the battery as structurally unconnected components.

The components (60), (70), (80) and (90) were assembled into the multicell battery
(5) as continuous strips, although they also could have been components which have
no structural connection with each other when assembled into successive multicell
batteries.

The cutting of the three electrically conductive plastic carrier strips plus the cutting
of any other continuous strips used in constructing the multicell battery (5) may be
the last step in the construction of a plurality of multicell batteries, thereby retaining the advantages of processing continuous strips rather than individual unconnected
pieces for as much of the assembly process as possible. It is not essential that the
cutting of the continuous, electrically conductive plastic carrier strip into structurally and electrically unconnected duplex electrodes be postponed until all other

steps in the assembly of multicell batteries are complete. The cutting of the strips may, for instance, be done immediately after the positive and negative electrodes are applied intermittently on opposite sides of the carrier strips and the unconnected duplex electrodes may then be assembled into multicell batteries.

If this sequence of steps is taken, then the assembly of a multicell battery after the cutting of the carrier strip comprises: placing at least one of the structurally and electrically unconnected duplex electrodes between an outer positive electrode and an outer negative electrode; placing an electrolyte impregnated separator between each positive and negative electrode; sealing around the perimeter of each electrolyte impregnated separator to produce a liquid impervious seal; sealing a liquid impervious layer around the electrodes and electrolyte impregnated separators; and, connecting to the outer positive electrode electrically conductive means which extend to the exterior of the liquid impervious layer and connecting to the outer negative electrode additional electrically conductive means which extend to the exterior of the liquid impervious layer.

The unconnected duplex electrodes could be assembled into multicell batteries as described above in a process in which some other component of the finally constructed batteries was used in the form of a continuous carrier strip during some or all of the assembly steps.

For instance, the outside layers could be continuous carrier strips and the duplex electrodes, electrolyte impregnated separators, and outer electrodes could then be placed along those continuous strips, with the cutting of those strips to produce structurally unconnected multicell batteries being postponed until after all other assembly steps have been concluded.

The electrically conductive plastic used in the continuous carrier strip (50) and also shown in items (80) and (90) in Figure 8.4c may be produced by casting, extrusion, calendaring, or other suitable techniques. The conductive plastics may be made, for example, from materials such as polymers loaded with electrically conductive particles and containing various stabilizers and/or plasticizers.

The conductive particles may be carbonaceous materials such as graphite or acetylene black, or metallic particles may also be used. Polymers which by themselves are sufficiently conductive may also be used. The conductive plastic, whether loaded or unloaded, must be made from a composition which is compatible with other components of the battery.

For batteries using Leclanché and moderately concentrated alkaline electrolytes, the conductive plastic may be made, for example, from materials such as polyacrylates, polyvinyl halides, polyvinylidene halides, polyacrylonitriles, copolymers of vinyl chloride and vinylidene chloride, polychloroprene, and butadiene-styrene or butadiene-acrylonitrile resins.

For batteries using strongly alkaline electrolytes, polyvinyl chloride and polyolefins such as polyethylene and polyisobutylene may be used in the preparation of the conductive plastic. For batteries using acid electrolytes such as sulfuric acid, polyvinyl halides, copolymers of vinyl chloride and vinylidiene chloride may be used.

Continuous Carrier Strip

In a process described by B.C. Bergum; U.S. Patent 3,694,268; September 26, 1972; assigned to ESB Incorporated a continuous strip of separator material is used as a carrier of positive and negative electrodes in the construction of multicell batteries. The positive and negative electrodes are first placed on opposite sides of the continuous carrier strip, and subsequently segments of the carrier strip are assembled into batteries, each segment having opposed positive and negative electrodes on the opposite sides.

Preferably the assembly of the segment into batteries occurs while the segments are structurally connected together as undivided parts of the continuous carrier strip, but alternatively the carrier strip may be cut into structurally unconnected segments before the segments are assembled into batteries.

Figure 8.5a is a schematic diagram showing a continuous strip of separator material (40) being passed by positive and negative electrode applicators (220) and (230), respectively, where the applicators place intermittent deposits of positive and negative electrodes (20) and (30), respectively, on opposite sides of the carrier strip from each other.

The separator (40) may be made from a wide variety of materials including the synthetic fibers and cellulosic materials which are conventional in battery construction as well as from woven or nonwoven fibrous materials such as polyester, nylon, polypropylene, polyethylene and glass. Each deposit of negative electrode (30) is substantially opposite a deposit of positive electrode (20).

The applicators (220) and (230) may be spaced apart as shown in Figure 8.5a so that one applicator first makes its deposit and later the other applicator makes the opposing deposit, or they may be spaced opposite one another so that they make their opposing deposits simultaneously. The applicators (220) and (230) shown in Figure 8.5a are intended to represent electrode applicators in general and are not intended to illustrate only one or more specific types of applicators.

Figure 8.5a also shows two steps preceeding the placement of electrodes onto the continuous strip of separator material. The first is the impregnation of patches of adhesive sealant (100) along the strip of separator material, each such impregnation resulting in a nonimpregnated area (42) inside the adhesive patch as shown in Figure 8.5b.

The second step is the application of electrolyte onto each area (42). The first step, the impregnation of adhesive sealant patches (100), is not essential for purposes of this process but is nevertheless desirable as a means for providing a liquid impervious seal around each cell in subsequently assembled batteries.

If desired, a deposit of gel-like material may be applied to one or both the electrodes in adhering or bonding to the strip of separator material (42) as well as to decrease the electrical resistance across the interface between each electrode and the separator. These gels may be simply viscous deposits of the electrolyte or they may include a wide variety of other materials such as polyvinyl alcohol, methylcellulose, and the many formulas of starches commonly known in the battery industry. The gels may be dried to form films or they may remain as gels.

The gels may or may not contain water-soluble salts. It is to be understood that this process contemplates the application of electrodes onto a continuous strip of separator material which has been coated or impregnated into or onto one or both sides with such gels.

As can be seen from Figure 8.5a, the application of the electrodes (20) and (30) along the continuous strip of separator material (40) results in a series of segments of separator material each of which has a positive and negative electrode on the opposite sides. Immediately after passing the two electrode applicators these segments are structurally connected to each other as parts of the continuous separator strip. Each of these segments is subsequently assembled into a battery.

If assembled into batteries while still part of the continuous carrier strip, these segments may simply be left structurally connected so as to form a tape of structurally connected batteries, or an additional and subsequent step may then be taken, that of cutting the segments and the batteries of which they are components into structurally unconnected members, retaining to the last the advantages of being able to process a continuous strip rather than a series of individual pieces.

The batteries which are structurally connected together by the continuous carrier strip may be electrically unconnected to each other, or they may be electrically connected in series or parallel, depending upon other aspects of the battery construction. See Figure 8.5b for dashed lines indicating where such cutting might occur.

Cutting of the separator strip into structurally unconnected segments may occur before those segments are assembled into batteries, however, retaining through at least a part of the battery production the advantages of being able to use the separator as a carrier or processing implement for the electrodes of both polarities. These structurally unconnected segments may, if desired, be assembled into batteries by being placed on continuous strips of other materials required in the battery.

Figures 8.5c and 8.5d illustrate a multicell battery (5) assembled with four continuous carrier strips of the type which were processed as shown in Figures 8.5a and 8.5b. Although shown in the drawing as being structurally unconnected to any other battery, the multicell battery (5) may be considered as being one of many batteries structurally connected together by the continuous strips of separator material, or it may be considered as being structurally unconnected to other batteries as a result of cutting the separator strips into unconnected segments.

Referring to Figure 8.5d, during the assembly of the multicell battery (5) intercell connectors (50) are placed between the continuous carrier strips. Later in the assembly a liquid impervious layer is placed around the electrodes, separators and intercell connectors. The impervious layer shown in Figure 8.5d comprises two pieces (80) and (90) both of which are electrically conductive plastic.

Also shown in Figure 8.5d are outer metal foils or sheets (60) and (70) which function both as vapor barriers to prevent evaporation of electrolyte from the battery and as current collecting means; these foils, which are optional and are not required by this process, may be laminated to the liquid impervious layers (80) and (90) if desired. As a final step in the assembly of the battery, the liquid impervious layer is sealed around the electrodes, separators and intercell connectors. This sealing may conveniently be done by using the adhesive impregnations (100) shown in Figures 8.5a,

FIGURE 8.5:   MULTICELL BATTERY CONSTRUCTION

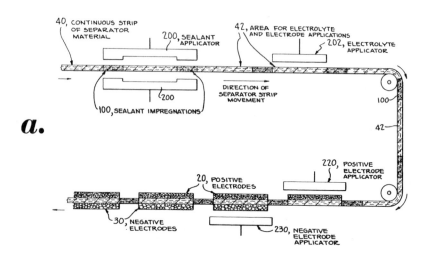

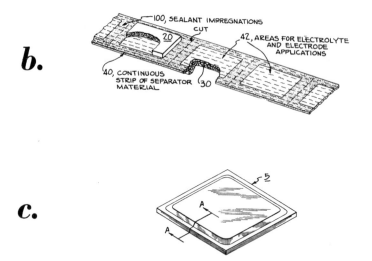

(a) Schematic of Process Employing Continuous Separator Strip

(b) Continuous Separator Strip with Electrodes in Place

(c) Oblique View of Multicell Battery

(continued)

FIGURE 8.5:   (continued)

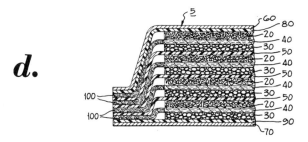

d.

Cross-Sectional View Along Line (A—A) of Figure 8.5c

Source: B.C. Bergum; U.S. Patent 3,694,268; September 26, 1972

8.5b and 8.5d as being present in the continuous strip of separator material, but alternative sealing constructions such as the use of adhesive patches deposited on the intercell connectors and impervious layer may also be used.

Subsequent to the assembly of the multicell batteries as described above, the batteries may be structurally unconnected from each other if desired by the additional step of cutting the separator strips (and any other members of the batteries assembled into the batteries while in the form of continuous strips) between pairs of opposed positive and negative electrodes.

Cuts may be made between each successive pair of opposed electrodes or between longer portions each of which connects several multicell batteries. In either case, the cutting may be described generally as occurring between pairs of opposed positive and negative electrodes to obtain structurally unconnected batteries.

The positive electrodes (20) may each comprise particles of electrochemically positive active material contained in and dispersed throughout a binder matrix. The positive active material conventionally is divided into tiny particles so as to increase the ratio of total surface area to weight in the active material and thereby increase the rate at which the electrochemical reactions can occur by increasing the surface areas where they occur.

The binder maintains the internal electronic conductivity and the structural integrity within the positive electrodes. Since electrolyte must have access to the surface of each active material particle, the binder matrix must be made sufficiently porous so that the electrolyte may diffuse throughout the electrode rapidly and thoroughly.

Preferably the pores in the electrode are produced by the evaporation of liquid during the construction of the electrode; the evaporating liquid may be part of a dispersion binder system in which the solid binder contained in the finally constructed electrode comprises tiny particles of binder material dispersed throughout and not dissolved in the liquid while the electrode is being constructed, or the evaporating liquid may be part of a solution binder system in which the solid binder contained in the finally constructed electrode is dissolved in the liquid which is later evaporated. The porosity of the positive electrodes may be increased as the discharge rate desired in the battery is increased. Electrodes may also be constructed using combinations of the dispersion and solution systems.

Alternatively, the pores might be produced by the dissolving of a solid which was present during construction of the electrode or by passing gases through or generating gases within the electrode at controlled rates during electrode construction. The positive electrodes (20) may, and preferably will, also contain amounts of a good electrical conductor such as carbon or graphite to improve the electrical conductivity between the active material particles, the positive active material particles themselves generally being relatively poor conductors of electricity.

The conductivity of the active material particles together with the conductivity of the binder itself will influence the amounts of conductors added to the electrode. The electrodes (20) may also contain if desired small amounts of additional ingredients used for such purposes as maintaining uniform dispersion of active material particles during electrode construction, aiding the diffusion of electrolyte through the pores of the finally constructed electrodes, controlling viscosity during processing, controlling surface tension, controlling pot life, or for other reasons.

When the positive electrodes (20) are applied to the separator they may be of liquid or slurry consistencies capable of being applied by painting, brushing or printing techniques and the electrode deposits may be as thin as one or a few thousandths of an inch.

The negative electrodes (30) may comprise metallized spray or vacuum deposition or may comprise tiny particles of metals contained in and dispersed throughout a binder matrix. Such metallized sprays or vacuum depositions and such binder mix formulations may all be considered as alternative constructions in which the negative electrodes comprise a plurality of individual, discrete particles of electrochemically active material.

If the negative electrodes utilize a binder matrix, in general the same considerations regarding that matrix apply to the negative electrodes as do for the positive electrodes except that no electrical conductor may be needed to achieve desired electrical conductivity between the active material particles since the negative active materials are generally better conductors than are the positive materials.

When the negative electrodes utilize a binder matrix, the binder system need not be the same as the one used in the positive electrodes, and even if it is the proportions of binder, active material particles, and other ingredients in the negative electrodes may have a different optimum than the proportions of analogous ingredients in the positive electrode.

The initial porosity of the negative electrodes may sometimes be less than that of the positive electrodes; since the electrode discharge reaction products are sometimes dissolved in the battery electrolyte, the porosity of those electrodes will sometimes increase as the battery is discharged. The porosity of the negative electrodes may be increased as the discharge rate desired in the battery is increased. It is apparent that electrodes which comprise particles of active material would be unable or poorly suited to be produced as continuous strips. They should therefore be deposited upon a substrate which, in the case of this process, is the continuous separator carrier strip. If segments of the separator having positive and negative electrodes deposited on the opposite sides are assembled into multicell batteries such as the one shown in Figure 8.5d, then an impervious intercell connector (50) must be placed between each consecutive pair of cells.

These intercell connectors (50) may take several different embodiments, but regardless of the specific embodiment the impervious intercell connector (50) must meet three essential requirements: it must be impervious to the electrolyte of the battery so that one cell may be sealed off from the next; it must provide some means by which electrical current may be conducted between the positive electrode in one cell and the negative electrode in the next cell; and it must not create any undesired reactions with the electrodes or other components of the battery.

The impervious intercell connector shown in Figure 8.5d may be a sheet or film of electrically conductive plastic. Alternative forms of the impervious intercell connector may be achieved with metal foils, conductive adhesives, or combinations of them, since with the use of metal foils it may be necessary or desirable to interpose a layer of conductive adhesive or other conductive polymer between the foil and the positive electrode (20) to prevent the foil from engaging in an undesired electrochemical reaction with the positive electrode or the electrolyte.

The impervious intercell connector (50) may also be a combination of an electrically nonconductive member such as plastic, with one or more members of electrically conductive material extending around the edge of or through the nonconductive member to conduct electrical current between the positive electrode in one cell and the negative electrode in the next consecutive cell.

Another essential of the multicell battery (5) is a liquid impervious layer comprising members (80) and (90) sealed around the electrodes and electrolyte impregnated separators as shown in Figure 8.5d. When a battery is in storage waiting to be placed into service there is an opportunity for liquids from the electrolyte to escape from the battery, leaving the battery incapable of performing as desired when later placed into use.

Also during discharge the battery may produce liquid by-products which are corrosive, poisonous, or otherwise harmful, and it is desirable to prevent these liquids from escaping from the battery. The liquid impervious layer provides means for preventing or minimizing the loss of these liquids.

The multicell battery (5) must also be provided with means for conducting electrical current between the positive electrode (20) in one cell and the exterior of the liquid impervious layer and additional means for conducting electrical current between the negative electrode (30) in the other end cell and the exterior of the liquid impervious layer.

This additional requirement of the battery may be met by the liquid impervious layer members (80) and (90) themselves by constructing those members from a conductive material such as an electrochemically inert, electrically conductive plastic, and such a construction is shown in Figure 8.5d. As an alternative to the conductive plastic, metals which are either themselves electrochemically nonreactive or are made so by appropriate conductive, nonreactive coatings may be used for the liquid impervious layer. Another alternative construction not illustrated in the drawings is to use a liquid impervious layer which is made from an electrically nonconductive material and then extend separate conductive means from the edge of the nonconductive, liquid impervious layer so that current may be withdrawn from the battery. Conductive materials such as coatings, adhesives, or inert metal depositions may also be used to collect current, and such materials may be applied directly onto the positive electrodes prior to cutting the continuous separator strip.

It is to be understood that all of these alternative constructions are encompassed by the general statement that a liquid impervious layer is sealed around the electrodes and electrolyte impregnated separators, that electrically conductive means are connected to the end positive electrode which extend to the exterior of the liquid impervious layer, and that additional electrically conductive means are connected to the end negative electrode which extend to the exterior of the liquid impervious layer.

Two additional components, members (60) and (70), are shown in Figure 8.5d and are illustrated because they may be used in the construction of the multicell battery produced by this process. Those members are metal foils or sheets, e.g., steel foil, which function both as vapor barriers to prevent evaporation of electrolyte from the battery and as current collecting means.

Where a nonmetallic, nonconductive vapor barrier is used instead of steel foil, additional means must be provided to conduct current from the exterior of the liquid impervious layer [members (80) and (90)] to the exterior of the vapor barrier. Where vapor barriers such as the members (60) and (70) shown in Figure 8.5d are used with the battery, they may be laminated to the liquid impervious layers (80) and (90) if desired.

Liquid impervious sealing means must be provided around the perimeter of each cell to prevent electrolyte loss from the battery and to prevent the electrolyte of one cell from migrating to another cell around the perimeter of an intercell connector. Adhesive impregnations (100) may serve as the needed liquid impervious sealing means. By being made from an electrically nonconductive adhesive, impregnations (100) also serve an additional purpose, that of preventing undesired electrical connections between electrically conductive members of the battery. Other means for providing the seals may also be used.

Using the same general production technique as noted above, W.J. Dermody and and J.E. Oltman; U.S. Patent 3,701,690; October 31, 1972; assigned to ESB Inc. describes a battery having a sealant impregnated into the separator. In the process, patches of adhesive are impregnated into battery separator material, each patch being in the form of a closed loop inside of which is an area of separator material not impregnated with the adhesive. Electrolyte is then impregnated into the area inside each patch of adhesive, after which the separator segments are assembled into batteries where the adhesive patches serve as seals.

Preferably the adhesive patches are nonconductive and are impregnated along continuous strips of separator material, and a further preference is to assemble the separator segments into the batteries which are structurally connected together by the continuous strips.

After assembly of the batteries the continuous strips may be cut between patches of adhesive to obtain structurally unconnected batteries.

The above multicell battery process has been modified by B.C. Bergum and K.H. Kenyon; U.S. Patent 3,674,565; July 4, 1972; assigned to ESB Incorporated to include a double adhesive liquid sealing window.

Around each cell of a multicell battery is a window, both sides of which are adhesive. The inside perimeter of each window, which overlaps a portion of the intercell connector between two adjacent cells, provides a first seal around a cell to prevent liquid from one cell from migrating either to another cell or to the exterior of the battery.

The outside perimeter of each window extends beyond the edges of the intercell connector to provide a second seal which also prevents liquids from migrating from the interior to the exterior of the battery. The windows may be electrical insulators which serve the additional function of keeping electrically conductive intercell connectors and/or liquid impervious layers from coming into electrical contact with each other.

Housing Design

G.E. Kaye; U.S. Patent 3,575,725; April 20, 1971; assigned to P.R. Mallory & Co., Inc. describes a battery case consisting of two hollow semicylindrical sections, to receive a string of cells and to receive and position a pair of terminals connected to the cells, which terminals are to be disposed at one end of the cylindrical housing, to be available for electrical coupling to an external prong plug.

The two semicylindrical sections of the housing are appropriately provided with a male rib or bead on one section and with a receiving flute or cavity on the other section, the rib and the cavity being so positioned that the two housing sections will be readily engaged for proper closure along the entire length. Such beads or ribs may be patterned to intricate configurations which will assure closely and properly engaging intricate structure surfaces, which may then be readily bonded by suitable agents, or by the application of ultrasonic energy to the seam edges of those structures.

Appropriate disposition of the ribs and cavities need not be relied upon as sole means for the actual holding purposes, but they do serve particularly to align the two half-structures in an automated process, which will permit further economy in the assembling and closure of the housing parts by the simple application of the bonding agent or ultrasonic energy.

One of the features of the process is that either half-section of the housing may be utilized as a carrier tray or manufacturing fixture for assembling the battery cells in the manufacture of the assembled battery in the housing, and for receiving the terminals connected to the battery for access to an external circuit through an appropriate external plug. After such assembly and disposition of the cells in one-half of the housing as such a carrier tray during manufacture assembly, the other half of the housing may then be positioned in proper place to close the housing as a structure, with assurance of proper fit and positioning of the two half-sections of the housing. Such assured proper fit permits automatic handling of the housing when so completely assembled and closed on the battery cells, so that bonding of the two half–sections of the housing by any automatic procedure may be resorted to with assurance that the housing structure is properly assembled.

As shown in Figure 8.6a, one half–section (10) of the housing consists of a hollow, semicylindrical shell (12) having a semicircular closed rear wall (14) and a semicircular open wall (16) at the front, which is provided with two semicircular grooves or half–openings (18) and (20), which will cooperate with similar half–openings in the other half–section to form two circular openings when two proper half–sections are later combined.

Those two openings (18) and (20) are provided to permit access of an external prong of a plug connected to an external circuit. The border edge or seam surface (24) on the one half shell (12) is provided with a concave groove or flute (26) that extends all around the two seam edge surfaces (24) of the half shell (12), and along a corresponding seam edge surface of the semicircular rear wall (14).

A semicylindrical space within the half shell (12) is subdivided into a main compartment (28) and two semicircular pockets (32) and (34), at the front end of the housing half shell (12). The main compartment (28) is to receive the battery cells. The two pockets (32) and (34) are respectively coaxially aligned with the two semicircular openings (18) and (20).

The two pockets (32) and (34) are separated by a wall (36) provided with a transverse flange (38) at the rear of pockets (32) and (34), which substantially blocks a major portion of each of the two pockets (32) and (34) to serve as a bracing wall for the two socket terminals (42) and (44), indicated in those pockets, as terminals of the string of battery cells. Wall (36) extends backward to and supports a second transverse flange (46), which serves as a front buttress or bracing wall for the front central coaxial button electrode (50) on the front battery cell (52). The top surface of that backwardly extending wall (36) is provided with an elongated flute (56).

The general disposition of the two transverse flange pieces (38) and (46) on that wall (36) may be better appreciated upon also looking at sectional Figures 8.6c and 8.6d, respectively. The socket terminal (44) is connected to front button (50) of cell (52) through a short flat conducting strip (45); and the socket terminal (42) is connected to the outer or wall electrode of rear cell (53), for example, through a long, flat conducting strip (54), wrapped in thin insulation, that fits snugly between the cells and the inner surface of half shell (12).

In Figure 8.6b is shown the second half shell (12–A) which is similar in all respects to the half shell (12) shown in Figure 8.6a, except that the half shell (12–A) in Figure 8.6b is provided with upstanding ribs (64) and (66) to fit into the grooves (26) and (56) of the half shell (12) in Figure 8.6a. The parts on the shell (12–A) of Figure 8.6b are otherwise numbered the same as those in Figure 8.6b, the rib (66) is provided to fit into the slot (56) of the half shell in Figure 8.6a.

Figure 8.6c shows the disposition of the bracing flange (38) for the terminals (42) and (44) in pockets (32) and (34), to show how the flange blocks part of each pocket to brace those terminals against the insertion force of a prong, (55), on an external plug (57). Figure 8.6d shows the bracing flange (46) which serves as a front buttress for the button electrode (50) on the front battery cell (52).

FIGURE 8.6:   BATTERY ASSEMBLY

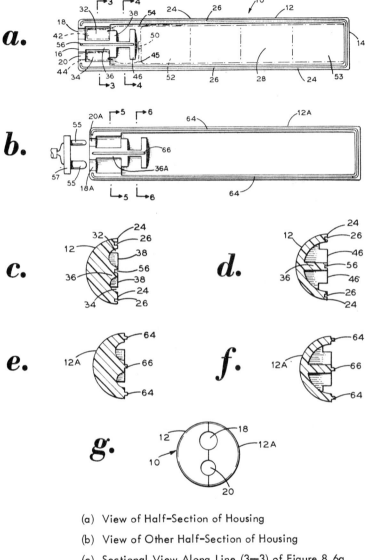

(a) View of Half—Section of Housing

(b) View of Other Half—Section of Housing

(c) Sectional View Along Line (3—3) of Figure 8.6a

(d) Sectional View Along Line (4—4) of Figure 8.6a

(e) Sectional View Along Line (5—5) of Figure 8.6b

(f) Sectional View Along Line (6—6) of Figure 8.6b

(g) Front View of Final Housing Assembly

Source: G.E. Kaye; U.S. Patent 3,575,725; April 20, 1971

Figures 8.6e and 8.6f are similar to Figures 8.6c and 8.6d, except that in Figures 8.6c and 8.6d, being sections of the half shell (12-A) in Figure 8.6b, ribs (64) and (66) are shown instead of the flutes (26) and (56) of the shell in Figure 8.6a.

Figure 8.6g shows the completed openings (18) and (20) to permit the prongs (55) on the external terminal plug (57) to pass through and enter the battery terminals (42) and (44). The front annular border faces of those terminals (42) and (44) are larger in diameter than the openings (18) and (20), so the terminals will be blocked by the borders of those two openings (18) and (20) when the terminal plug is moved to disconnect the prongs from those terminals.

The casing thus serves as a carrying tray or manufacturing fixture to hold the cells and their terminals and terminal conductors in place and in correct alignment, until the mating half shell is applied, with proper mating assured by the cofitting rib and flute arrangement. Suitable bonding procedure may then be employed, by selected bonding material or by ultrasonic bonding between the mated edges. The housing, when sealed, becomes a unitary structure that holds the cells rigidly in place, in all positions of the housing and provides proper insulation for the terminals and the cells.

Container Assembly

A process described by R.E. Brindley; U.S. Patent 3,650,841; March 21, 1972; assigned to Union Carbide Corporation relates to a multiple cell galvanic battery and in particular to a container assembly for such a battery. Standard galvanic dry cells usually produce an open circuit voltage of about 1.5 v. When a higher voltage is required, it is customary to combine a multiplicity of these cells to form a battery having the required voltage. The cells are usually disposed in a container and connected in series, parallel, or series parallel with external terminals attached to the container and making contact with the cells. Therefore, a 6 v. battery, commonly used in portable lanterns, would contain four 1.5 v. cells connected in series.

The individual galvanic cells within the multiple cell battery conventionally may be connected electrically by soldering leads to the appropriate cell electrodes. Soldering the connections is a costly step, however, in assembling the battery. The soldering must be done manually and consistent quality connections are hard to achieve. Also, if the batteries are subjected to severe shock and/or vibration, either in shipment or in usage, the soldered joints may be damaged or sometimes completely destroyed.

Another more recent type of multiple cell battery has eliminated the necessity for soldered connections to electrically connect the cells. The battery utilizes pressure contact conducting strips to achieve proper intercell electrical connection. These strips are usually held in place by spring means which provide the necessary pressure to insure proper electrical contact between the individual cells and the strips.

This construction has a lower assembly cost and is more shock and vibration resistant than the soldered connection type. Military usage, however, has put more stringent demands on the ability of a multiple cell battery to withstand shock and vibration. It is not uncommon for batteries to undergo severe shock and vibration during usage or shipment, especially in the extraordinary modes of transporting, delivering and use of military equipment. This severe shock and/or vibration may result in damage to the conventional multiple cell battery.

Another problem with conventional multiple cell batteries is the possible leakage of corrosive exudate from the individual cells. Due to the close proximity of the cells to one another, the exudate leakage from one cell may promote leakage in the outer cells by externally corroding the outer casing, usually a zinc can, of an adjacent cell.

This process provides a multiple cell galvanic battery which comprises a container for the individual galvanic cells, intercell connectors electrically connecting the individual galvanic cells, and a cover for the container which is so assembled as to provide adequate mechanical pressure to insure good electrical contact between the cell electrodes, intercell connectors and battery terminals.

Features of the container assembly which result in a multiple cell battery of improved construction include a container which may be rectangular, cylindrical, modified cylindrical, or modified rectangular, with partition members disposed within the container forming individual compartments for the galvanic cells, a cover with webs which fit over the partition members on the open end of the container so as to seal off the individual compartments and a bond between the cover and container which maintains sufficient mechanical pressure between the intercell connectors, cell electrodes and battery terminals to insure the maintenance of good electrical contact between these members without any additional spring means within the battery.

Referring to Figures 8.7a, 8.7b, 8.7c and 8.7d, a battery according to this process, denoted generally at (10), comprises a container (12), having a base (14) and side walls (16), partition members (18) integrally formed with the base (14) and side walls (16), within the container (12) and forming four individual compartments in each one of which is disposed an individual galvanic cell (20), a cover (21) integrally formed with webs (24) and side walls (26), intercell connectors (28), (28a), (28b), electrically connecting the positive electrode (30) of one cell (20) to the negative electrode (32) of an adjacent cell (20) and battery terminals (34), (35) which pass through the cover (21) and provide external terminal connection to the battery assembly (10).

The bond (36) between the cover (21) and the container (12) maintains sufficient mechanical pressure between the intercell connectors (28), (28a), (28b), cell electrodes (20), (32) and the battery terminals (34), (35) to insure good electrical contact. The bond (36) is formed along the periphery of the cover side walls (26) and container side walls (16) and along the intersection of the cover webs (24) and the container partition members (18).

The intercell connectors (28), (28a), (28b) are fastened to the cover (21) by sprues (38) and pass through notches (40) in the cover webs (24) to make electrical connection with the individual galvanic cells (20). In the arrangement shown in Figures 8.7a through 8.7c, (13), the intercell connectors (28), (28a), (28b) connect the positive electrode (30) of each cell (20) to the negative electrode (32) of an adjacent cell (20) in series relationship. The battery terminals (34), (35) are sealed to the cover (21) by an integral seal (42) on the inside of the cover (21).

A boss (44) is integrally formed with the cover (21) to prevent cell movement along an axis parallel to the container side walls (16). The boss (44) provides added protection against damage when the battery (10) is subjected to severe shock or vibration. Alignment means (46) are integrally formed with the cover webs (24) to insure proper alignment between the cover webs (24) and the container partition members

FIGURE 8.7:   MULTIPLE CELL BATTERY

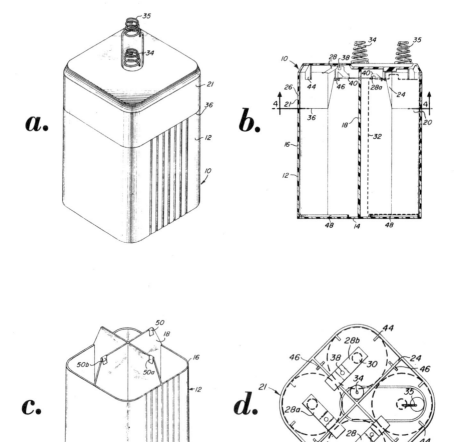

(a) Isometric View of Battery

(b) Vertical Section

(c) Isometric View of Battery Container

(d) Horizontal Section Taken Along Line (4—4) of Figure 8.7b

(continued)

FIGURE 8.7:   (continued)

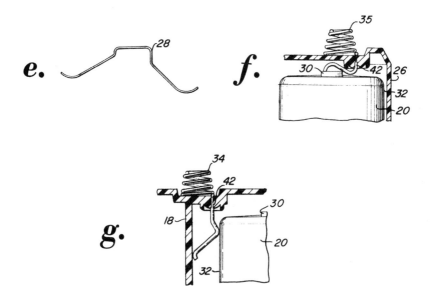

(e) Enlarged View of Intercell Connector

(f) Enlarged View of Juncture of Positive Terminal and Cover

(g) Enlarged View of Juncture of Negative Terminal and Cover

Source: R.E. Brindley; U.S. Patent 3,650,841; March 21, 1972

(18) during formation of the bond (36). A circular ridge (48) in cooperation with the base (14) forms a crater-like structure on top of which a galvanic cell (20) rests. Any expansion of the cell (20) due to intercell gases will be taken up by the expansion into the center of the crater.

An important feature of this process is the bond between the cover side walls and the container side walls. The bond must have sufficient strength to keep the necessary contact pressure between the intercell connectors, the cells, and the battery terminals. The intercell connectors and the portion of the battery terminals within the container are made of a flexible conductive metal. The intercell connectors rest upon the electrodes and because of the flexibility of these connectors, pressure must be exerted to force the cover side walls against the container side walls for the bond to be formed.

In the preferred form of the process the bond is ultrasonically formed. The bond by compressing the cover against the container, creates pressure between the intercell connectors, battery terminals, and the cells, thereby providing good electrical contact. The bond between the cover webs and container partition members serves a dual purpose, i.e., [1] the bond cooperates in providing the necessary mechanical pressure for achieving good electrical contact between the intercell connectors, the

cells and battery terminals and [2] the bond in cooperation with the webs and partition members, forms individual compartments within which the individual galvanic
cells are disposed. These compartments serve the purpose of containing any exudate
leakage from the individual cells and prohibiting the exudate leakage from coming
into contact with adjacent cells and corroding them during use of the battery.

The intercell connectors are arranged in the cover in such a way as to enable placement of the cover upon the container in any one of four positions, each 90° apart,
without impairing the electrical contact between the cells, intercell connectors and
battery terminals. Also, due to the internal connector arrangement, all the cells
are placed within the container compartments with the positive electrode at the open
end of the container.

The cover and container materials of construction should be impervious to the constituents of the cells and electrically nonconductive. The material used must be at
least as shock and vibration resistant as the rest of the battery. The container need
not be rigid, since the galvanic dry cells, which are rigid in construction, make the
battery assemblage rigid when the cells are disposed with the container.

The cover with cover webs, alignment means and boss is integrally formed by injection molding a suitable plastic such as polystyrene, polyolefins, acrylo-butadiene-
styrene, polyvinyl chloride, polycarbonates, etc. The container with partition members and circular ridges is also integrally formed by injection molding a suitable plastic. In the preferred embodiment of the process the cover and container are molded
of a rubber modified polystyrene.

Example: Batteries made according to this process and conventional batteries were
made and tested. The conventional batteries were of the type which utilize pressure
contact conducting strips for electrical connection between cells, which strips were
held in place by a spring means to insure electrical contact.

All the batteries were subjected to the same test conditions, i.e., consecutive free
fall drops of 3 feet, 4 feet, 5 feet and 6 feet respectively until the battery went open
circuit. The voltage of the batteries was measured after each drop at the specified
height level. The table below is a compilation of data taken from the above described tests. The batteries made according to this process are denoted generally as
Type A and the conventional batteries are denoted generally as Type B. A consider-
ation of the data presented shows the batteries made according to this process to be
superior to conventional batteries in resistance to severe shock.

Consecutive Free Fall Drop Tests

	Battery No.	Initial volts	Voltage subsequent to—			
			3′ drop	4′ drop	5′ drop	6′ drop
Type A....	1	6.2	6.2	6.2	6.2	6.2
	2	6.4	6.4	6.4	6.4	6.4
	3	6.35	6.35	6.35	6.35	6.35
	4	6.35	6.35	0		
	5	6.45	6.45	6.45	6.45	6.45
Type B......	6	6.4	0			
	7	6.4	0			
	8	6.4	0			
	9	6.4	6.3	0		
	10	6.4	6.3	0		

Replaceable Cell Partitions

A process described by S. Quisling; U.S. Patent 3,635,766; January 18, 1972 involves multicell electrochemical devices which comprise nonconductive, replaceable cell partitions formed in a closed configuration, such as cylindrical, and disposed substantially in series, such as concentrically. The cell partitions are coated on both their inside and outside surfaces with conductive material to provide an electrode in each of two adjacent cells separated by the partition.

The cell partitions are set in a nonconductive, yieldable, sealing material forming the bottom of the cells. The electrode-forming cell partitions are united by a bridging element which may be in the form of a removable cover for the battery to facilitate removal and replacement of the electrode forming cell partitions when the electrode material thereon has been substantially depleted. Accordingly, the electrodes can be readily replaced with an electrode refill unit rather than recharging the battery or replacing the worn out battery in its entirety.

Figures 8.8a and 8.8b show an example of a multicell electrochemical battery (10). The battery (10) has a nonconductive cylindrical container (11) made of plastic or other suitable nonconductive material. Sealing means (12) shown in the form of a nonconductive, yieldable, matrix material such as tar or the like is provided in the bottom of the container (11). A circular cover (13) of substantially rigid nonconductive material is removably secured on the container (11) by peripheral wingnuts (14).

The cover has a concentric series of nonporous, nonconductive, cylindrical partitions (15) depending therefrom and extending into the yieldable sealing material (12) to provide a series of concentrically spaced individual annular cells (16). While the cell partitions (15) are shown in a cylindrical configuration, it is understood that they may be of any other closed geometric configuration such as triangular, rectangular or polygonal. The central element (17) may be rod shaped as shown.

In Figures 8.8a and 8.8b, the inwardly facing surface of each of the partitions (15) is coated with a conductive material such as lead peroxide to form an annular positive electrode (18) in each of the cells (16). The outwardly facing surface of each cylindrical partition (15) is coated with a conductive material having a different electrochemical value than the positive electrode material, such as lead, to form an annular negative electrode (19) in each cell.

An electrolyte (20) such as a sulfuric acid solution is provided in each cell. The electrolyte in the cells may be replenished through capped inlets (21). The conductive materials forming the electrodes (18) and (19) may be formed on the partitions (15) by dipping, brushing, spraying, pasting, laminating, blowing, or any other coating method suitable for the particular materials involved. Each of the positive electrodes (18) is connected to a positive terminal (22) by an electrical conductor (23) and each of the negative electrodes (19) is similarly connected to a negative terminal (23) by an electrical conductor (25). The conductors (23) and (25) are preferably imbedded in the nonconductive cover (13) as shown in Figure 8.8b.

The cover (13) provides a bridging unit connecting the electrode forming partitions (15). When the battery has lost its charge or when the electrodes have deteriorated to a point such that the battery cannot be recharged, the electrodes may be removed

FIGURE 8.8:   MULTICELL BATTERY

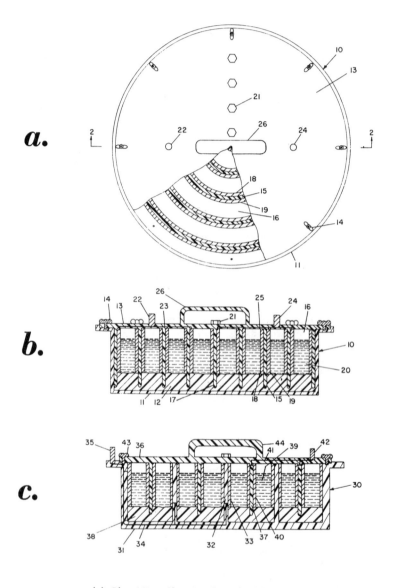

(a) Plan View Showing Detail of Concentric Cells

(b) Sectional View Along Line (2—2) of Figure 8.8a

(c) Sectional View of Modified Battery Design

Source: S. Quisling; U.S. Patent 3,635,766; January 18, 1972

as a unit by releasing wingnuts (14) and lifting the cover off with handle (26). A new or recoated electrode refill unit may then be inserted whereby the coated partitions (15) engage the sealing material (12) to form the individual sealed annular cells.

Another example shown at (30) in Figure 8.8c is particularly well suited for a primary type cell in which only the negative electrode material is consumed during operation. Unlike the common dry cell where the entire battery must be replaced, in this example the depleted negative electrode can be readily replaced as a unit.

In this form, the battery (30) has a nonconductive cylindrical container (31). The container has upstanding, cylindrical, concentrically disposed, cell partitions (32) formed integral with the bottom of the container. Both sides of these upstanding partitions (32) are coated with a conductive material such as copper or the like to form a positive electrode (33). The positive electrodes (33) are connected by an electrical conductor (34) embedded in the container material to a positive terminal (35) mounted on the edge of the container.

The nonconductive cover (36) has a series of depending cylindrical cell partitions (37) that are alternately interspaced between the anode-forming partitions (32) and extend into the yieldable sealing material (38) to provide a series of concentrically spaced individual annular cells (39). Both sides of partitions (37) are coated with a conductive material such as zinc to form an annular negative electrode (40) in each of the cells (39). The negative electrode (14) are connected by an electrical conductor (41) to a negative terminal (42) in the cover.

Accordingly, when the negative electrode material has been consumed by reaction with the electrolyte, the battery may be immediately rejuvenated by merely releasing the wingnuts (43) and lifting handle (44) to remove the cover which forms a bridging unit for the negative electrode forming partitions (37) and replacing same with a new or recoated refill unit.

Intercell Connectors for Stacked Cells

A process described by E.R. Cich and R.C. Ivey; U.S. Patent 3,615,867; Oct. 26, 1971; assigned to ESB Incorporated relates to batteries made up of individual cells stacked one upon another and connected in series in order to provide a battery for use in various electronic equipment. In the past, two common methods of interconnecting the cells have been used; namely, the single strap and the cup-type interconnector.

In the single strap connection a single piece of metal strip is welded to one electrode of a first cell and to the electrode of opposite polarity of a second cell which is stacked on top of the first. In the cup-type of interconnector, a metal cup is welded to the one electrode of the first cell and the second cell is then forced down into the cup and spotwelded at one or more points so that a series connection is made between the cells. Neither of these methods have been extremely satisfactory since they have not overcome various problems encountered in stacking a group of cells on top of each other. In the case of the single strap method the individual cells tend to move about and become loose, or are subject to shock and vibration which tend to loosen the stack. Also, these multicell stacks tend to buckle or lean over and therefore one cell does not maintain a fixed position relative to another.

In those situations using a cup intercell connector, air pockets exist in spaces which result due to lack of perfect mating between cell and cup. When the stack of cells is potted or encapsulated in a potting resin, these air pockets contain trapped air which can be detrimental to battery structure when subject to high g forces. This is undesirable since such batteries are used in instruments subjected to such forces.

This process has a primary goal of overcoming the above problems which have existed in the past in assembling multicell, stacked batteries for use in electronic equipment. This process provides a battery where the individual cells are joined together by means of an intercell connector which comprises a flat section having several extensions projecting beyond the edges of the flat portion. The flat portion of the connector is welded to one electrode of a cell and on top of this is placed another cell.

The extensions or ears which project beyond the flat portions of a connector are then bent upwards and are welded to the other cell thereby locking the two cells together. As a result, an excellent mechanical and electrical intercell connection is made which does not provide spaces for entrapping air when the battery stack is potted or encapsulated in a resin.

In Figure 8.9a, there is shown an intercell connector used in this process. The connector (20) is shown as a flat, annular piece of metal having three extensions or ears (21) projecting from the edge of the flat section. The intercell connector is preferably made of cold, rolled steel which if desired can be coated with such materials as cadmium, nickel, gold and chromium in order to protect against corrosion or to improve the appearance of the connector and battery.

Also, the metal of the intercell connector should be compatible with the metal of the individual cells in order to reduce any form of corrosion due to dissimilar metal effects. In practice, the flat section of the connector is first welded to one electrode of one cell. Thereafter another cell is placed on top of the connector and the ears are bent upwards and welded to the electrode of opposite polarity of the stacked cell. Thus a mechanically rigid connection is made between the cells as well as a good electrical connection.

The connector in Figure 8.9a is shown as having an annular center although the connector could be solid if desired. Normally, the annular connector is used since it facilitates the welding procedure by providing the open area in the center for one of the welding probes. However, with large cells the solid connector may be used since the cell electrode to which the flat center of the connector is to be welded is large enough that the welding probe that must contact the cell electrode can do so at a point on the electrode periphery.

Other modifications of the connector are readily apparent such as punching a small hole in the flat section of the solid connector to provide a place for the welding probe to contact the cell electrode. With each type of connector, however, a positive mechanical and electrical connection between cells is made.

Although the annular connector appears to provide a space in its center to trap air, this has not been found to be very significant and both the annular and the solid connectors are decided improvements over the conventional cup connectors. Figure 8.9b illustrates the manner in which one cell is connected to another using the intercell connector of Figure 8.9a.

FIGURE 8.9: BATTERY CONSTRUCTION USING INTERCELL CONNECTORS

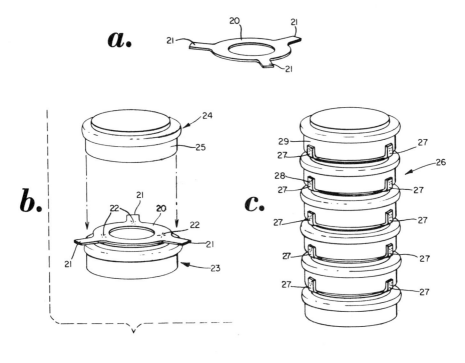

(a) Interconnector

(b) Single Cell Design

(c) Stacked Multicell Battery

Source: E.R. Cich and R.C. Ivey; U.S. Patent 3,615,867; October 26, 1971

In this figure, the connector (20) has been spot welded at points (22) to the top positive electrode of cell (23). The connector ears (21) are shown extending out over the edges of the cell (23). A second cell (24) is about to be placed on top of the connector and after it has been in position the ears are bent upwards and welded to the sides of the negative electrode can (25) of the second cell. In practice, in assembling a battery a number of flat cells are stacked one on top of another with the intercell connector of Figure 8.9a placed between each pair of the cells and with each connector welded to the top electrode of each cell except for the uppermost one.

The stack of cells is placed within a cylindrical tube which has three vertical slots through and out from which extend the ears of the connectors. The stack of cells is held rigid within the tube by pressure at each end of the stack and the ears are then bent upward to make contact directly with the surface of the electrode can of the individual cell which has been placed on top of the connector. The ears are all then welded to the respective electrode cans in order to interconnect the cells to each other. After this welding takes place, the stack of cells can be removed from the tube and a rigid single battery structure is thereby formed. Depending on the height of the cylindrical tube, a variety of relatively rigid and straight cell stacks can be welded to various heights without buckling.

In Figure 8.9c there is shown a stacked multicell battery which has been assembled as described in Figure 8.9b using intercell connectors of the type shown in Figures 8.9a and 8.9b. In viewing the figure there can be seen the connector ears (27) that have been bent upwards and spotwelded to the cell stacked next above as illustrated at (28). Cell (29) is the uppermost cell and its positive electrode terminal is the positive terminal for the battery.

This stacked column of cells is not subject to buckling and the individual cells will not move but will remain fixed to each other. These features make this battery stack desirable for use in electronic equipment which has somewhat stringent space displacement requirements. It is customary practice to pot or encapsulate a battery stack of the type shown in Figure 8.9c in a potting resin. When this is done to the stack of Figure 8.9c there are no air pockets formed in the resin as often happens when the cup-type intercell connectors are used. This is another important advantage of using the intercell connector described here in multicell battery stacks.

OTHER CELLS AND BATTERIES

Disposable Cell with Encapsulated Electrolyte

In a process described by T.F. Bolles; U.S. Patent 3,653,972; April 4, 1972; assigned to Minnesota Mining and Manufacturing Company the reserve cells comprise a casing containing a cathode, an anode, a multiplicity of capsules ranging in diameter substantially from 10 to 4,000 microns filled with an aqueous electrolyte and means for rupturing capsules from the exterior of the casing. The capsule shells are composed of a material which melts between 40° and 110°C. and which is immiscible in the aqueous electrolyte when both are in the liquid form.

Ordinarily, but not always, the reserve cells also have a nonconducting porous separator between the cathode and the anode which prevents their direct contact and which can act as a wick to draw electrolyte into contact with the electrodes. The capsules can be ruptured from the exterior of the casing by either pressure or by heat, usually the former. Capsule rupture by heat is envisioned in reserve cells used in integral fire alarm reserve cell devices and other heat-activated systems.

In Figure 8.10a, (11) denotes a casing which is preferably of a plastic material having a chamber opening on one side designed to receive the elements of the cell and provided with means for conveying the current generated by the cell to contact points where it can be utilized by external appliances.

Within the casing are a cathode (12), an anode (13) and a nonconducting porous separator (14) (e.g., of a suitable paper, other nonwoven webbing, porous plastic sheeting or a similar material), a current collector (15) adjacent the cathode (although this may not always be required), capsules (16) containing electrolyte as liquid fill therein, a thin impermeable elastic cover (17) (which can be of an elastomeric or deformable polymeric material) and a retaining ring (18).

FIGURE 8.10: DISPOSABLE CELL WITH ENCAPSULATED ELECTROLYTE

a.

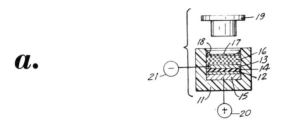

Sectional View of Basic Unit

b.

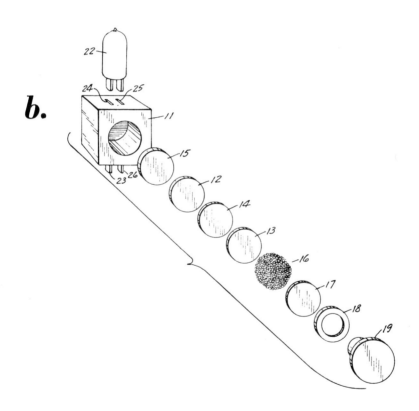

Exploded View of Cell Adapted for Use with Photo Flash Bulb

Source: T.F. Bolles; U.S. Patent 3,653,972; April 4, 1972

The cover (17) and the retaining ring permit pressure rupture of the capsules from out-side the casing while preventing leakage of the electrolyte. The cell can be sealed by known means such as heat or an adhesive. The plunger (19) provides a means for rupturing the capsules thereby activating the cell. The leads (20) and (21) conduct the electrical current produced by the cell to an external circuit. No separator be-tween the electrodes is actually required if they do not touch, but may be desired for other reasons, e.g., to provide a wick for distribution of the electrolyte.

Figure 8.10b illustrates a similar cell adapted to be used at the power source for ig-niting a photo flash bulb (22). Elements (11) through (19) are as in Figure 8.10a, (23) and (26) are conductor prongs through which the circuit can be completed (and which can, for example, be plugged into a camera) and (24) and (25) are electrical connection sites adapted to receive a flash bulb base. Thus, the reserve cell is placed in series with (23) and (24) while (25) and (26) are connected directly by a conductor. The reserve cell is activated and a flash bulb is plugged into sites (24) and (25). When the circuit is completed through (23) and (26), e.g., simultaneously with the release of a camera shutter, the flash bulb is ignited.

Threaded socket, or other types of electrical connections could, of course, be used in place of the prong connectors shown. One cell is normally sufficient to ignite several flash bulbs and will remain active for some time (e.g., several hours) after it has been activated or until it is exhausted. Thus the cells can be readily adapted for use with flash cubes. The following are some examples of electrochemical sys-tems suitable for use in the reserve cells of the process together with the approximate voltages generated.

[1]	Anode	Zinc
	Cathode	PbO_2
	Electrolyte	Aqueous H_2SO_4
	Voltage	2.4
[2]	Anode	Zinc
	Cathode	MnO_2
	Electrolyte	Aqueous NH_4Cl, $ZnCl$
	Voltage	1.5
[3]	Anode	Zinc
	Cathode	MnO_2 + NH_4Cl, $ZnCl$
	Electrolyte Solvent	H_2O
	Voltage	1.5
[4]	Anode	Zinc
	Cathode	Ag_2O
	Electrolyte	Aqueous KOH solution
	Voltage	1.9
[5]	Anode	Lead
	Cathode	PbO_2
	Electrolyte	Aqueous sulfuric acid
	Voltage	2.0
[6]	Anode	Magnesium
	Cathode	AgCl
	Electrolyte	H_2O or aqueous NaCl solution
	Voltage	1.6

(continued)

[7]   Anode   Zinc
        Cathode   HgO
        Electrolyte   40% aqueous KOH solution
        Voltage   1.3

A preferred electrochemical system is the zinc-lead dioxide–aqueous sulfuric acid system. Such a cell can be quite small (due to high energy density and amperage), is composed of available and inexpensive material and is easily manufactured. Finely divided lead dioxide is advantageously used to obtain high current density. A suitable lead dioxide cathode can be prepared by overcoating a chemically inert conductor sheet (which acts as a current conductor) with finely divided lead dioxide in a water-soluble binder, such as polyvinyl alcohol.

The chemically inert conductor can be prepared on a polymeric film base, e.g., a polyester such as polyethylene terephthalate, by first applying a metal layer (such as by vapor deposition or by lamination of a foil) then coating the metal layer with finely divided carbon, such as acetylene black or graphite, in a water-insoluble binder, such as a styrene–butadiene copolymer.

In such a cell the aqueous electrolyte contacts the lead dioxide particles in the water-soluble layer as soon as it is released from the capsules but does not readily penetrate the adjacent water-insoluble binder. Such a cell furnishing a total charge of 26 amp.-sec. requires only about 26.4 mg. of H_2SO_4 (258 mg. of 1 molar aqueous solution), 30 mg. of PbO_2 and 8 mg. of zinc.

The electrolytes are generally aqueous in order to generate useful voltages. Aqueous fills are quite difficult to encapsulate, although it has been found that they can be prepared by the process described in U.S. Patent 3,423,489. Preferably these are in about the 300 to 3,000 micron range, for reasons of ease of capsule preparation and use. As previously stated, the capsule shells should be composed of a material which melts between 40° and 110°C. and which is immiscible in the aqueous electrolyte fill when both are in the liquid form.

These restrictions relate to the process of encapsulation (U.S. Patent 3,423,489). Among the encapsulating materials are waxes, fats, proteins, carbohydrates, gellable colloid materials such as gelatin and agar-agar, low polymers, and the like. Specific encapsulating materials include paraffin waxes having melting points of 48°, 55° and 83°C., microcrystalline waxes, etc. These can be used alone or together with other waxes and materials. A useful encapsulating composition is the following (the parts being given by weight).

	Parts by Weight
Paraffin wax, MP 55°C.	44.95
Paraffin wax, MP 83°C.	44.95
Copolymer of ethylene and ethyl acrylate (Dow EA-2018a)	10
Butylated hydroxy toluene (antioxidant)	0.1

Capsules as described here are essentially impossible to fracture by jarring but are easily broken by compression or by heat.

The capsules can be utilized as such or can be first coated on a web. The latter is advantageous in the manufacturing process since it means that all components of the cell can be added to the casing in sheet form (cut into circular disks or other desired shapes). The sheet carrying the capsules can be placed between the electrodes, if desired, thus doubling as a separator.

Casing for Seawater Activated Battery

K.R. Jones; U.S. Patent 3,655,455; April 11, 1972; assigned to Globe-Union Inc. describes a method of encasing batteries where the battery electrodes are assembled with spacers maintaining the electrodes in a spaced relationship to define an electrolyte cavity.

In practicing the method, the electrode assembly is immersed in a body of molten casing material whereby the casing material solidifies to form a web between the electrodes and define the electrolyte cavities. The supports for the assembly during immersion extend across the electrodes and are coated with casing material so that upon removal of the supports transverse tubular manifolds are defined. A single port to the manifold and the plurality of electrolyte cavities is also provided upon removal of the supports.

These batteries employ an anode of magnesium and a cathode of silver chloride in a battery cell in which the electrolyte is, for example, seawater. The cell is activated by immersion of the unit into seawater, provision being made to permit the seawater to enter the cell and to act as an electrolyte and also to pass through the cell to remove products of the battery action and, in some cases, to provide control of thermal conditions within the battery.

Referring to Figure 8.11a, there is shown a battery construction (9) with a modified casing (10). As shown in Figures 8.11b and 8.11c, the battery comprises an electrode assembly (11), which includes a plurality of cathodes (12), anodes (14) and partitions (16), the partitions separating adjacent anodes (14) and cathodes (12). The partitions (16) are formed of impermeable, electrically conductive material, so that they serve to prevent electrolyte from flowing between the various cells of the battery and also to electrically connect the several cells of the battery in series.

A plurality of small insulating spacers (18) operate to maintain a given separation between the cathode (12) and anode (14) of each respective cell, and each of the cathodes (12) is provided with a plurality of embossed or protuberant portions (20) for the purpose of increasing electrolyte circulation in the nonreactive areas between the cathode and the spacer for better temperature stability and in the reactive space between anode and cathode to maintain the cell areas substantially free from undissolved salts which may be generated during the battery reaction. An end plate (13) is disposed in contact with the endmost cathode (12a) and serves as an electrical connector to which a lead may be attached.

The electrode assembly (11) is preferably secured together, at least temporarily, with a strip of insulating tape (17), shown in Figure 8.11e and partially shown in Figure 8.11c, which insures that the components of the electrode assembly remain in fixed relationship, until the casing (10) is formed. The casing (10) which surrounds the electrode assembly (11) is formed by dipping the assembly into a supply of molten casing material, which is preferably a plastic compound, and preferably one which

FIGURE 8.11:   BATTERY ASSEMBLY

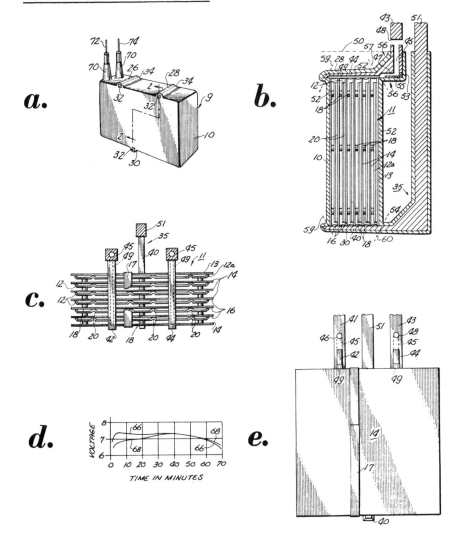

(a) Perspective View of Battery

(b) Cross-Sectional View Along Line (2—2) of Figure 8.11a

(c) Top View of Assembly Prior to Dipping in Casing Material

(d) Voltage-Time Characteristics of Battery

(e) Side View of Electrode Assembly

Source: K.R. Jones; U.S. Patent 3,655,455; April 11, 1972

is relatively rigid at room temperature but fluid at temperatures not greatly in excess of ambient ranges. One satisfactory material which is employed is cellulose acetate butyrate. A thin inner surface film of compound congeals on contact with the relatively cool electrode assembly (11), and an additional thickness sets on the outer surface upon removing the assembly from the compound supply and cooling the unit.

As shown in Figure 8.11a, the basic body of the casing (10) also includes integral hoods (26), (28) and (30), which are disposed over the openings in the casing (10) communicating with edges of the electrode assembly. Each of the hoods (26), (28) and (30) are open at both ends to form ports (32) and (34) through which electrolyte may pass into or from the interior of the casing (10). The passageways of conduits, defined by hoods (26) and (28), comprise small manifolds and extend between the ports (32) and (34).

The manifolds define openings at the edges of all of the cells of the battery, and therefore permit some intercell leakage to occur. However, the presence of the hoods over these areas limit the cross-sectional area over which such leakage current may flow, which thereby maintains the resistance of the leakage path, at a relatively high level and consequently minimizes energy wasted by this leakage. A reinforcing stem (70) for the terminal wires (72) and (74) is also formed integrally with the casing (10) and the hoods (26), (28) and (30).

Figures 8.11b, 8.11c and 8.11e illustrate portions of apparatus (35), employed for supporting and submerging an electrode assembly (11) into the molten casing material. The jig (35) is provided with a support bracket (51), with a centrally disposed bottom arm (40) adapted to support the weight of the assembled battery during dipping. Two or more bottom arms (40) may be employed, if desired.

Two upper arms (42) and (44), secured to supporting members (41) and (43), straddle the lower arm (40) to hold the assembly in position on the lower arm (40). Each of the arms (40), (42) and (44) extends entirely across one side of the assembly (11), and engages an edge of each of the electrodes. The supporting members (41) and (43) for the upper arms (42) and (44) are provided with vents (46) and (48) connected through bores (45) and (47) to grooves (49) extending along the lower surface of each of the upper arms (42) and (44).

The air trapped in the electrode assembly (11) during dipping is permitted to expand and escape through the vents (46) and (48), which remain above the level of the casing material at all times. Thus a uniform, smooth casing results which is free of blisters or bubbles. The ends of the bores (45) and (47) are preferably closed by plugs (53) and (55) to prevent the casing material from entering the bores which would seal the bores and prevent the venting of air from the battery. The support members (41) and (43) for the upper arms (42) and (44) are preferably adjustably mounted by a clamp on the support bracket (51) to permit a degree of adjustability so that the jig (35) may be used on various sizes of batteries and to clamp the assemblies in place for dipping.

During dipping, the jig (35) and assembly (11) are dipped into a supply of molten casing material, the level of which is indicated at (50) in Figure 8.11b. The entire electrode assembly and jig assembly are therefore coated with casing material up to the level (50), and when the jig and the electrode assembly are withdrawn, a coating (52) surrounds the assembly (11) and that portion of the jig (35) which was submerged in the casing material. The casing material is then cut in the area (54) adjacent the lower arm (40), and preferably along the line (60), and in the area (56), adjacent each of the upper arms (42) and (44), and preferably along the line (57), whereby the completed battery, together with its surrounding casing, may easily be slid longitudinally off of the arms (40), (42) and (44), to free the battery from the jig.

The apertures left by the removed jig arms comprise the vent holes (34). The arms (40), (42) and (44) have tapering thicknesses to facilitate removal of the battery from jig (35). Thereafter the coating (52) of casing material adjacent the tip end of the arms (40), (42) and (44) is cut back flush with the side walls of the casing along the lines (59), to form the ports (32) on the other side of the battery (see Figure 8.11a). This process results in a battery casing where all portions are integrally formed, including the hoods (26), (28) and (30) and the reinforcing stem (70).

Referring to the graph illustrated in Figure 8.11d, two voltage-time characteristic curves of a battery constructed in accordance with the process are illustrated. In both curves, the current is maintained constant. The design maximum voltage of the battery is 7.5 v. and the design minimum is 6.25 v. The two curves differ in that the curve (66) represents the operation of the battery with a relatively warm electrolyte, while the curve (68) represents the operation of the battery when a cold electrolyte is used. In the curve (66) the voltage of the battery rises from zero to a maximum voltage within a very short time, and then falls at a relatively constant rate until the minimum voltage of 6.25 v. is reached.

In the other characteristic curve (68), which applies for a cold electrolyte, the output voltage quickly rises above the minimum voltage and then continues to rise over a substantial portion of the life of the battery, and eventually becomes equal to output voltage indicated by the warm electrolyte curve. This mode of operation contrasts sharply with that of previously known batteries in which the output voltage for a cold electrolyte condition remains substantially below the warm electrolyte voltage over the entire life of the battery.

It is believed that the improved characteristic is due to warming the electrolyte more effectively by virtue of the decreased cross-sectional area of the ports (32) and (34), so that the heat generated within the battery during use cannot be dissipated as effectively. An additional advantage is that the reduction in intercell leakage improves the life and capacity of the battery.

Replaceable Unit for Electric Powered Toys

A process described by R.C. Hill and M.J. Pollack; U.S. Patent 3,556,858; January 19, 1971; assigned to Honeywell, Inc. involves a toy powered by electric current where the current is derived from an electrochemical cell which is replenishable after discharge by replacement of the spent components.

A cavity is provided in the toy housing which may be constructed of plastic or other appropriate material. The cavity portion of the toy is generally sealed so that it may serve as a container for a liquid electrolyte. In the preferred form, the anode of the electrochemical cell is attached to the inside wall of the cavity and is not designed to be removed in the course of the normal operation of the toy. The remaining elements of the cell, however, are packaged in such a way that the resulting package is a replaceable unit which may be inserted into the toy cavity to energize the cell upon addition of a liquid solvent and may be removed when the cell is discharged, to be replaced by a new unit. The replaceable unit may comprise of a porous cup which also serves as a separator and electrolyte salt deposited on the cup or placed within the cup, a cathode material, and a cathode collector with a connection for the external load. The other connection with the external load is made at the anode.

The concept for a toy battery with replaceable parts is made practical by development of new battery chemistry based on nontoxic materials. It would otherwise be too dangerous for use by children. Special care must be exercised to select materials which are not excessively toxic and which do not react with the other components of the battery to produce toxic substances. For the purpose of this discussion, the term "nontoxic" shall mean any substance which exhibits no more than moderate toxity as defined and classified in the second edition of "Dangerous Properties of Industrial Materials," by N.I. Sax and published by the Reinhold Publishing Corporation.

The anode of the electrochemical cell is made of magnesium. The cathode is selected from a group consisting of ferric citrate, ferric acetate, cobalt citrate and cobalt acetate. It will be noted that all of the above cathode materials are nontoxic. The electrolyte comprises a salt of a light metal such as KCl and/or $MgSO_4$ in a nontoxic solvent, preferably water. A porous separator is provided to physically separate the anode from the cathode.

Spiral Wound Plate Type Rechargeable Battery

A process described by R.K. Sugalski; U.S. Patent 3,503,806; March 31, 1970; assigned to General Electric Company involves sealed batteries of the spiral wound plate type. In the process, a cell construction is utilized in which the negative plate edges extend longitudinally beyond the positive plate edges in one direction and the positive plate edges extend longitudinally beyond the negative plate edges in the opposite direction.

Terminal straps are attached to the extending edges of the plates at a plurality of points. The plates may be spirally wound and the negative plate connected to a conductive casing through one terminal strap and the positive plate connected to terminal rivet through the remaining terminal strap. As a separate or combined feature a resilient washer having a durometer rating the range of from 15 to 95 may be positioned adjacent either or both ends of the plates.

As shown in Figure 8.12a, a prior art battery of the type to which the process relates has been conventionally constructed of a nickel-plated steel casing (5) in which is contained a battery coil (6). The battery coil (6) consists of a positive plate (6a) and a negative plate (6b) separated by separator insulative layers (6c) which are wound together to form the battery coil (6).

A tab (7a) connected to or integral with the negative plate (6b) is secured to the bottom of the case (5) and another tab (7b) connected to or integral with the positive plate (6a) is secured to a positive terminal (8) located in a cover member (9) of insulative material. The cover member (9) has been sealed to the casing (5) by crimping upper end thereof about the cover as shown. As shown in Figure 8.12b, the construction and also the overall appearance of the battery has been improved by

FIGURE 8.12:   SPIRAL WOUND PLATE TYPE BATTERY

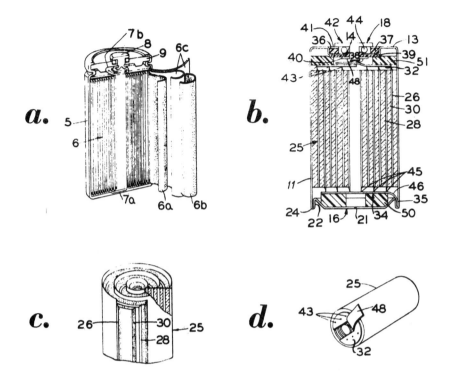

(a) Prior Art Battery

(b) Cross-Sectional View of Battery

(c) Perspective View of Battery Coil

(d) Perspective View of Battery Coil with Connector Terminal Strip

Source: R.K. Sugalski; U.S. Patent 3,503,806; March 31, 1970

elimination of the insulative cover member and the crimping of the case. This is accomplished by constructing the battery with the case of the battery comprising a generally tubular casing (11) closed at its upper end by an integral recessed end wall
(13) having a central opening (14). The lower end of the casing (11) is closed by
closure means (16).

The central opening (14) of integral end wall (13) is closed by a positive terminal
and a resealable vent means (18) which means operates to relieve excess gas pressure in accordance with the principles set forth in U.S. Patent 3,320,097. The closure means (16) for the open end of the case is a cap which has a generally dish

shaped bottom (21), as viewed in Figure 8.12b, and an integral peripheral flange (22) extending generally at right angles. The flange (22) closely contacts the interior lower side wall of the casing (11) and is sealingly secured to the casing by a peripheral weld (24) which shall be described in greater detail hereinafter.

The battery coil (25) comprises a positive plate (26) and a negative plate (28) spaced apart by separator insulative means (30). Connector terminal strap (32) for the positive plate (26) is connected to the positive terminal means (18) in the top of the casing (11) and a connector terminal strap (34) for the negative plate (28) is entrapped between the cap and the interior side wall of the casing (11), at a joint (35). The casing (11) is provided with a suitable electrolyte.

The end wall (13) of the casing is provided with a depressed area, or recess (36) surrounding the central opening (14). The casing (11), to withstand the pressures encountered during operation of the battery and to withstand the corrosive effects of the electrolyte contained therein, is composed of a nickel-plated steel as in the prior art.

However, since the crimping of the case to the insulative cover of the prior art shown in Figure 8.12a, is no longer required, the material used in manufacture of the case may be of thinner, e.g., 0.012", material as compared to the thicker, e.g., 0.019" material customarily used. The crimped seal to the insulative cover made it necessary to anneal the steel so as to get a proper ductility to the steel to form the crimp. Annealed steel is no longer required in construction of batteries in accordance with the process.

The resealable vent and positive terminal means (18) is comprised of outer and inner insulative washers (37) and (39), respectively on either side of the recessed portion (36) of the end wall (13) and surrounding the central opening (14). A headed rivet member (38), acting in conjunction with a washer member (40) secures the two insulative washers (37) and (39) concentric to the central opening (14) of the casing, the head (42) of the rivet (38) serving as the positive terminal of the battery.

An O-ring (44) is provided in the space between the rivet head (42) and the outer support washer (37) and operates as a resealable vent in accordance with the principles set forth in U.S. Patent 3,320,097, by forming a seal between the underside of rivet head (42) and bottom of recessed portion (36). Suitable slots (41) are provided in the inner and outer insulative washers (39) and (37) to provide venting routes for the gasses developed during operation of the battery. While the resealable vent means operates in accordance with the same principles as the above patent, it will be noted that this process construction of the vent means eliminates the circumferential crimp which poses a potential leak problem.

The battery coil (25), shown in greater detail in Figure 8.12c, is wound so that the positive plate (26) is offset to the top of the battery as viewed in the drawing and the negative plate (28) is offset to the bottom of the battery, the separator layers (30) extending the distance between the plates from the upper edge of the negative plates to the lower edge of the positive plates.

This construction offers several advantages, the primary one being the ease of connection of the connector leads to the respective battery plates. The prior art construction connected a tab (7a) or (7b), usually a single tab for each plate, or was made an integral part of the plate. With the construction in accordance with the process, terminal straps (32) and (34) may be connected at respective ends of the coil (25) after winding thereof because only one of the plates is exposed at either edge. Not only can the terminal straps (32) and (34) to the battery electrodes be connected after coiling, but the making of multiple connections to the turns of the battery plates with a plurality of welds (43) and (45) greatly increases the rate at which the battery may be discharged by providing ohmic connections at several points along the length of each of the plates.

By connection of the terminal strap (32) over the end of the coil (25) as best viewed in Figure 8.12d, the coil (25) is given much greater strength and rigidity. Connector strap (34) may be similar in construction to that of strap (32). Since the positive plates (26) only extend upwardly toward the positive terminal strap (32) and the negative plates (28) only extend downwardly toward the bottom of the case, insulating spacer members required at the ends of the coil in the prior art construction may be eliminated.

The elimination of these insulating spacers not only eliminates the cost of the same, but also provides more space within the casing for the electrolyte, active plate material and the recombination of the gases which are formed during operation of the battery. The construction of case (11) with the integral top wall (13) recessed to receive the positive terminal means (18) also provides extra space within the battery casing (for any given size of battery) since the relatively thick insulative cover (9) of the prior art construction has now been eliminated.

Thus, the construction offers more space within the casing (11) than the prior art construction. Although the releasable vent means (18) is provided on the casing, it is desirable that the gases produced during the operation do not escape to the atmosphere since this, in effect, reduces the amount of electrolyte remaining in the cell, hence reducing the total capacity of the battery.

Where the battery is intended to be used under conditions of extreme shock and vibrations, it is desirable that the battery coil (25) be restrained against movement within the case by use of spacers (50) and (51) which are formed of a material having sufficient resiliency to permit clamping of the battery coil between the ends of the case (11). The material preferably has a durometer hardness of from 15 to 95, with 40 to 60 being preferred for most applications.

Any resilient material which is chemically inert toward the electrolyte used in the cell construction may be utilized. Rubbers having good acid and base resistance are ethylene-propylene rubbers, neoprene and fluorinated rubbers. Where a material in the form of a solid spacer is too hard for use, it is anticipated that it may be used as a foam. For example, it is anticipated that foamed rubbers may be utilized in applications where extreme shocks or vibrations may be encountered. Polystyrene, polyethylene and polypropylene are examples of materials that may be utilized effectively as foams.

The battery may be constructed by forming the primary body (11) and the cap member (21) out of suitable steel. No annealing of the case material is now required, and as pointed out above, thinner material may now be utilized because the crimping operation has been eliminated. The positive terminal and resealable vent means (18) may then be assembled, surrounding the central opening (14) of the recessed end wall (13) of the casing (11) by placing the respective insulative washers (37) and (39) on either side of the central opening and securing the rivet (38), O-ring (44) and washer (40) together. The battery coil (25), being wound in the offset manner as aforestated, may be provided with the connector terminal straps (32) (as shown in Figure 8.12d) and (34) on opposite ends of the coil. The positive terminal strap (32) is provided with a tab portion (48) while the negative tap strap (34) is provided with the tab end (46).

The coil (25) is then inserted into the open end of the casing (11), and using a properly elongated welding rod (not shown), the central portion (48) of the tap strap (32) is connected to the positive terminal means (18) by welding. The tab end (46) of the negative tap strap (34) is folded over the open outer end edge of the casing prior to placement of the closure means (16) in the open end of the casing. The casing may then be filled with the battery electrolyte, and the closure means (16) placed in the open end of the casing entrapping the end (46) of the negative terminal tap strap (34). A peripheral weld (24) is made between the edge of the main body (11) of the casing and the edge of the peripheral flange (22).

While the use of resilient spacers such as (50) and (51) has been discussed with particular reference to the cell construction shown in Figure 8.12b, it is recognized that the use of resilient spacers to prevent damage to cells of the type described due to extreme shocks or vibrations may be applied to cells of conventional construction.

For example, it is anticipated that resilient spacers could be conveniently mounted in compression between the battery coil (6) and the cover member (9) shown in Figure 8.12a. (Also, a washer could be mounted between the casing (5) and the lower end of the battery coil). With reference to the cell construction of Figure 8.12b, it is apparent that either spacer (50) or (51) or both could be omitted.

In related work R.J. Lehnen and R.K. Sugalski; U.S. Patent 3,684,583; August 15, 1972; assigned to General Electric Company describe a rechargeable cell holder. The rechargeable cell comprises an inner cell having a metallic casing with positive and negative electrodes and an electrolyte within the metal casing. An external nonmetallic cylindrical case surrounds the metal casing. The nonmetallic case has external dimensions, i.e., diameter and length which conform to conventional predetermined cell sizes such as C size and D size cells.

The nonmetallic case is provided with internal means to center the inner cell. An electrode connector coaxially positioned on the internal cell protrudes through a centrally formed aperture on one end wall of the nonmetallic case. An enlarged second electrode connector which is attached to the opposite end of the internal cell has an exterior diameter substantially equal to the internal diameter of the nonmetallic casing. These elements cooperate to position the internal cell within the nonmetallic casing.

Thus, a single size internal cell can be used to construct a series of various external sizes of cells such as the C and D size cells which are commonly used in flashlights and the like. By constructing the outer case of a nonmetallic plastic material which can be molded, a substantial gain can be realized both in economics and in dimensional stability and accuracy of the final product thereby enabling the use of automatic equipment to produce the cell. Complete design details are provided.

Multilayer Separator for Secondary Alkaline Battery

In a process described by J.L. Devitt and D.H. McClelland; U.S. Patent 3,669,746;
June 13, 1972; assigned to The Gates Rubber Company a sealed or resealably safety-
valved alkaline zinc electrode-containing battery cell is provided employing a multilayer separator material which is composed of the following:

[1] A first layer in close engagement with the zinc electrode, microscopically homogeneous, nonreticulated, highly absorbing and
retentive of electrolyte, and providing a uniformly wetted interface with the zinc electrode substantially lacking occluded voids;

[2] A second layer, contiguous to the opposite polarity electrode, which
is composed of the same material as the first layer, or of a more
porous, reticulated and less retentive material than the first
layer; and optionally,

[3] A third layer, interposed between the first and second layers, of
a semipermeable membranous layer. The net effect of this multilayer separator is to provide a variance in separator material
characteristics normally to and transversely between the electrodes.

Referring to Figure 8.13, the inner cell components comprised of positive electrode
(4), negative electrode (2), and separator (6), saturated to near dampness with 35%
KOH, are constrained within cylindrical cell casing (8) having a safety valved top
(10). The separator (6) is composed of three layers (24), (26) and (28). Layer (28)
is microporous, highly uniform strip of high grade filter paper having a permeability
to air of about 4.7 cubic feet per minute per square foot of separator per 1/2" water
pressure differential.

Layer (24) is contiguous to the nickel electrode and is a single layer of a relatively
nonretentive, highly porous strip of Pellon. Separator layers (24) and (28) are sandwiched about a single layer of cellophane (26).

The multilayer separator strip (6) and the nickel plate (4) are spirally wound on
a machine under a pressure of about 300 psi to give a tightly wound concentric configuration. Figure 8.13b is a cutoff diagrammatic view of a partial section taken
along section (2—2) of Figure 8.13a. The cylindrical axial void (12) represents the
space vacated by the mandrel on which the stacked cell contents are wound. In general, the stacking pressure is firm, sufficient to minimize the occurrence of voids.

The stack pressure will depend on the cell configuration, particularly the type of
electrode plates used. Preferably at least about 100 and more preferably from 250
to 1,000 psi are employed. It has been found that this minimum stack pressure is
highly beneficial in reducing dendritic growth, apparently because firm stack pressure reduces greatly the presence of pockets and voids within the separator and along
the zinc plate/separator interface.

The spirally wound package in Figure 8.13b is tightly constrained to prevent unwind-
ing or slippage and is then inserted into the steel cylindrical container (8). The
spiral wound components may alternatively be wound and compressed into oval, rectangular or other shapes to accommodate the ultimate shape desired. The can is
preferably electrically insulated in some way, coated with a plastic layer, polyvinyl

FIGURE 8.13:   SEPARATORS FOR SECONDARY ALKALINE BATTERIES

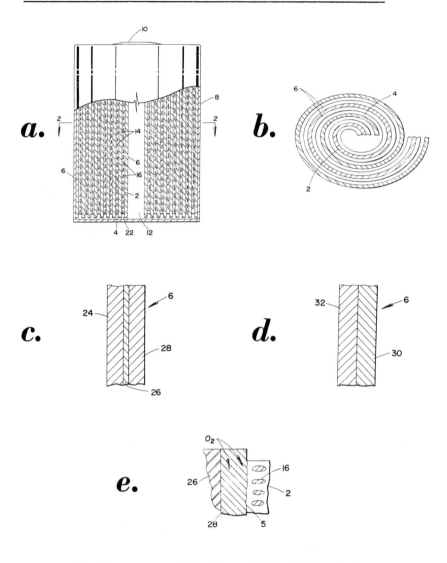

(a)   Cutaway View of Spirally Wound Nickel/Zinc Cell

(b)   Cross-Sectional View Along Line (2—2) of Figure 8.13a

(c)   Enlarged View of Three Layer Separator

(d)   Enlarged View of Two Layer Separator

(e)   Enlarged View of Separator/Zinc Interface

Source: J.L. Devitt and D.H. McClelland; U.S. Patent 3,669,746; June 13, 1972

chloride, polytetrafluoroethylene or some other insulating material. Alternatively, the entire can (8) may be made of a suitable electrically inactive material, such as plastic. Metal conductive tabs for current collection respectively are attached to and extend from the anode to the steel can (8) and from the cathode to the under side of the top (10), where they are separated by some suitable hermetically sealing insulator material such as rubber or plastic. Thirty-five percent KOH is added under vacuum until the cell stack is very slightly damp, making sure that no free electrolyte is present. The top is preferably secured to the can by crimping its edges over the mouth of the can.

The top (10) preferably contains a suitable safety valve for discharging excessive gas pressure above a predetermined minimum value which may build up in the cell (e.g., as a result of an excessive charge) but such a buildup is rare and hermetically sealed cells may be in some instances be useable, though not preferred for safety reasons. Paths for oxygen transfer and recombination are provided by space (22) at the bottom of the cell, a like space near the top of the cell and the cylindrical axial void (12). The void space (22) is further beneficial in preventing the localized buildup of electrolyte and formation of zincate pools.

In Figures 8.13c and 8.13d certain preferred separators are depicted. In Figure 8.13c, any of strips (28), (24) and (26) may be composed of a number of the same or like strips. However, it is generally preferred to use only one strip of semipermeable membrane (26) to present a barrier to zinc growth yet allow adequate conduction of ions. In Figure 8.13d, the two layer separator is shown where the first layer (32) is a microporous filter paper layer which closely engages the zinc electrode, and a strip of Pellon (30) separates the filter paper layer from the positive electrode. Each of these layers (30) and (32) may be composed of a plurality of contiguous strips of the same or like character.

In Figure 8.13e, the interface (5) separating zinc electrode (2) from first separator layer (28) is depicted where the view is blown up to about 100 times its actual size. The separator strip (28) intimately follows the contour of the zinc plate (2) and no voids are visibly present. The filter paper strip (28) and cellophane membrane (26) extend beyond the edge of zinc plate (2), discouraging dendritic growth around the top and bottom of the separator layers. Oxygen recombination path is shown. Normally the cellophane layer (26) is essentially impermeable to oxygen transfer. The following examples illustrate the process.

Example 1:   A sealed pillbox cell contains a single conventional sintered nickel plaque cathode and an anode consisting of an expanded mesh of metal pasted with a pliant mixture of zinc powder, zinc oxide and mercuric oxide. The separator consists of three layers. Adjacent the zinc plate is a single layer of a fibrous mat of vegetable parchment paper, microscopically uniform.

The remaining two layers consist of Viskon 3011, which is a bonded cellulosic nonwoven material, substantially more porous and permeable to air than the paper layer adjacent the anode. The cell is compressed to 2/3 psi and repeatedly discharged (50% level) and charged. The cycle life is determined to be 122 cycles.

Example 2:   The same type of cell is constructed as in Example 1 except that two cathodes lie on either side of the single anode and the separators are composed of two layers of high grade uniform filter paper (permeability to air 7.0 cfm) enclosing a membranous layer of regenerated cellulose (cellophane). The mid life voltage is about 1.6 and cycle life found to be 223 cycles.

Example 3:   The same cell as used in Example 2 is employed except that two anodes and one cathode are used, the cell is pressurized to 500 psi stack pressure and the depth of discharge is between 50 and 80% of capacity. The cell life is 220 cycles.

Example 4:   As a comparison, a cell very similar to that of Example 2 is constructed with the separator composed of four layers of high grade filter paper. The depth of discharge is 25%. The cycle life is only 13 cycles. This example demonstrates the advantages of transverse variance in separator materials, typified by the separators of Example 1, 2 and 3.

High Performance Carbon Electrode for Organic Depolarized Cells

F.A. Hinson, Jr.; U.S. Patent 3,637,436; January 25, 1972; assigned to Ashland Oil, Inc. describes a cathode mixture for a cell which contains a furnace carbon black having a surface area (iodine adsorption) of about 600 to 1,300 m.^{2}/g., a structure level as measured by oil factor of about 250 to 450 ml./100 g. of carbon, a pH of about 6 to 10 and an ash content of 0.5 to 10%.

There is considerable interest in organic depolarized cells. Such cells differ from Le Clanche cells in that the cathode depolarizer is an organic substance. Thus, for instance, it has been proposed to produce organic depolarized cells having a magnesium anode and a cathode which is a mixture of carbon black, magnesium perchlorate and an organic depolarizer such as meta-dinitrobenzene which is reduced, 12 electrons are transferred:

In contrast, for each molecule of MnO_2 reduced in the Le Clanche system, only one electron is transferred. The potential advantages of the organic depolarized cell are evident. Although it has been reported in the prior art that high structure is an essential property of a black suited for use in cells, the art still lacks a comprehensive understanding of the exact combination of properties needed to produce suitable carbon blacks for cells. Two conclusions follow: structure is not the sole property which determines the suitability of a black for cell applications; and a black suitable for use in one type of cell is not necessarily useful in other types of cells.

Support for these conclusions is provided by experimental evidence which shows that the same high structure acetylene-type blacks which have been successfully employed in Le Clanche cells are not suitable for organic depolarized cells. In this process, it has been found that with a particular class of reactor, it is possible to produce pronounced increases in surface area as well as a high level of structure. The applicable class of reactors is characterized by the ability to carry out a furnace carbon black process and by having two distinct chambers or zones separated by a restriction or choke having an opening which has substantially less cross-sectional area than either zone. The reactor in which it is particularly preferred to carry out the process is quite similar to that shown in U.S. Patents 3,060,003 and 3,222,131, but is slightly modified in respect to burner design and provision of means for injection of alkaline earth metal. In all other respects, the reactor is capable of operating in the same manner as the one shown in the patents.

The reactor shown in the patents is adapted to direct a linear flow of a burning mixture of fuel and a combustion-supporting oxygen-bearing gas through a first zone toward a second zone. At the same time, a second flow comprising feedstock is directed along the axis of the first zone toward the second zone, the flow being introduced into the first zone in the form of a conical spray, the angle of which is preselected to produce a carbon black of preselected structure. The two flows are intimately mixed and enter the second zone. The feedstock is thermally decomposed to carbon black, the carbon black being recovered from the second zone.

Referring to Figure 8.14, numeral (1) denotes a generally tubular reactor which is divided as shown into a first chamber or zone (2), a second chamber or zone (3) and a quench chamber or zone (4) having quench ports (5). As illustrated, the quench zone constitutes merely an extension of the second zone and is of substantially similar configuration. The first zone, however, is of greater diameter and shorter length than the second zone. For optimum results, moreover, it is preferred that the diameter of the first zone be greater than its length.

First zone (2) is provided with an inlet opening through which injector assembly (6) projects, while quench zone (4) is provided with an outlet opening for withdrawal of reaction products. Positioned in the inlet end of the second zone is a replaceable choke ring (7) of a high temperature refractory material having an orifice (8), the length, diameter and overall shape of which may vary. Each of the zones and their inlet and outlet openings is formed by a high temperature refractory liner backed up by a castable refractory insulation, the entire reactor in turn having an outer steel shell or casing.

Injector assembly (6) comprises substantially tubular members (9), (10) and (11), members (9) and (10) supporting a heat resistant ring member (12) between their inner ends. Fixed to the end of member (11) and positioned within the first zone is a circular deflector (13) having a diameter substantially equivalent to that of member (9).

The position of deflector (13) within the first zone may be adjusted by means not shown, so as to provide a circumferential orifice (14) of desired width, the orifice being formed on the one side by ring member (12) and on the other side by the deflector. As are all surfaces that are subjected to the high combustion and decomposition temperatures, the deflector is constructed of a high temperature refractory material, the inner surface of which is further provided with a heat resistant stainless steel insert (15).

Extending through tubular member (11) is a hydrocarbon feedstock conduit (16) provided at its inner end with a nozzle or injector (17) adapted to inject the feedstock into zone (2) in the form of a vaporized or atomized spray. Connected to zone (2) through tubular member (11) is a source of oxygen-bearing combustion-supporting gas, referred to as axial air, which serves to support in part the combustion of the fuel and/or feedstock.

In like manner, a source of an oxygen-bearing combustion-supporting gas, referred to as process air, for supporting combustion of the fuel in zone (2) is connected to zone (2) through circumferential orifice (14) and conduit (10). Communicating with zone (2) through orifice (14) is a source of fuel for providing the heat to sustain thermal cracking of the feedstock. The particular means for injecting the fuel into zone (2) may take various forms, a particularly effective arrangement comprising a conduit (19) connecting the source of fuel to the interior of enclosed space (27) between tubular members (9) and (10).

Ring (12) constitutes one end of space (27) and it is provided with a plurality of gas jets (20) which are secured at angularly spaced intervals around the front face of ring (12). The jets communicate with the enclosed space (27). The jets are directed outwardly so as to project the fuel toward the circumferential surface of zone (2).

An alkali metal compound injector pipe (29) is provided. See Figure 8.14b, the pipe (29) has been omitted from Figure 8.14a to simplify the drawing. Pipe (29) extends alongside the feedstock pipe (16) in tubular member (11) and has its outlet adjacent feedstock spray nozzle (17).

FIGURE 8.14:   CATHODE MIXTURE CONTAINING HIGH STRUCTURE, HIGH
SURFACE AREA CARBON BLACK

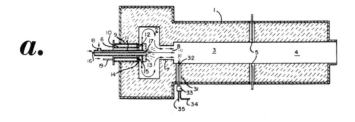

a.

Sectional View of Reactor

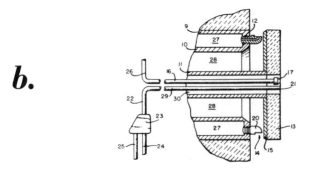

b.

Enlarged View of Reactor Detail

(continued)

FIGURE 8.14:   (continued)

C.

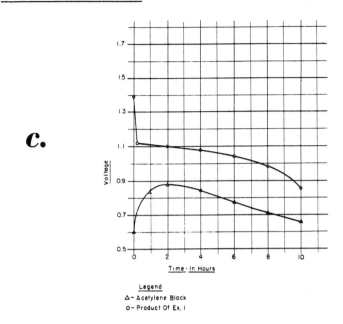

Legend
Δ– Acetylene Block
o– Product Of Ex. I

Discharge Curve for Organic Depolarized System

Source: F.A. Hinson, Jr.; U.S. Patent 3,637,436; January 25, 1972

Pipe (29) is connected through supply pipe (22) with an atomizer (23) which in turn is connected through pipe (25) with a source of alkaline earth metal solution under pressure, and through pipe (25), with a source of high pressure air for atomizing the alkaline earth metal solution.

In accordance with the process, the reactor is also provided with a racial port (31) through the metal casing and refractory lining of the furnace opening into the second zone (3) immediately downstream of restriction or choke (7). In port (31) is an alkali metal injection pipe (32) having its inner end slightly within the inner end of port (31) and having its outer end connected to atomizer (33). The latter, in turn, is connected through pipe (35) with a source of alkaline earth metal solution under pressure and through pipe (34) with a source of high pressure air for atomizing the alkaline earth metal solution.

During the operation of the above described reactor, a continuous stream of process air is injected into circumferential orifice (14) and flows radially outward passing the jets (20), at which point it is at its maximum velocity and minimum static pressure. Simultaneously, a stream of hydrocarbon fuel is injected into orifice (14) through jets (20) resulting in a thorough and rapid mixing thereof in the process airstream. The fuel-air mixture is ignited as it passes into zone (2), the burning mixture and its products of combustion flowing radially outward from the axis thereof as a uniformly expanding disc shaped stream. It then follows a flow pattern as determined by the configuration of zone (2) and as shown by the arrows in the drawing, flowing substantially parallel to the circumferential surface of the zone towards the opposite end thereof where it is directed radially inward toward the axis of the zone and orifice (8).

As hydrocarbon fuel and process air are introduced into the reactor, hydrocarbon feedstock is injected into zone (2) through injector nozzle (17) in the form of a vaporized or atomized spray cone. Alkaline earth metal solution is introduced to the interior of the reactor through injection pipe (29) and/or injection pipe (32) and is rapidly dispersed in the reaction mass. The temperature of the feedstock is rapidly raised as it approaches orifice (8) and it is thoroughly mixed with and dispersed in the hot combustion gases resulting from the burning of the hydrocarbon fuel.

The mixture of combustion products, alkaline earth metal [when pipe (29) is used] and feedstock passes through orifice (8) into zone (3), cracking of the feedstock being terminated in zone (4) by quenching with water or other suitable cooling medium introduced through quench ports (5). The cooled reaction gas with entrained carbon black then exits from zone (4) for subsequent separation and collection of carbon black.

The following examples illustrate the process. The dimensions of the reactor used in the examples are as follows: First Zone — diameter 36", length 15"; Air Inlet Annulus — diameter 12", width 3/4"; Restriction Ring — internal diameter 7", length 9"; Second Zone — diameter 15", length 10', 6"; Additive Injector 32 — distance of pipe from downstream face of restriction 1/2" approximately; and Quench Sprays — distance from additive injector 32 is approximately 10', 5 1/2". A hydrocarbon feedstock having approximately the following analysis is employed in the examples.

Gravity	2.5 API at 60°F.
Viscosity	43.6 SU sec. at 210°F.
Conradson Carbon	6.42 %
Correl. Index	116
Distillation at 760 mm.	
Initial BP	463°F.
5%	632
10%	674
20%	700
30%	704
40%	721
50%	739

Oil began to crack.

Natural gas having a Btu rating of about 1,050 Btu/ft.^{3} net is employed as fuel with ambient air, as the combustion-supporting gas. The pipe (32) for injecting the additive into the second zone of the reactor was a 1/4" IPS stainless steel pipe extending to within 1" of the wall of the zone. The oil absorption tests employed in the examples are conducted in accordance with the procedure set forth in U.S. Patent 3,222,131. Where iodine numbers are referred to in the examples they were determined by the following procedures: 0.3333 g. of carbon black are placed in a flask; 10 ml. of 0.479 N iodine solution and 40 ml. of water are added to the carbon black, the mixture is shaken 30 seconds, then allowed to stand 5 minutes. The mixture is then filtered and 20 ml. of the filtrate are titrated with standardized sodium thiosulfate solution to the usual iodine endpoint using starch solution indicator. A blank solution of 10 ml. 0.479 N iodine solution and 40 ml. of distilled water is prepared and 20 ml. of this blank solution is titrated with the same sodium thiosulfate solution as used with the filtrate. The iodine number is computed as follows: iodine number = 3 x normality of sodium thiosulfate x 317.26 x (B − S). B is the number of ml. of sodium thiosulfate required to titrate the blank and S is the number of ml. of sodium thiosulfate required to titrate the filtrate from the sample.

Example 1: In this example, the reactor was operated at a combustion air rate of 190,000 scfh, a natural gas rate of 12,260 scfh, a feedstock rate of 140.75 gph (metered at 60°F.), a feedstock preheat temperature of 450°F. (at the reactor), and a feedstock spray angle of 120° (solid cone spray). Eight gallons per hour of $CaCl_2$ solution (35 lb./100 gal.) was atomized with 500 scfh of air and introduced through injector pipe (32). Under these conditions, about 54% of the feedstock was burned and a temperature of about 3200°F. was attained in the reactor. The product had an ash content of about 2.5%. The yield was 2.02 pounds of carbon black per gallon.

Example 2: A cathode mix for an organic depolarized cell was prepared in accordance with the following procedure. Grind 100 g. meta-dinitrobenzene and 4.5 g. barium chromate together with a small portion of a 50 g. sample of the above acetylene carbon black until free of lumps of m-DNB. Blend the above mixture and the remaining carbon in a blender for 5 minutes. Add 225 ml. of electrolyte which consists of a 2.5 N magnesium perchlorate solution containing 0.2 g. of lithium chromate/liter. The mixture is kneaded until a moist crumb is obtained.

The moist crumb was placed in a cylindrical container of plexiglass having a magnesium anode plate at one end. A conventional coated paper separator is placed between the anode plate and the cathode mix. It serves as a conducting medium when wetted by the electrolyte solution. At the other end of the container is a carbon rod which is embedded in the cathode mix and extends through the end wall to the positive terminal of the cell. It serves as a collector of electric current from the anode and also is sufficiently porous to permit the escape of gas from the container which, except for the porosity of the carbon collector, is tightly sealed.

Example 3: Example 2 is repeated except that the carbon black produced in accordance with Example 1 is substituted for the acetylene black.

Example 4: The cells prepared in accordance with Examples 2 and 3 were discharged at a 16 2/3 ohm drain rate which was selected at a reasonable simulation of the type of load to which such cells might be subjected in normal use. The performance of the cells is illustrated graphically in Figure 8.14c and in the following table.

Cell	No Load Voltage	Time to 0.9 v.(min.)	Power to 0.9 v. (watt-hour)	Amp.-hr. to 0.9 v.
Example 2, Acetylene black	1.60	—*	—*	—*
Example 7, Furnace carbon black of this process	1.825	582	0.615	0.6

*Cell with acetylene black never attained 0.9 v.

As may be seen from the graph in Figure 8.14c and the data in the table, the furnace carbon black produced in accordance with the process proved superior to the acetylene black.

Organic Depolarized Cell

A process described by D.M. Larsen and M.J. Terlecke; U.S. Patent 3,622,392; November 23, 1971; assigned to ESB Incorporated provides a rechargeable cell having the same outer physical dimensions as the conventional primary cell. One of the basic elements that enables this cell to be recharged is an organic depolarizer that is used in place of the inorganic depolarizer in use today. The basic purpose of having a rechargeable cell is also achieved in part by the fact that the improved construction allows for venting of the gasses formed during charge.

The cell construction basically comprises a zinc can electrode, a laminated separator which lines substantially the entire vertical walls of the zinc can, a porous carbon rod in the center of the zinc can, insulation about the carbon rod in the area known as the airspace of the cell, an azodicarbonamide depolarizer mix, laminated insulation at the bottom of the zinc can, a sealant and a vent washer on top of the sealant to allow the venting of gasses during storage and operation.

Referring to Figure 8.15, a detailed elevational and partially diagrammatic drawing in section of a cell of this process is outlined. The cell is generally designated as (20) with an outer metal jacket denoted by (21). This outer jacket is lined in the interior by an insulator (22). The jacket (21) and the insulating liner (22) are shown folded over at the base of the cell thereby retaining a metal disk (23) clamped against the negative electrode can (24) which usually is made of zinc.

The metal disk (23) serves as a contact point for the negative electrode to the outside circuit. A metal cap (28) is shown at the top of the cell insulated from the metal jacket (21) by the insulator (22) and serves as the contact point for a porous carbon rod (25) shown centrally located in the cell.

A separator (26) lines the inner wall of the negative can (24) and separates the can from the organic depolarizer mix (27) which surrounds the porous carbon rod (25) and fills the annular area between the rod and the separator (26) up to a top-insulating material located at (37) on top of the depolarizer mix. The insulator (37) is preferably paraffined kraft paper although a laminated construction of paper and nonconductive plastic could be used. This insulator aids in preventing dendritic growths developing from the top of the depolarizer mix (27) to the zinc can.

The carbon rod (25) and the depolarizer mix (27) are insulated at the bottom of the cell from the negative electrode can (24) by a bottom-insulating cup (29), which preferably is a lamination of paraffined kraft paper (30) and any nonconductive, inert plastic material (35) such as polyethylene.

This insulator also prevents dendrite growths from the zinc can to contact the depolarizer mix or the carbon rod. Instead of the laminated construction of the insulator (29), a disk of plastic material could be used or a coating of plastic could be allowed to solidify on the bottom of the can (24). Resting on the crimped top edge of the negative can (24) and covering the open end of the can is an insulating element (31) referred to as a seal washer which can be made of a strong kraft paper.

FIGURE 8.15:   ORGANIC DEPOLARIZER CELL CONSTRUCTION

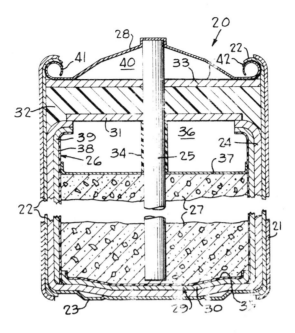

Source: D.M. Larsen and M.J. Terlecke; U.S. Patent 3,622,392; Nov. 23, 1971

A sealing compound (32) seals the interior of the cell and is supported by the crimped edge of the negative can and the washer (31) so that a maximum airspace (36) is provided. Between this sealing compound and the metal terminal cap (28) is an insulator (33) called a cap washer which serves to prevent the metal cap (28) from becoming embedded in the sealant (32) when the cap is forced downward under the crimped edges of the casing (21) and the insulating liner (22). If the cap were allowed to become embedded in the sealant, the area (40) would be effectively sealed so that gasses formed in the cell could not be vented as described below and a dangerous pressure would thereby be able to build up in the cell.

Although the cap washer (33) prevents the cap (28) from becoming embedded in the sealant (32), there remains another possibility that the area (40) could be sealed and thereby prevent the cell from being vented. When the cap (28) is forced down onto the washer (33) and sealant (32), some of the sealant can be, and often is, forced up along the paper insulator (22) and acts as a seal between the insulator and the rolled edge (41) of the cap. As a result the area (40) is sealed and the cell cannot be vented.

Therefore, there is provided on top of the cap washer (33) a second, thin vent washer (42). This washer is larger in diameter than the inner diameter of the cell (20) and its edge extends up around and thereby lines the rolled edge (41) of the cap (28). The washer is thin in order to be capable of being folded over the edge (41) and also must be tough enough so that the sealant (32) cannot break through it and seal the

area between the edge (41) of the cap and the insulator (22). The vent washer therefore performs the extremely useful and important function of being a vent path for
gasses to escape to the atmosphere. A material that has been successfully used for
this washer (42) is a 5 mils thick laminated construction of kraft paper-polyethylene
kraft paper. A thin layer of paraffined kraft paper, or wax paper or cellophane or
other plastic would also be suitable for use in this washer. Also, with the proper
choice of material and thickness, it is contemplated that one thin washer can be used
in place of the two separate washers (33) and (42).

Venting of gasses from the cell results from the gasses entering the porous carbon rod
(25) and collecting in the area (40) under the cap (28) and the escaping from beneath
the cap in the vicinity where the casing (21) and insulating liner (22) are crimped
over the rolled edge (41) of the cap (28). As mentioned above the vent path around
the rolled edge (41) is provided by the vent washer (42).

The carbon rod is constructed to have a controlled porosity to permit gasses to flow
up the rod for venting purposes and still prevent electrolyte from entering and wetting
the rod and corroding the cap (28). Surrounding the carbon rod (25) in the airspace
(36) is another insulator (34). With the use of this particular cell construction and
an organic depolarizer mix the basic purposes of this process are achieved.

The insulator (34) which surrounds the carbon rod in the airspace (36) is impermeable
to electrolyte and together with the separator (26) restricts the growth of zinc dendrite from the zinc can to the carbon rod and thereby aids in preventing a short circuit between the zinc can and the carbon rod. A suitable material to be used as this
insulator (34) has been found to be polyethylene. The separator (26) lines the entire
inner wall of the zinc can including the portion in the airspace (36) and is not folded
over the top of the depolarizer mix as in the conventional cell.

A preferred separator material comprises using a semipermeable barrier material in
combination with an absorbent material, such as a thermoplastic resin which serves
as a continuous elastomeric binder matrix for a gelling agent such as a starch-wheat
flour mixture, carboxymethylcellulose, etc. In particular, a separator used successfully in a cell of this process consists of a laminated construction of a layer of cellophane with a layer of an ethylene/vinyl acetate copolymer, such as Elvax, having
therein a gelling agent such as starch-wheat flour mixture.

The cellophane layer (38) being the semipermeable barrier, is in contact with the depolarizer mix while the Elvax layer (39) is placed against the zinc can. Another
laminated separator which has been used consists of a layer of victory paper on a
layer of cellophane with the cellophane again in contact with the depolarizer mix
while the paper lines the zinc can. To have the carbon rod and zinc can insulated
from each other in the airspace (36) in such a way as to prevent a short circuit between these two elements greatly aids in enabling the cell to be recharged numerous
times. The laminated separator also prevents zinc dendrite growths from growing from
the side walls of the zinc can to the depolarizer after numerous cycling.

The depolarizer mix used in the cell contains a substituted or an unsubstituted azodicarbonamide compound of the type described in U.S. Patent 3,357,865 and which
is present in the depolarizer mix in an amount ranging from about 5 to about 60%
by weight of the total mix. These compounds may be generally represented by the
formula shown on the following page.

$$R_1 \diagdown N-\overset{\overset{\displaystyle O}{\|}}{C}-N{=}N-\overset{\overset{\displaystyle O}{\|}}{C}-N \diagup R_1$$

In the above formula R_1, R_2, R_3 and R_4 may be hydrogen, alkyl of 1 to 8 carbon atoms, mono- and dicarbocyclic aryl or substituted aryl, cyclo-alkyl, aralkyl, alkoxyalkyl, cyanoalkyl, haloalkyl, nitroalkyl, alkenyl and R_1 and R_2 and/or R_3 and R_4, when alkyl, may be joined together through a nitrogen, sulfur or oxygen linkage to form a heterocyclic ring.

The preferred azodicarbonamide compounds are those in which the nitrogen atoms carry an alkyl radical of 1 to 4 carbon atoms. In particular, a di-N-butyl azodicarbonamide compound has been successfully used in the depolarizer mix in amounts ranging from between 15 to 3% by weight of the depolarizer mix.

COMPANY INDEX

The company names listed below are given exactly as they appear in the patents, despite name changes, mergers and acquisitions which have, at times, resulted in the revision of a company name.

INVENTOR INDEX

U.S. PATENT NUMBER INDEX

NOTICE

Nothing contained in this Review shall be construed to constitute a permission or recommendation to practice any invention covered by any patent without a license from the patent owners. Further, neither the author nor the publisher assumes any liability with respect to the use of, or for damages resulting from the use of, any information, apparatus, method or process described in this Review.

BATTERY MATERIALS 1970

by P. Conrad

Electronics Materials Review No. 10

Demands for new types of power systems for earth and space vehicles are putting new battery systems in the limelight. More compact power systems for space vehicles operating under exotic conditions of temperature, pressure and weightlessness are providing opportunities for commercial application of new battery designs. In addition, pressures for environmental improvement on earth have focussed attention on new types of batteries for automotive propulsion. It may be possible that new battery systems will replace, or at least supplement, the internal combustion engine as man tries to reduce air pollution and to improve the quality of the atmosphere.

This latest in the Electronics Materials Review series is concerned with reporting on the search for technological advances in battery materials. It is based on the U.S. patent literature and is written from the practical, processing point of view, giving detailed information regarding operating conditions for these manufacturing processes.

The great quantity of material provided by this review is indicated by the Table of Contents shown below. The numbers in ( ) indicate the number of processes given for producing each of the indicated battery materials. This book will serve as a guide to a productive industry.

171 pages

CAPSULE TECHNOLOGY
AND MICROENCAPSULATION 1972
by M. Gutcho

A review of 210 processes collected from recent U.S. patents. The book deals with the manufacture and use of macrocapsules, microcapsules, and other methods of enclosing a liquid or solid in a polymer substance.

The most modern methods are described. These include simple and complex coacervation, physical processes involving centrifugal and electrostatic methods, colloidal wall formation, chemical synthesis in situ, interfacial polycondensation reactions, and many more.

A much abridged table of contents is given here. Numbers in ( ) indicate nos. of processes. Chapter headings are given, followed by examples of important subtitles.

372 pages

ELECTRONICS INDUSTRY OF JAPAN 1972

The electronics industry has been Japan's fastest growing industrial sector and one of the country's most important export earners. Over the past thirteen years the growth has averaged 24 percent per year to reach a value of production of 3,400 billion yen in 1970.

"Electronics Industry of Japan" is aimed at giving you a concise but detailed picture of the present state of this important industry, describing 854 Japanese electronics firms. The book is divided into three sections.

The **first section** gives details of 145 electronics companies listed on eight Japanese stock exchanges—Tokyo, Osaka, Nagoya, Fukuoka, Hiroshima, Kyoto, Niigata and Sapporo. The following information is given on each company:

> **Full name**
> **Address**
> **Date of establishment**
> **Capital**
> **Total assets, sales and profits for 1968, 1969 and 1970**
> **Principal executives**
> **Number of employees**
> **Major stockholders**
> **Bankers**
> **Technical agreements**
> **Products**

The **second section** gives more limited information on a further 714 unlisted companies. Details given include:

> **Full name and address**
> **President**
> **Capital**
> **Number of employees**
> **Main products**

The **third section** comprises a statistical analysis of the present state of the industry and its historical development. Figures on production, trade, research and development, and employees are included.

The book is completed by an alphabetical list of registered trade names.

To anyone concerned with electronics in any part of the world, a knowledge of the important Japanese industry is essential. This significant new guide will give you all the pertinent information in one volume, providing both a convenient source of reference and a valuable marketing aid.

This guide was prepared in Japan by one of our associates, in order to give you on-the-spot coverage.

147 pages

SEALING AND POTTING COMPOUNDS 1972

by J. A. Szilard

Sealing and potting compounds are used to fill and seal cavities between components, to coat and seal porous surfaces, to completely cover or encapsulate electrical and other devices. The purpose of all these treatments is to protect against the ingress of liquids or gases. In most cases the desired protection is against the penetration of moisture. Many "telechelic" polymers have been developed for sealant purposes—the properties required for integrated circuit sealants are naturally quite different from the properties of a composition used to seal the soil around an oil well. This book describes 166 sealant manufacturing processes with strong emphasis on applications. Numbers in parentheses indicate the number of processes for each topic.

1. ACRYLICS (11)
   Building Construction and Automotive Use (3)
   Soil Treatment
   Anaerobic Sealants for Closely Facing Surfaces (6)
   Potting Compound for Electrical Use
2. BITUMENS (11)
   Building Construction (5)
   Highways and Runways (5)
   Soil Treatment
3. BUTADIENES (18)
   General and Building (6)
   Automotive (5)
   Electrical Potting (5)
   Sealant for Shoemaking
   Container Sealant
4. CHLOROPRENE (3)
   General Uses
   Pipe Joint Sealants (2)
5. EPOXIES (23)
   General Uses (9)
   Electrical and Electronic (11)
   Soil Treatment (3)
6. ESTERS AND AMIDES (23)
   General Uses (3)
   Electrical Potting (6)
   Soil Treatment (10)
   Aircraft Construction (2)
   Patent Leather
   Container Sealing
7. FLUOROCARBON RESINS (6)
   General Use (2)
   Liquid Oxygen Apparatus

Shafts and Stuffing Boxes
   Carton Sealants
   Extreme Temperature Uses
8. SILICONES (13)
   General Uses (5)
   Electrical and Electronic (7)
   Bouncing Putty
9. SULFIDES AND MERCAPTANS (29)
   General Uses (10)
   Building Construction (10)
   Aircraft Construction (6)
   Appliance Base Coatings (2)
   Automotive Use
10. URETHANES (19)
   Building Construction (7)
   Soil Treatment (3)
   Electrical and Electronic (6)
   Automotive (3)
11. VINYLS (6)
   Building Construction (2)
   Soil Treatment (3)
   Paper Sealing
12. OTHERS (14)
   General Uses (2)
   Soil Treatment (7)
   Automotive
   Electrical
   Sealing Wax
   Lubricating Sealants (2)

To indicate the wealth of information found in each chapter, a breakdown of Chapter 8 on Silicones follows:
   Diorganopolysiloxane-Bis (alkoxysilyl) hydrocarbon Sealant
   Hydroxyendblocked Organosiloxane and Polyvinyl Alkoxysilane
   Joint-Sealing Polysiloxane
   Curable Organopolysiloxanes (2)
   Silica-Reinforced, Transparent Composition
   Silicone-Based Electrical Contact Sealants
   Highly Filled Vinyl Polysiloxanes
   One Component Rubbery Siloxane Sealant
   Encapsulants for Printed Circuits
   Heat Resistant Carborane-Substituted Silicones
   Potting Compounds for High Frequency Circuits
   Bouncing Putty (Boron Polysiloxane)

280 pages